Building Research Estab

BRE Building Elements

Floors and flooring

Performance, diagnosis, maintenance, repair and the avoidance of defects

P W Pye, MRIC, CChem

H W Harrison, ISO, Dip Arch, RIBA

Prices for all available BRE publications can be obtained from:
Construction Research Communications Ltd
151 Rosebery Avenue
London, EC1R 4QX
Telephone
0171 505 6622
Facsimile
0171 505 6606
E-mail
crc@construct.emap.co.uk

BR 332
ISBN 1 86081 173 6

First published 1997

Published by Construction Research Communications Ltd by permission of Building Research Establishment Ltd.

Contents

Preface

It has been said that most problems with floors occur because people insist on walking on them, pushing trolleys over them, placing large objects on them and dropping things on them – if only they were ceilings they would never wear out! A small witticism that reflects the way some people, including professionals in the construction industry, see floors. Or, rather, don't see them! After all, what is there to a floor: floorboards nailed to joists. What can go wrong with that? And if it happens to form a ceiling, even better. But the facts belie this perception.

BRE's figures on faults in buildings of all types (given in greater detail in the introductory chapter which follows) show that a substantial number concern floors. Despite the advice that has been available to the industry from the 1920s, faults in flooring, such as cracking, detachment and entrapped water, recur frequently. If some of the errors appear elementary, this only reflects what happens in the design office and on site. All that we can do is show to those who work in the flooring industry what is being done incorrectly and how to take corrective action – preferably before faults or defects lead to costly damage.

This book describes the materials and products, methods and criteria which are used in the construction of floors and flooring. It draws the reader's attention to those elements and practices which ensure good performance or lead to faults and failure. There is sufficient discussion of the underlying structure to enable an understanding of the behaviour of the whole floor without going very far into engineering design principles. It does not purport, though, to be a book of construction practice; nor does it provide the reader with the information necessary to design a floor, but, mainly through lists and comprehensive illustration, shows him or her what to look for as good and bad features of floors and flooring. It also offers sources of further information and advice.

Readership

Floors and flooring is addressed primarily to building surveyors and other professionals performing similar functions, such as architects and builders, who maintain, repair, extend and renew the national building stock. Lecturers and other educators in the building field will also find it to be a useful adjunct to their course material.

Scope of the book

Although books on flooring are few, there is no shortage of industry guidance on floors and flooring. The problem is that people do not use the guidance that exists.

To try to remedy that situation, the contents of this book are configured so that the principles, features and functions of floors and flooring are described first (Chapter 1). There needs to be sufficient discussion of principles to impart understanding of the reason for certain practices; without that understanding, practitioners will have difficulty following correct procedures – or until they make the mistakes, or overlook precautions, as previous generations have done. The criteria presented in Chapter 1 are then related to the different types of floors and their finishes (Chapters 2–8).

The text concentrates on those aspects of construction which, in the experience of BRE, lead to the greatest number of problems or greatest potential expense if carried out unsatisfactorily. It follows that these problems will be picked up most frequently by maintenance surveyors and others carrying out remedial work on floors. Although most of the information relates to older buildings, surveyors may be called upon to inspect buildings built in relatively recent years. It is therefore appropriate also to include much material concerning observations by BRE of new buildings under construction in the period 1985–95.

Many of the difficulties which are referred to BRE for advice stem from too hasty assumptions about the causes of particular defects. Very often the symptoms are treated, not the causes, and the defects recur. It is to be hoped that this book will encourage a systematic approach to the diagnosis of floor and flooring defects.

The case studies provided in some of the chapters are selected from the files of BRE Advisory Service and the former Housing Defects Prevention Unit, and represent the most frequent kinds of problems on which BRE is consulted. They are not meant to be comprehensive in scope since the factors affecting individual sites are many and varied.

As has already been said, this

book is not a textbook on building construction. Hence, the drawings are not working drawings but merely show either those aspects to which the particular attention of readers needs to be drawn or simply provide typical details to support text.

Other more specific aspects of the subject not deemed to be relevant to this book are mentioned briefly in appropriate chapters – usually in the introductory paragraphs.

Passive fire protection measures are those features of the fabric, such as structural frames, walls and floors, that are incorporated into building design to ensure an acceptable level of safety. These measures, so far as they affect floors, are dealt with in outline in this book. Measures which are brought into action on the occurrence of a fire, such as fire detectors, sprinklers and smoke exhaust systems are referred to as active fire protection, and are not dealt with in this book.

The standard headings within the chapters are repeated only where there is a need to refer the reader back to earlier statements or where there is something relevant to add to what has gone before. We have assumed that readers will know many of the more common abbreviations used in the industry – DPC, PVC etc – and we have declined to spell them out.

This book deals with all kinds of floorings (ie floor coverings), including both in situ and manufactured products, and these are covered in detail in Chapters 5–8. A classification of floorings for use internationally has been published by the European Union of Agrément, and this is examined in more detail later in the book.

Ceilings, whether suspended or applied directly to soffits, are treated as integral parts of the element of floors since many aspects of performance, such as fire and sound, affect all parts of both floors and ceilings. Ceilings are mainly dealt with in Chapter 2.

Ramps, landings and stair treads are included as elements of floors, but not staircase enclosures. There is an argument for dealing with staircases in conjunction with walls since, in most cases, it is necessary to consider the enclosures for stairs in conjunction with stair flights; enclosures for staircases, and such matters as protected shafts, are therefore considered as elements of walls.

Much that is relevant to ordinary floor finishes applies also to stair tread finishes.

The weatherproofing aspects of balconies, insofar as they are similar to those of roofs, are dealt with in the book on roofs. Thresholds are handled as parts of external walls.

In many places through the book we have quoted British Standards and codes of practice which have been withdrawn; however they would have been current at the time a particular floor or flooring was laid. We have done this deliberately since they often gave better specifications than those now current. Indeed, in some cases, standards and codes have been withdrawn and not replaced. Copies of old standards and codes are often retained by BRE for use in disputes; they can also be seen in the British Library.

In the United Kingdom, there are three different sets of building regulations: the Building Regulations 1991 which apply to England and Wales; the Building Standards (Scotland) Regulations 1990; and the Building Regulations (Northern Ireland) 1994. There are many common provisions between the three sets, but there are also major differences. Although the book has been written against the background of the building regulations for England and Wales, this is simply because it is in England and Wales that most BRE site inspections have been carried out. The fact that the majority of references to building regulations are to those for England and Wales should not make the book inapplicable to Scotland and Northern Ireland.

We have deliberately not provided more than an outline of the major points which specifiers will need to take into account in the cleaning and maintenance of floorings since this is not a topic which has been studied in depth by BRE or its predecessor, the Building Research Station; in any case there is suitable literature available from industry sources and other publishers.

Some important definitions

Some of the more general terms used in floors and ceilings will be found in Section 1.3, Subsection 1.3.3 of BS 6100:Part 1 (Glossary of building and civil engineering terms: General and miscellaneous).

Since the book is mainly about the problems that can arise in floors, two words, 'fault' and 'defect', need precise definition. Fault describes a departure from good practice in design or execution of design; it is used for any departure from requirements specified in building regulations, British Standards and codes of practice, and the published recommendations of authoritative organisations. A defect – a shortfall in performance – is the product of a fault, but while such a consequence cannot always be predicted with certainty, all faults should be seen as having the potential for leading to defects. The word failure has occasionally been used to signify the more serious defects.

By 'floor' we mean the whole of the horizontal elements of a building (excluding roofs but including ceilings); 'flooring' refers simply to the finish of the upper surface of the floor.

Where the term 'investigator' has been used, it covers a variety of roles including a member of BRE's Advisory Service, a BRE researcher or a consultant working under contract to BRE.

'Topping' has been used to describe an in situ material laid to provide good abrasion resistance and to provide the wearing surface. It has also been used to refer to a cementitious mix used to lock together components of a structural concrete floor, such as hollow pots, where it is called a structural topping.

'Underlay' is a layer used between the structural deck or slab and the flooring, either prefabricated (eg plywood or hardboard) or a thick in situ layer (eg mastic asphalt or aggregate filled latex cement). 'Underlayment' is in situ material used to smooth or level the base prior to laying the flooring. (The term underlay seems to have become restricted to preformed substrates and the term underlayment to those formed in situ, and we have adopted this distinction. However, in other industry publications and documents the former term will be found to apply to both applications.)

'Wear' is the progressive loss from the surface of a material or component brought about by mechanical action.

Acknowledgements

Photographs which do not bear an attribution have been provided from our own collections or from the BRE Photographic Archive, a unique collection dating from the early 1920s.

To the following colleagues – many from the BRE Scottish Laboratory and the Fire Research Station – and former colleagues who have suggested material for this book or commented on drafts or both, we offer our thanks:

D R Addison, Dr B R Anderson, R W Berry, Dr A B Birtles, Dr P Bonfield, P J Buller, Dr R N Butlin, A H Cockram, Dr J P Cornish, S A Covington, R N Cox, Dr R C deVekey, Diane M Currie, R J Currie, Maggie Davidson, Dr J M W Dinwoodie, Dr B R Ellis, Dr V Enjily, Dr L C Fothergill, E Grant, G J Griffin, C J Grimwood, J H Hunt, Dr P J Littlefair, T I Longworth, K W Maun, N O Milbank, Dr D B Moore, W A Morris, Dr R M Moss, F Nowak, E F O'Sullivan, R E H Read, J F Reid, M R Richardson, D Seller, A J Stevens, P M Trotman, C H C Turner and Dr T J S Yates, all of BRE.

We have also drawn upon some notes prepared by the late Dr Frank Harper and the late Wilfred Warlow. In addition, we acknowledge the contributions of the original, though anonymous, authors of *Principles of modern building*, Volume 2, from which several passages have been adapted and updated.

PWP
HWH
August 1997

Chapter 0 Introduction

The majority of floors, of whatever kind of structure or surface finish, perform well (Figure 0.1). However, there is evidence from BRE sources of all kinds that avoidable defects in floors and flooring occur too often. These sources include surveys of housing, of both new construction and rehabilitation, evidence from the United Kingdom house condition surveys (undertaken every five years), and also, particularly for building types other than housing, from the past and current commissions of BRE Advisory Service.

Records of failures and faults in buildings

BRE Advisory Service records

An examination of BRE Advisory Service records of enquiries received by letter or by telephone, and of investigations, involving site visits, laboratory analysis where required, and reports, shows how the pattern of defects in floors and flooring brought to the notice of BRE has changed over the years. Of course not all requests for investigations can be acted upon; indeed, it is estimated that around half of all requests to BRE for investigations on flooring have had to be turned down for resource reasons, and these therefore do not show in the records.

An analysis of enquiries received from 1925 to 1949 shows that 1 in 10 of all enquiries received by BRE Advisory Service related to floors and flooring, with those on finishes outnumbering those on the structure by three to one. In the years since 1949, enquiries on floors and flooring have continued to form a substantial portion of all enquiries received, with site visits having to be made in many cases.

To give an example, from February 1970 to May 1974 a total of 69 investigations were carried out on floors and flooring (Figure 0.2). These did not include investigations arising from fires.

In 1970–74, housing and offices each comprised 1 in 12 investigations; 1 in 10 for each of laboratories, schools and colleges, commercial buildings and hospitals; and around 1 in 3 for factories.

In 1987–89, 1 in 7 were in housing; 1 in 40 schools and colleges; 1 in 6 in each of factories and offices; 1 in 5 for commercial and public buildings (which includes, for the first time significantly, 1 in 28 for swimming pools); and getting on for 1 in 3 in

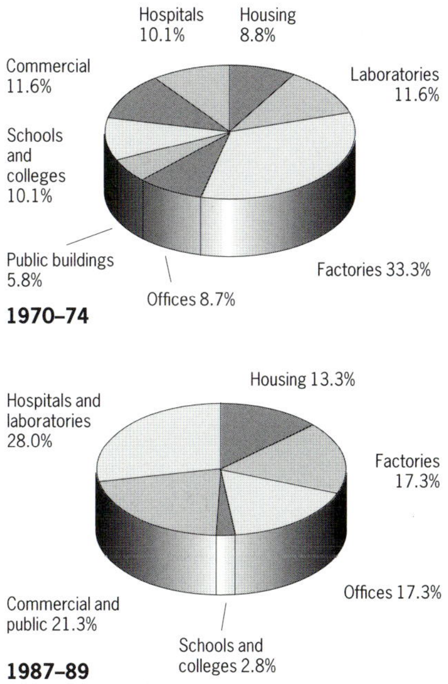

Figure 0.2
BRE investigations of floors and flooring by building type

Figure 0.1
A terrazzo tile floor like this, often specified for its hard wearing and easy-to-clean qualities, should last for many years

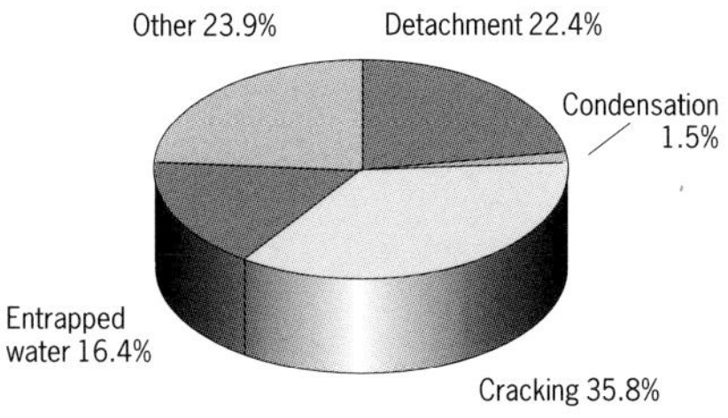

1970–74

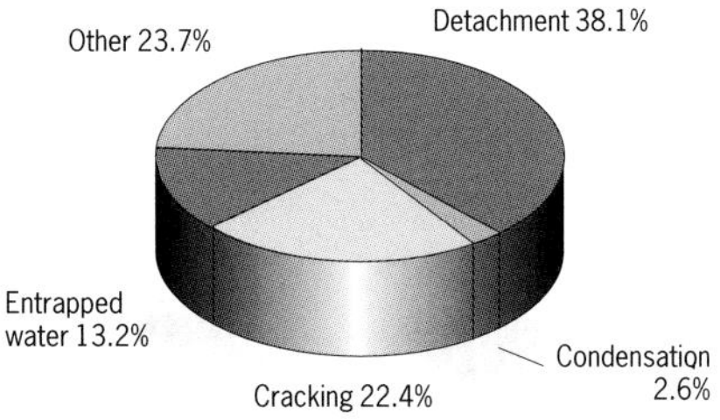

1987–89

Figure 0.3
BRE investigations of floors and flooring by type of defect

hospitals and laboratories. These show very big changes from earlier years.

In 1970–74, detachment of the finish was involved in 1 in 5 cases; cracking in nearly 1 in 3; entrapped water in nearly 1 in 6; and the remainder of 1 in 4 was ascribed to a variety of other causes (Figure 0.3). There was just one case of condensation.

The figures for 1987–89 show that of the 550 investigations carried out, 76 were on the element of floors and flooring. About one quarter of the 76 involved the floor as a whole, and three quarters involved only the finishes. Detachment of the finish was involved in 1 in 3 cases; cracking in nearly 1 in 4; entrapped water in nearly 1 in 8; and the remainder of 1 in 4 was due to other causes. There were two cases of condensation.

Comparison between the two periods by defect shows that, proportionately, the incidence of detachments has risen while that of cracking has fallen. The industry seems to be learning the lessons of entrapped water only slowly. However, some care must be exercised in interpreting these figures for BRE investigations since they are probably not completely representative of the true pattern of failures in the industry. For instance, entrapment of water leading to failure is fairly well documented, so most of these failures can be dealt with by professionals and contractors without specialist knowledge. Even so, the number of practitioners who are unaware of the long drying times for thick concrete bases to dry sufficiently to receive moisture sensitive floorings is surprising. Detachment of finishes is more variable and less well documented, so it may well be that it is for this reason that more cases are referred to BRE for specialist advice.

A further reason why care should be exercised in interpreting the figures is that relatively few cases referred to BRE are concerned with domestic buildings. While most of the cases in domestic buildings do not need specialist advice, failures in hospitals, shopping malls, offices, schools and factories may well need specialist advice.

In 1970–74, just over one quarter of all failures were attributed by BRE to the designer, just under half to site construction or its supervision, 1 in 12 to material defects, and about 1 in 10 to other causes (Figure 0.4).

In 1987–89, nearly two thirds of the failures were ascribed to be the responsibility of those on site, 1 in 8 to the designer, 1 in 7 to the material, 1 in 12 to unexpected user requirements, and the remainder to other agencies.

Before the early 1980s there had been a significant increase in the number of cases of breakdown of sand and cement floor screeds covered by relatively thin floorings. Responsibility for these failures could be ascribed approximately equally to design and specification and to workmanship. However, since that time there has been a dramatic fall in their incidence, largely as a result of the use of the BRE screed tester (see Chapter 4.1).

BRE defects database records

A comparatively large amount of detailed information on faults which occur in floors and flooring in housing is available in the BRE

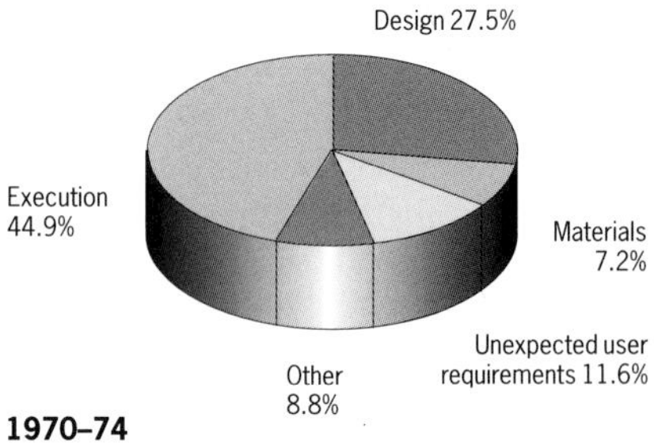

1970–74

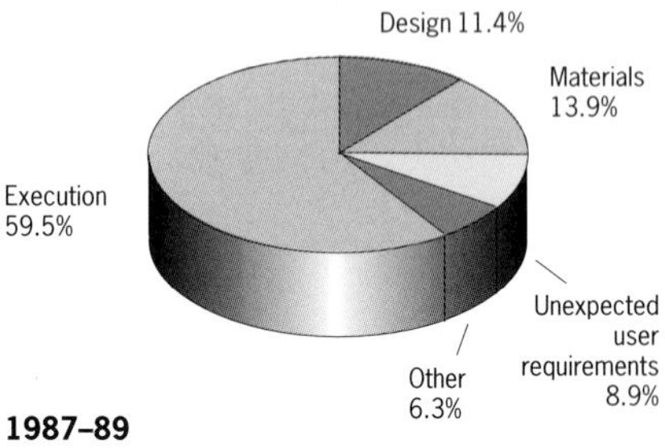

1987–89

Figure 0.4
BRE investigations of floors and flooring by responsibility for defect

Quality in Housing database. The database records items of non-compliance with requirements whatever their origin, whether building regulations, codes of practice, British and industry standards or other authoritative requirements. It also records actual inspections by BRE or by external consultants working under BRE supervision. Although the data were gathered over the years 1980–90, there are indications that the mistakes made then are recurring in current construction. It is possible to see with some confidence exactly where mistakes were being made and by whom, and therefore who needs further guidance, and also what form that guidance should take. The data have been analysed, according to dates of original construction of dwellings, broadly into two categories: those dwellings being built new at the time of inspection and those being refurbished. Only faults which occur frequently in the database have been listed. Faults which occur infrequently have been omitted.

Of the 1,073 faults concerned with floors and flooring in the database, which form about one quarter of the total of faults of all kinds, one third occur in the floor element itself, and the remainder, two thirds, at the junctions with other elements; for instance separating and external walls (Figure 0.5).

Figure 0.6 shows the distribution of faults by types of floor. The difference in figures for suspended timber floors reflects the relatively few suspended timber ground floors existing in the 1980s. These data are not representative of Scottish practice as comparatively few Scottish sites were examined.

From an examination of performance attribute affected (Figure 0.7), the greater proportion of items relating to strength and stability is most likely accounted for by lateral restraint problems which did not arise in older construction. As might be expected, sound and fire figure more extensively in older properties. It is difficult to account for the greater proportion of safety and habitability items in older construction.

Although it was only comparatively infrequently that building regulations were invoked for older properties, the database included records of when construction was unlikely to have complied with building regulations – in this case the earlier deemed-to-satisfy construction rather than provisions of later Approved Documents – in force at the time

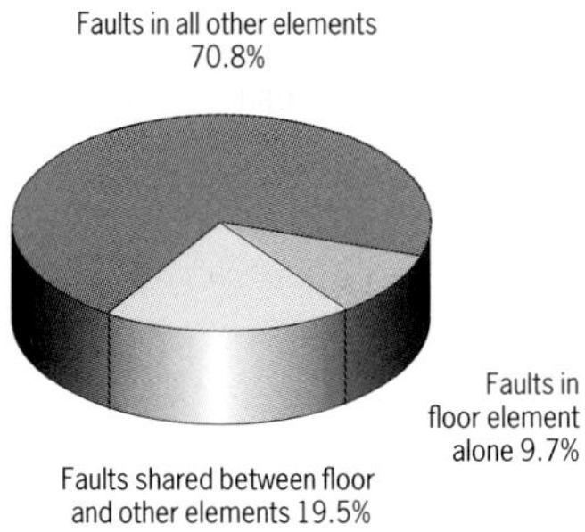

Figure 0.5
Faults in floors and flooring compared with faults in all other elements (from BRE Housing Defects Prevention Unit database)

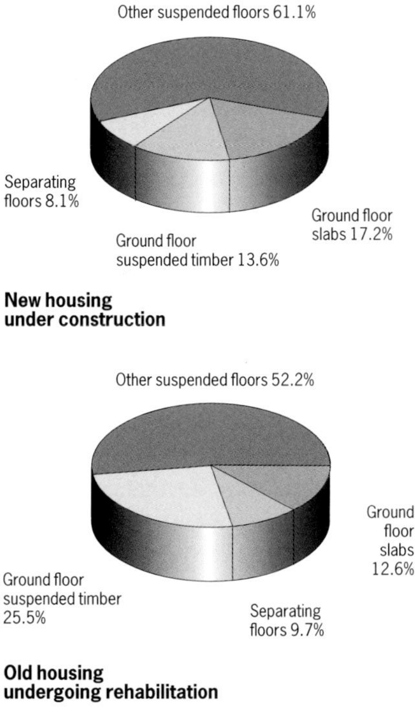

Figure 0.6
Faults in floors and flooring by types of floor (from BRE Housing Defects Prevention Unit database)

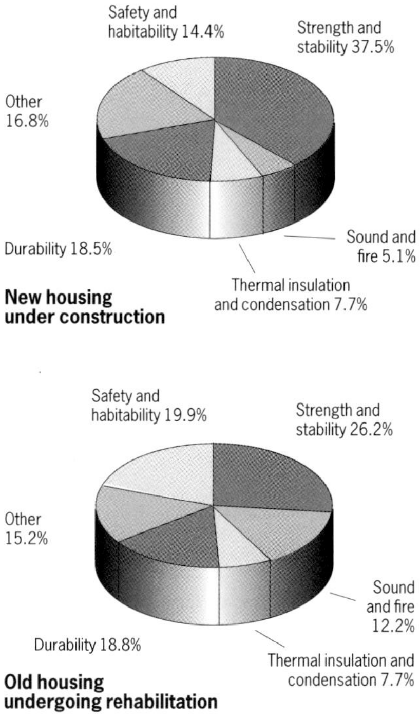

Figure 0.7
Faults in floors and flooring by performance attribute affected (from BRE Housing Defects Prevention Unit database)

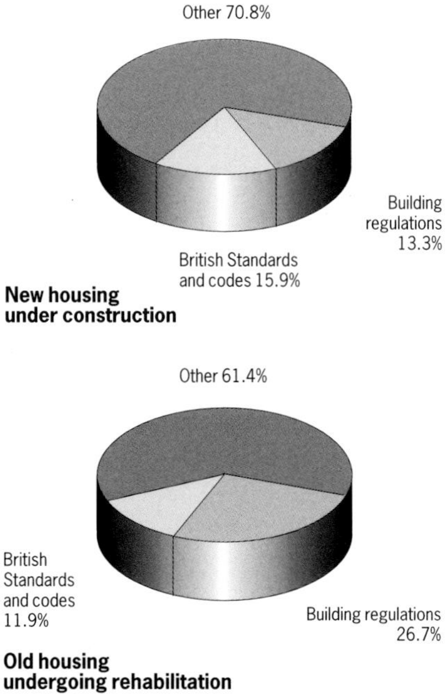

Figure 0.8

Faults in floors and flooring by authority contravened (from BRE Housing Defects Prevention Unit database). 'Other' includes Agrément certificates, industry requirements, client requirements etc, each of which forms a small proportion of the total

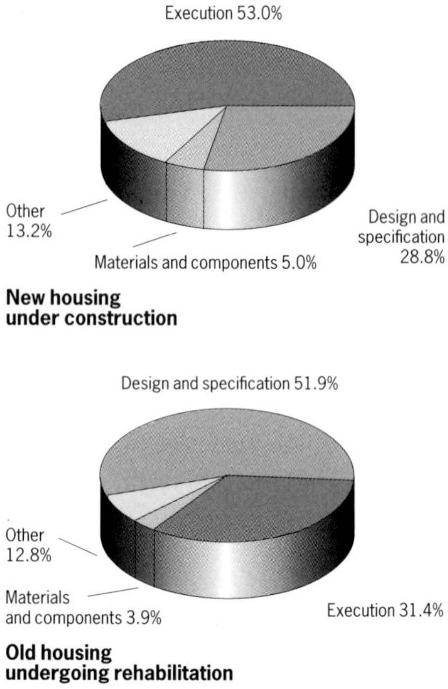

Figure 0.9

Faults in floors and flooring by responsibility for defect (from BRE Housing Defects Prevention Unit database)

the work was being carried out (Figure 0.8). There is, however, less excuse for non-compliance with building regulations in new construction. Approximately one eighth of all infringements of such requirements for newly built floors, as with other elements, have proved to be (in the view of BRE investigators) infringements of the building regulations (or strictly speaking, the provisions of deemed-to-satisfy construction) in force at the time the inspections took place. Where the provisions of the more recent Approved Documents were not followed, too high a proportion of the chosen solutions were considered not to comply with the underlying functional requirements.

There are considerable differences between the figures in the database for old and new dwellings with respect to design and site responsibility (Figure 0.9). Both new and old construction, though, show very small numbers relating to material and component quality – it is what people do with components and materials that is more important than their intrinsic quality.

For faults with their origin in execution it is estimated that a large majority were due to lack of care rather than to lack of knowledge.

Although in the site observations carried out by BRE there has been normally no weighting for the degree of importance of the consequences of a failure when it occurs, many failures in floors were potentially of a serious nature.

In common with faults in other elements, the majority of faults in floors and flooring of housing originate in practices acquired before the 1939–45 war. Of the total faults recorded, only about 1 in 8 of new-build, and 1 in 16 of older dwellings under rehabilitation, relate to practices introduced since the war. However, a different scenario is painted by evidence from a new-build survey of highly insulated dwellings (BRE Information Paper IP 3/93[1]); it points to misunderstanding of the serious consequences resulting from the lack of adequate ventilation and the need for elimination of cold bridges, in spite of the provisions in standards and in building regulations.

Summaries of the actual faults found are given in the appropriate sections of the chapters which follow.

House condition surveys

The 1991 English, 1993 Welsh, 1991 Scottish and 1991 Northern Ireland house condition surveys[2–5], which are the latest available, provide a limited amount of information about the numbers of dwellings having floors of different construction, and the proportions of those which show faults. Special analyses were carried out for this book on the data for England: the results follow. Although similar information was collected for the remainder of the UK, these have not been analysed here.

First, an important qualification must be made. The categories used by the *English House Condition Survey* comprised 'solid' and 'the remainder' (ie all other types of floor). Solid includes those floors made of concrete, whether suspended or groundbearing, and whether screeded or boarded; the remainder consists almost entirely of suspended timber joisted floors.

Ground floors

Taking ground floors first, of the nearly 16 million houses in England, of which nearly 2 million are bungalows, approximately 1 in 5 are of suspended timber construction over the whole of their area, and just under one third have suspended timber over a portion of their area and solid over the remainder; just over two fifths have solid floors over the whole of their area. These are figures for the whole stock. As might be expected, the vast majority of ground floor flats, of which there are over half a million, have solid floors. The few suspended examples, approximately 58,000, probably represent conversions.

There are differences in the proportions over the various age categories. Of the just over 4 million houses in the pre-1919 category, about half have some solid and some suspended floors, about one quarter have all solid, and the remainder all suspended. In the 1919–44 category, of the 3 million houses, more than 1 in 3 has suspended timber ground floors over the whole of its area, 1 in 7 has solid over its whole area, and the remainder, and therefore the majority, has part solid, part suspended. In the 1945–64 category, of nearly 3.5 million houses, about two thirds have solid ground floors, and of nearly 5 million in the post-1965 category, about three fifths have solid ground floors.

The number of floors having faults was also recorded. Of all ground floors, approximately 1 in 8 has faults and, in these instances, surface faults outnumber structural faults by nearly 10 to 1.

There is no comparable information for the rest of the UK, although for Scotland the proportion of suspended timber ground floors is expected to be very high.

Information from Housing Association Property Mutual (HAPM) indicates that in the early 1990s, of dwellings under construction and being monitored by HAPM, just under 1 in 50 had timber suspended ground floors, 1 in 10 ground supported slabs and a similar proportion were rafts, 1 in 3 precast concrete, 1 in 6 beam and block, and the remainder a mixture of beam and block and precast concrete.

Upper floors

As might be expected, the vast majority of houses have upper floors of suspended timber, although around 2 million in the total stock are of solid construction over the whole or part of their upper floors. A proportion of these will be of non-traditional system built construction in concrete or steel and concrete.

With regard to faults in upper floors, the picture is quite different from ground floors. Of those where it was possible to make an effective observation, which was in about one quarter of all houses inspected, nearly all the floors had faults, albeit about half of them being minor in character. Over all age categories, surface faults still outnumber structural faults by 9 to 1, but in pre-1919 houses, the figure is nearly 5 to 1. As might be expected, there is about twice the likelihood of finding structural faults in the upper floors of pre-1919 houses as for those in later age categories; this translates to about 1 in 27 of the just under 4 million houses (ignoring 78,000 bungalows) in this age category.

So far as recently built dwellings are concerned, information from HAPM indicates that in the early 1990s about half of the housing association dwellings monitored by it had timber suspended upper floors, 1 in 3 concrete, and the remainder a mixture of suspended timber and precast concrete.

Thoroughness of surveys

The work of the former Defects Prevention Unit at BRE showed clearly that insufficient time was spent on initial surveys carried out before houses were refurbished. Consequently, many faults were unidentified until work was well under way. Amongst the features most commonly requiring additional expenditure were suspended timber ground floors, affecting one third of the schemes examined.

Construction Quality Forum

The Construction Quality Forum Database, managed by BRE for the industry, contains few references to floors, though non-residential items are more numerous than residential. In the non-residential category, about 1 in 4 of all defects are related to the presence of moisture in floors and about the same proportion to inadequate preparation or insufficient attention to workmanship items during laying.

Of the 38 floor related items in the non-residential category for which costs were available, 14 cost between £1,000 and £5,000 to remedy, and 13 cost more than £5,000.

BRE publications on floors and flooring

Principles of modern building

Principles of modern building, Volume 2, Floors and roofs[6], was first published in 1962. In many respects it proved to be a milestone in the application of science to building construction. The chapters on floors are now showing their age, not only because the examples date from the 1950s and all the dimensions in the text are imperial, but also because performance requirements such as thermal insulation are minimal in today's terms. However, since the book deals with principles, in some respects it is less out of date than might be imagined.

Those general principles expressed in *Principles of modern building*, and which have stood the test of time, form the basis of the introductory sentences of some of the following chapters.

BRE Digests, Information Papers and Reports

Various BRE publications listed in the references and further reading lists at the end of the book provide a fairly comprehensive coverage of floors and flooring, although, since they have been published over a time span of many years, they can have received very little cross-referencing between them all. All these publications have been drawn upon to a considerable extent in this book, and it therefore provides a key to most BRE publications relevant to floors.

A number of BRE reports, particularly, contain sections relevant to floors and flooring, which may not be apparent from their titles. Those relevant to floors of housing include: *Assessing traditional housing for rehabilitation* (7), *Surveyor's check list for rehabilitation of traditional housing* (8), and *Quality in traditional housing*, Volume 2, an aid to design (9), and Volume 3, an aid to site inspection (10). Information on the floors of non-traditional housing is available in the many BRE Reports on particular systems.

Figure 0.10
Medieval ceramic tiles in need of conservation

Changes in construction practice over the years

Historical notes

Floors

The Romans used both suspended and groundbearing floors on a considerable scale. Perhaps one of the best known types was the hypocaust where closely spaced piers carried flagged or tiled decks, forming a plenum for the transmission of flue gases for room heating. The Romans also used timber floors, indicated by the sockets in walls for carrying joists.

In medieval times, ground floors in many dwellings consisted of rammed earth or clay. Upper floors in domestic construction were almost exclusively timber boards on timber joists. Ground floors of larger buildings were often covered with stone flags or ceramic tiles (Figure 0.10), and upper floors had stone or brick vaults (Figure 0.11).

The upper storeys of medieval military architecture mostly had stone vaulted floors or massive adzed timber beams carrying adzed timber boards. The ground floors often consisted of rammed earth which could not easily be cleaned, and which were renewed every ten or twenty years. During the middle ages there existed an official, commonly called the 'saltpetre man', who had the power to dig up the floors in buildings in order to obtain saltpetre for making gunpowder. The saltpetre content came from food scraps cast onto the floor.

In older buildings the design of floors was often carried out empirically using a great variety of materials and forms. From the earliest times, beams providing the primary means of support would be of massive single sections of timber. To give one example, the main beams or binders providing support to the bridging joists of the floor in the Hall of the Middle Temple in London are approximately 450 × 330 mm, spanning around 12 m, with bearings of around 600 mm. It used to be, and probably still for the most part is, virtually impossible to obtain an accurate measure of the carrying capacity of such beams of natural materials without testing to virtual destruction, which is clearly rarely acceptable. So judgement based on careful inspection needs to be used in assessing current serviceability.

As time progressed, so materials began to be modified – bolted composite or flitched timber beams,

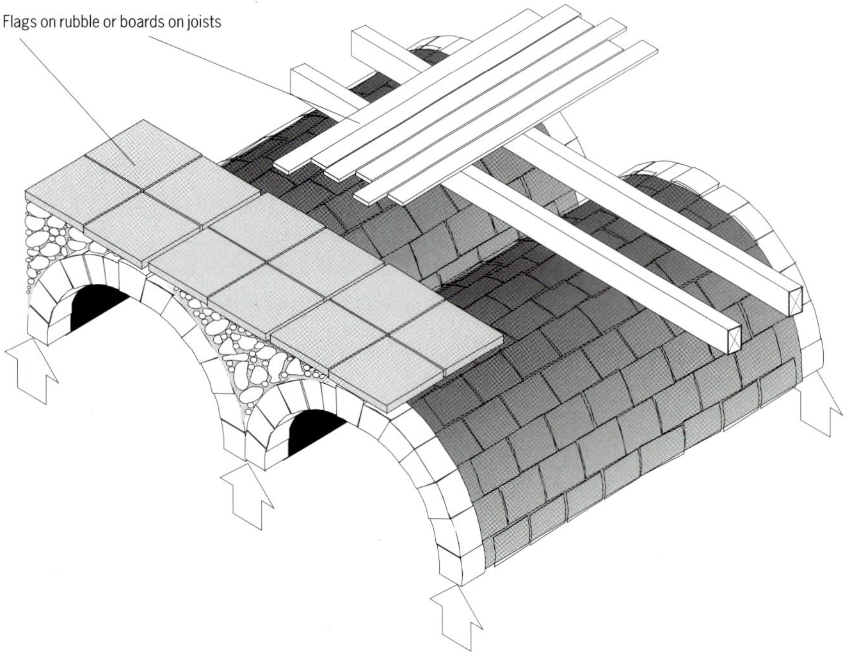

Figure 0.11
Barrel vaults have supported floors in many castles and cathedral crypts

steel flitched timber girders, trussed timber girders with wrought iron or steel ties, cast iron, wrought iron, steel and, in recent years, materials such as aluminium and laminated wood. This book is not the place to discuss design criteria in detail, suffice to say that as the material from which floors were constructed became more and more consistent, so it was possible to refine structural design calculations. With that refinement comes also a greater degree of reliance on scientific principles to assist in carrying out reassessments of the serviceability of existing construction.

The sixteenth to the nineteenth centuries saw only gradual improvements in techniques; for example increased spans for timber floors and of brick arches and barrels. But Victorian times saw massive changes. Until quite late in the nineteenth century, cast and wrought iron remained supreme for structural purposes. Then came the introduction of steel, the first steel joists being rolled in 1885 and the first steel framed building in London, the Ritz Hotel, was begun in 1904.

Portland cement had been invented in the 1820s, but its use was by no means widespread in reinforced concrete until the later years of the nineteenth century and the early years of the twentieth century. At about this time, floors in housing commonly consisted of suspended timber on sleeper walls, with kitchens and outhouses on the ground floor often having quarries laid directly on the earth or on a bed of ashes (Figure 0.12). Concrete received a big boost after the 1914–18 war, with the Ministry of Health allowing its use for housing with gravel or clinker aggregates, but not allowing the use of coke breeze. Nevertheless, a number of buildings were built with coke breeze aggregate concrete floors with disastrous effects on the integrity of steel joists incorporated within them. Concrete floors, widespread for public and commercial buildings, were, however, in the housing field, mainly confined to systems.

Flooring

The variety of floorings and floor coverings which has been available in recent times has been largely devised and produced in the last hundred years. Before the Industrial Revolution only a few floor coverings were available, and these were mainly confined to wood covered with rush mats or carpets, or were of stone flags, clay tiles and bricks often set on the bare earth. The use of stone has been common in public and ecclesiastical buildings throughout the civilised world from very early times. Marble goes back at least to Biblical times; in the Book of Esther it is mentioned that it was used in the palace of Shusan: '... *upon a pavement of red, and blue, and white, and black marble*'.

The Romans brought with them to Britain various types of flooring, including the jointless Ruberatio, prepared from crushed tiles and stone in a cementitious matrix, rammed and smoothed. Vitruvius describes several varieties of decorative floorings of coloured tile or mosaic (Figure 0.13), the art of which was virtually lost when the Romans left, such as the *opus tesselatum* which consisted of small cubes laid to form elaborate designs and the opus sectile of somewhat larger interlocking shapes. Vitruvius also describes the construction of the base upon which the flooring was to

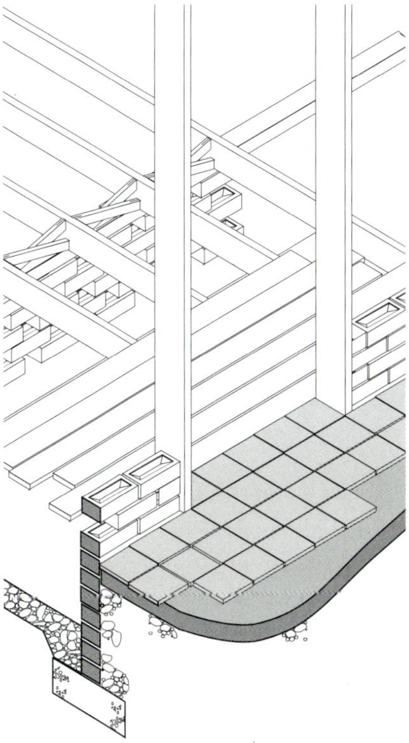

Figure 0.12
A suspended timber floor adjoining quarries laid on an ash bed

Figure 0.13
A mosaic floor from Bignor Roman Villa, Sussex (Photograph by permission of A Selkirk, *Current Archaeology*)

be laid so that defects could be avoided. *'In the case of upper floors, great attention must be given, lest any wall in the storey below is built right up to the pavement; it is rather to stop short, and have the joists carried free above it. For when the wall is taken up solid, if the flooring above dries or sags as it settles, the wall being of solid structure, necessarily causes cracks, right and left of it, in the pavements above.'* He goes on: *'When it is laid, and the proper fall is adjusted, it is to be rubbed down so that, if the pavement is of marble, no projecting edges may arise in the diamonds or triangles or squares or hexagons, but the adjustment of the joints is to be level one with another. If it is mosaic, the edges of the tesseræ are all to be level. For when the edges are not even, the rubbing down will be imperfect.'* Now as then.

In the middle of the reign of Henry III plain clay tiles were introduced and, towards the end of his reign, decorated tiles also, mostly imported from the Low Countries. The mosaic tile floors of Fountains Abbey date from 1220 to 1247. Tiles with an inlaid design of pale slip poured into stamped impressions on the unfired clay body, dried, scraped clean to reveal clear arrises, glazed and then fired all together in one process were known as encaustic tiles. They are still made by essentially the same hand process (Figure 7.7 in Chapter 7.1). Cheaper varieties of floor tiles were made by printing a design in pale slip on the flat body of the tile. Such designs self-evidently were not so hardwearing as the inlaid variety. The tiles were made in large numbers in north Warwickshire and Nottinghamshire.

The main developments in tile, mosaic and tessellated floorings came in the nineteenth century. British Patent 8042/1839, granted to Singer and Pether, for producing materials for mosaic work, stated: *'Our invention consists in the mode of producing small rectilinear pieces of pottery, porcelain or other plastic material, for the purpose of making mosaic work, and in the manner of combining them by calcareous or other cements, so as to produce slabs with ornamental devices'.* Such floors, as for example that designed by Barry and laid by the patentees in the Reform Club, London, could be laid by tradesmen who were far less skilled than formerly.

Tapestry carpets for use on floors were introduced probably in the fourteenth century, though were rare until the next century, and not common until the eighteenth. Before the fourteenth century they were probably too expensive to use on a wide scale as floorings and they were much more commonly used as wall hangings. Rush matting, as opposed to loose rushes, was not used in Britain until the reign of Henry VIII, though the ancient Egyptians used woven rush mats for flooring. Pile carpets were introduced from the Middle East after around 1500, at first as a table decoration before they transferred to the floor. Manufacture was introduced to the British Isles by the Earl of Ormonde in the early sixteenth century who brought from Flanders artificers whom he employed at Kilkenny. In 1756, the Royal Society of Arts offered a premium for the development of carpet manufacture from the hand knotting which was then common, and this stimulated the development of the industry in the UK.

The eighteenth century saw the development of 'floorcloth', a jute canvas sized and hand painted with several layers of linseed oil, and the whole overprinted with a coloured pattern (British Patent No 336/1763). This material was the forerunner of linoleum and probably made a relatively hygienic floor covering available to the majority of the population. Linoleum was first patented by Frederick Walton in 1860, but, by the time his patent lapsed in 1876, various other people made similar materials under different names.

Coating of fabrics by rubber was first applied to flooring in the middle of the nineteenth century (British Patent 9987/1843). Then came a rubber and cork mixture (British Patent 10,054/1844) which was given the name of Kamptulicon. It was used by Barry in the Houses of Parliament, though gradually fell into disuse with the development of linoleum.

Felts saturated with bitumen first appeared on the UK market in 1910 as a substitute for linoleum, and proved remarkably resistant to wear

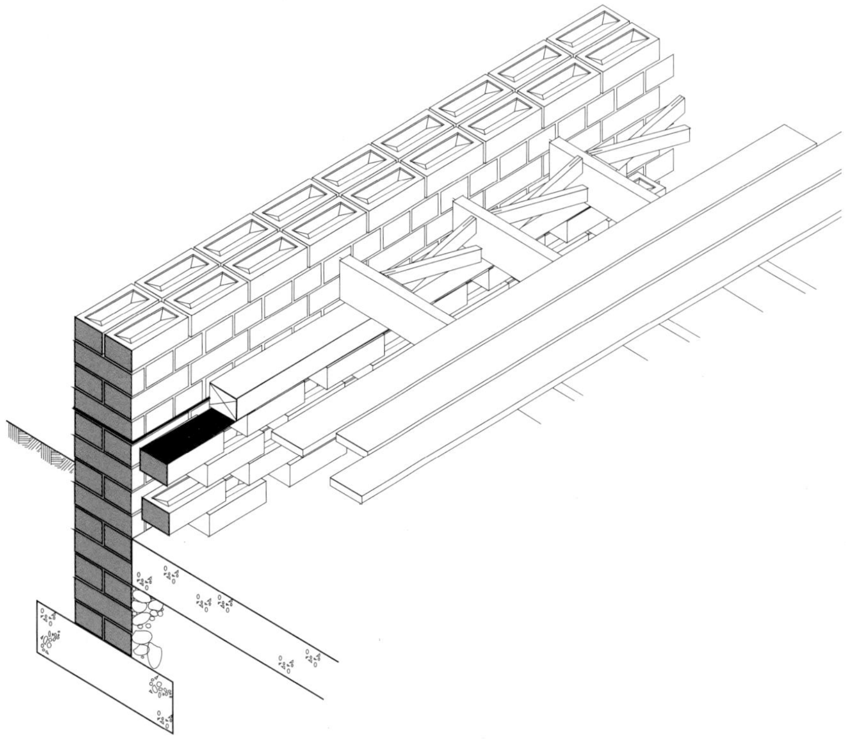

Figure 0.14
A typical suspended timber ground floor

and tear under foot.

Probably the first of the jointless floors to be developed since Roman times was the so-called grip, which was a composition of lime and ashes laid damp to a thickness of 100 to 125 mm and beaten flat with a wooden spade The finished floor was said to be relatively durable. Some of these old in situ floors contained other kinds of aggregates; for example crushed sea shells. Another traditional floor was one of puddled clay, trodden or beaten, which was also said to provide a relatively durable surface. In areas receiving hard wear, such as thresholds, rough stone flags would be set into the clay.

The reaction between magnesia and magnesium chloride was discovered in 1867, but it was 20 years before the discovery was applied to flooring following the discovery of large deposits of magnesia. The first magnesite floor in Britain was probably laid in the 1890s following its introduction to Austria and Germany.

Calcium sulfate flooring, using finely ground catalysed anhydrite with an aggregate of crushed rock anhydrite, has been used since the 1920s in the UK, though not on a substantial scale. It has mainly been laid in thicknesses of around 15 mm.

Summary of main changes in common practices since the end of the 1939–45 war

Many domestic scale buildings built during the interwar period had suspended timber floors (Figure 0.14), but a change to solid floors on a major scale took place in England following the 1939–45 war, though not to anything like the same extent in Scotland, Wales and Northern Ireland. One of the English New Towns was estimated to have 97% of new houses with groundbearing concrete floors. At this time a variety of floor finishes was used, the most common being magnesite, pitchmastic, mastic asphalt and linoleum laid directly onto concrete bases. However, following the introduction of thermoplastic tiles in about 1946, the use of all except linoleum had disappeared from domestic construction by around 1960. Because thermoplastic tiles and the solvent bitumen adhesives used to fix them were moderately tolerant of moisture rising from below, it was common to lay this type of flooring directly onto a concrete base, usually 4 inches (100 mm) thick, without any DPM or screed. Except on very wet sites, this construction worked well. The provision of DPMs below concrete slabs did not become common until sheet polyethylene became generally available in a suitable size and thickness in about 1960. Prior to that date, where DPMs were required to protect moisture sensitive floorings, they were formed by applying various brush or hot applied tar and bitumen based products to the top of the base concrete and covering it with a screed or board based products. The use of pitchmastic and mastic asphalt as both screed and surface DPM was also common.

In mining areas the use of colliery shale, or in other areas of pyritic shale, as hardcore under many of these solid concrete floors led to a number of major failures in slabs – of swelling due to sulfate attack. To give one example, in one of the New Towns in the late 1970s, over 600 houses were estimated to have problems caused by heaving of the ground floor slabs[11]: in some cases the amount of heave was of the order of 150 mm. The cost of remedial work was put at over £2 million (at 1980 prices).

Another significant change has been in the number of granolithic floors used – or perhaps more correctly stated as a reduction in the number of granolithic flooring cases investigated by BRE. In 1970–74, roughly one quarter of all cases involved granolithic flooring (Figure 0.15). The main problems seen were cracking, curling and lifting of separate bonded and unbonded toppings caused by drying shrinkage. Such floors are now little used having been replaced by medium or high strength concrete bases power floated and power trowelled to achieve the required abrasion resistance, or by cementitious

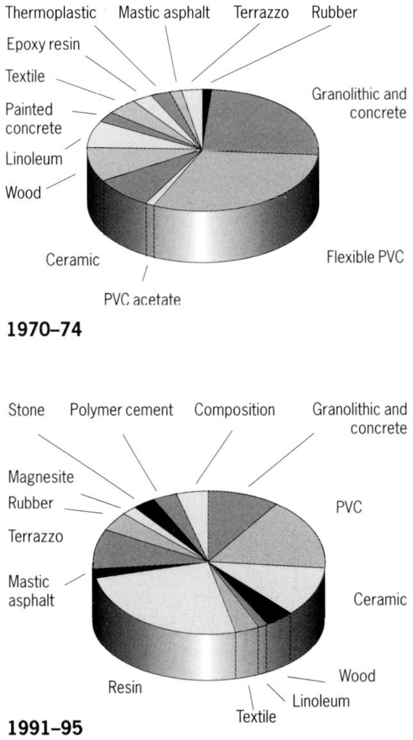

Figure 0.15
Flooring materials used in cases investigated by BRE

polymer screeds or resin screeds where wearing screeds are required.

Since 1980, a substantial proportion of the flooring problems investigated by BRE Advisory Service has been in health service buildings, mainly hospitals (Figure 0.16). A variety of problems have occurred including excess moisture, rippling in sheets and tiles, plasticiser migration into adhesives, shrinkage caused by unsuitable cleaning fluids, ceramic tile failures in kitchens, as well as the failure of anti-static floorings in operating theatres.

More recently, increases in both performance requirements and very large areas for floorings found in more sophisticated buildings (such as shopping malls) and other heavily used public access areas (like airport concourses and railway stations) have led to a number of problems. Failures in these floors can lead to enormous expense in rectification or replacement.

Probably the most serious and extensive failures, costing several millions of pounds for replacement and the disruption involved, have been the crushing and breakdown under the effects of traffic of sand and cement floor screeds covered by thin floorings. This problem rarely occurred before 1965, but, with the introduction of thin flexible PVC flooring, a change in the way floor screeds were laid occurred. To achieve the required tolerance in levels, and the relatively flat closed-in surface required by these floorings, screeds began to be laid with significantly less mixing water – the semi-dry screed had arrived. Low moisture content screed material has poor workability, and such mixes were often poorly compacted on the subfloor leading to low strength and subsequent breakdown in service. In addition, it was found that normal drum type concrete mixers did not thoroughly mix low moisture content screed material.

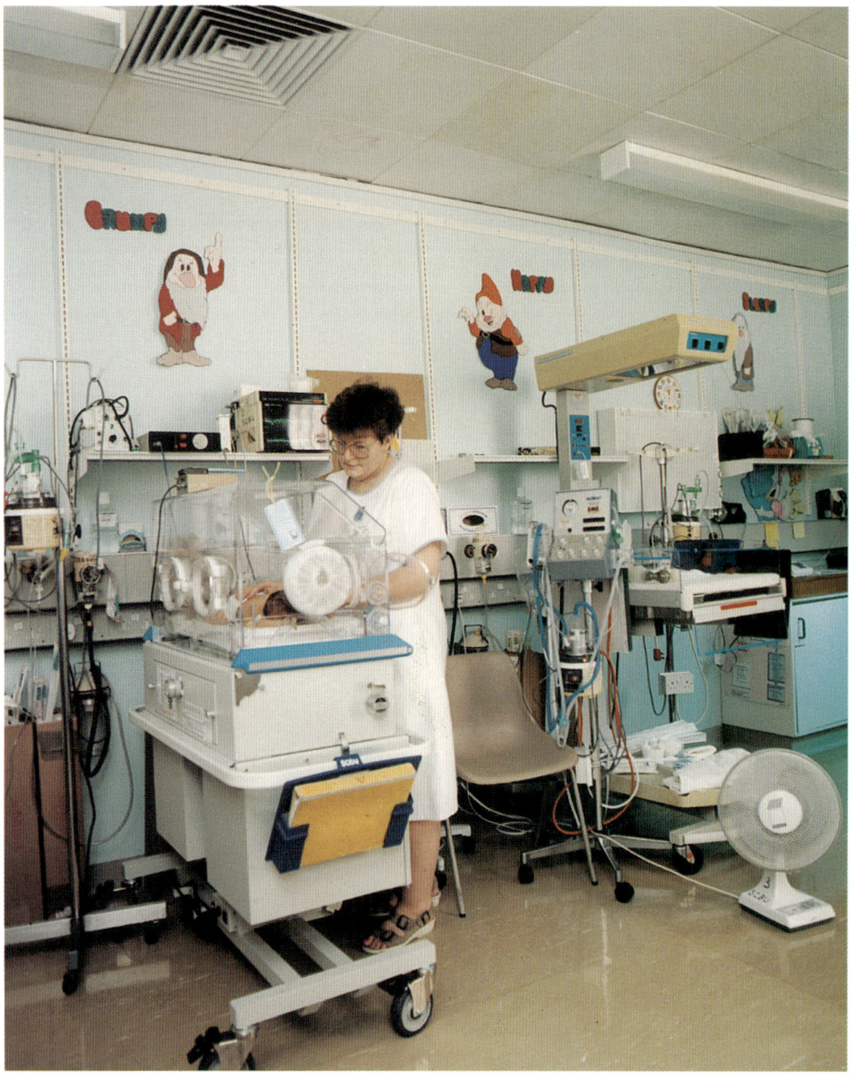

Figure 0.16
Hospitals and health service buildings are the source of many problems with floors and their coverings

To enable screeds to be assessed for their ability to carry traffic before they were covered with floorings, the BRE screed tester was developed. This instrument immediately gained widespread acceptance. It is currently incorporated into all British Standards concerned with floor screeds to measure what previously was called the soundness test, but will in future be called the in situ crushing resistance test. The change of name for the test is required as the method is likely to be accepted throughout Europe and the term soundness is not easily translated into other languages. The BRE screed tester is described in detail in a feature panel at the end of Chapter 4.1.

In situ thermosetting resin floorings (epoxy, polyurethanes etc) were introduced in the early 1960s. These materials have slowly gained ground, mainly in the many special industrial buildings used for food provisioning, pharmaceuticals and electronics, largely at the expense of ceramic tiles and to a lesser extent mastic asphalt. Resin floorings have brought their own types of failures which have required research to resolve. Problems have included detachment from the substrate, chemical attack, colour changes, surface bloom and the unique problems of blistering caused by osmosis.

If BRE records are considered to show a representative reflection of the overall quality currently being achieved in the industry, the element of floors and flooring compares very favourably with other elements in achieved quality. Nevertheless, many of the faults that do occur are avoidable, and, moreover, occur in sufficient numbers to merit the issue of further guidance and advice.

Chapter 1 The basic functions of all floors

The range of functions to be taken into account in the design of floors is not as extensive as in the case of those elements forming part of the external envelope of a building. Nevertheless there are several, such as control of noise and wear, which can be of crucial importance for floors (Figure 1.1).

The range of functions is considered in general in this main chapter as they apply to all floors and ceilings. In the later chapters on the different floors and floor coverings, each type is considered in relation to these functions in greater detail.

Figure 1.1
A highly polished decorative floor in a shopping mall. Such floors need to withstand onerous conditions of service for many years. Commercial pressures mean that periods out-of-service (eg for maintenance) must be minimal

Chapter 1.1 Strength and stability

As was noted in the companion book on roofs and roofing, and as *Principles of modern building*[6] pointed out, there are three primary structural systems that can be adopted for spanning across space: the chain or catenary, the arch and the beam.

The characteristic feature of the first of these, the chain, is that all its parts are in tension; of the arch, that they are all in compression; and of the beam, that the top fibres are in compression and the bottom in tension.

Unlike roofs, however, it is the last two methods rather than the first which are usually employed for floors. The requirement for uninterrupted clear space above the deck, coupled with inherent flexibility in the support system, effectively rules out the catenary for use in floors.

The **arch**, in the form of the barrel vault, has been used in Britain since Roman times for spanning comparatively large spaces such as the crypts of major churches to carry the floors of the naves and chancels above. As in the case of roof vaults, the materials, whether brick or stone, from which the vault is constructed need to be suitable to resist the large compressive forces which may be experienced. Later medieval church crypts had more complex intersecting vaults. In later times still, the shallow domed floor was introduced, and, in the absence of ring beams to resolve the horizontal thrust, massive abutments were provided. In parallel with the later barrels came the rather more slender ribs of the intersecting voussoir arches, infilled with stone slabs. Such arches stand up well given that the abutments do not move. Pippard and Chitty quote Hooke as saying: *'As the continuous flexible arch hangs downward so will the contiguous rigid stand up inverted'*[12].

The **beam**, however, provides the most versatile solutions for the whole range of spans for floors, with solid sections used over the shorter spans, and progressively lighter and structurally more efficient solutions as the span increases: from perforation or hollow coring of beams and infill decks to web construction over the largest spans. As in the case of roofs, much attention has been given to the development of the most efficient sections of beams, often using a combination of different profiles, sections and reinforcement techniques to exploit the characteristics of the chosen materials.

Again, as in the case of roofs, the examination of the efficiencies and economics of each combination is totally beyond the scope of this book. All that can be done is to provide information relating to the very commonest of solutions, with perhaps a brief mention of a few of the more infrequently found solutions.

It is, of course, rare that floors can be considered separately from the remainder of the structure since they usually contribute significantly to the overall strength and stability of the structure (Figure 1.2). Nevertheless, it is the correct assessment of the residual strength and serviceability of floors which is vital to their

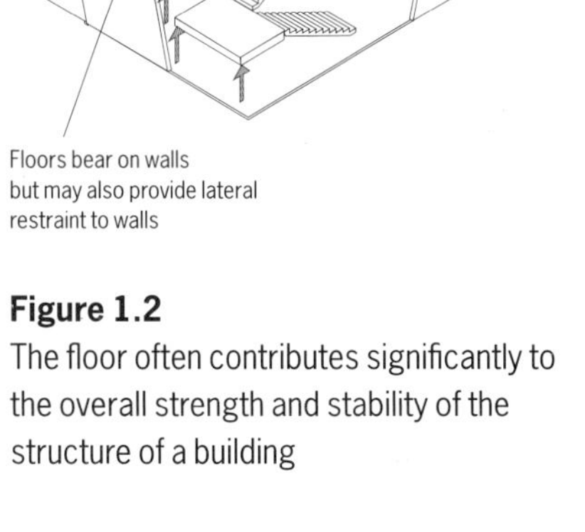

Figure 1.2
The floor often contributes significantly to the overall strength and stability of the structure of a building

continued safe use in buildings.

Even smaller domestic scale buildings of loadbearing brick often rely on the floors to provide lateral restraint to the external walls. All that this chapter can do is to draw attention to some of the more important considerations in relation to the contribution which floors make to the structure as a whole, but not to provide sufficient information to allow the structural design of floors to be carried out.

The most straightforward structural design is to be found in the ground supported slab, although even in this case it is possible, but rare, to find errors in specification or construction which will lead to failure. Floors spanning between supporting walls or beams are another matter entirely.

However, all floors need to be sufficiently strong to carry the self weight of the structure together with imposed loads; for example those due to furniture (Figure 1.3), equipment or the occupants of the building.

Current requirements as far as the structure of floors is concerned are embodied in the various national building regulations[13–15]. Taking Requirement A1 of Schedule 1 of the Building Regulations 1991 as an example:

'*(1) The building shall be constructed so that the combined dead, imposed and wind loads are sustained and transmitted by it to the ground*

(a) safely; and

(b) without causing such deflection or deformation of any part of the building, or such movement of the ground, as will impair the stability of any part of another building

(2) In assessing whether a building complies with sub paragraph (1) regard shall be had to the imposed and wind loads to which it is likely to be subjected in the ordinary course of its use for the purpose for which it is intended.'

Structural design of floors for buildings in this category is covered by the main British Standard codes of practice for the various materials:

- steel to BS 5950:Parts 1–9[16]
- concrete to BS 8110:Parts 1–3[17]
- timber to BS 5268:various Parts[18]

Loadings

As *Principles of modern building* has noted, the main function of a floor is to support the loads placed on it during the life of the building. With some types of building, loadings are easy to predict and unlikely to increase with time. With others, particularly with industrial buildings, change of occupancy or developments in production may require floor strengths to be increased from those originally considered adequate. Strengthening an existing floor is normally inconvenient, difficult and expensive. On the other hand, design for improbably high loadings to allow for change will result in excessive cost of the floor construction itself, of the supporting framework or walling, and of the foundations.

Design values of floor loads for current use are specified in BS 6399:Part 1[19]. This code applies to existing buildings where a change of use is contemplated; also alterations and additions to existing buildings.

Dead loads include the weight of the structure of the floor, calculated from the values given in BS 648[20], or, better still, from known actual weights. The load calculations should include all tanks etc filled to capacity – in fact all permanent construction including services of a permanent nature.

Imposed loads are all those loads arising from the use of the building including moveable partitions but excluding partitions of a permanent nature; wind loads are also excluded. There are reduction factors to take account of the likelihood that all floors will not be loaded to the maximum values simultaneously, but these are not always applicable.

The structural codes over the years have given uniformly distributed loads and point loads for residential buildings; institutional buildings; offices, banks etc; halls, libraries and theatres etc; shops; workshops and factories; warehouses; and garages and car parks. The values for uniformly distributed loads, for example, have

Figure 1.3
Office furniture being weighed in the BRE office floor loading survey carried out in 1966. Overloading is only an issue where bulk storage exceeds the original design loads

Figure 1.4
Steel wheels cause severe wear on floorings

ranged from 1.5 kN/m^2 for bedrooms and 2 kN/m^2 for most other domestic scale accommodation excluding circulation areas, up to 7.5 kN/m^2 for boiler rooms and such like. Ceiling structures with access provide for a uniformly distributed load of 0.25 kN/m^2 with an additional point load, though ceilings without access do not need such allowance. The code was revised in 1996.

In the case of warehouses, the minimum loads given in the code may not be adequate for all circumstances. A BRE report in 1987 concluded that there was such a large variability in possible loads in warehouses that the use of a single minimum imposed load in design could result in a significant proportion of warehouses being either excessively expensive or structurally inadequate. Steel stockholders were reported to load their warehouses to 80–90 kN/m^2 for considerable periods of time and this caused settlements in ground supported slabs (*Floor loading in warehouses: a review* [21]). This Report also pointed out that it was common practice at the end of last century to design warehouses for 13.4 kN/m^2, but after 1909 the value was lowered to 10.7 kN/m^2.

In addition, the loads from forklift or other kinds of mechanical handling vehicles, especially those with solid wheels, can be considerable. As one example, turret trucks with solid wheels, when stationary and fully loaded, give contact pressures of up to 12 500 kN/m^2, which has obvious implications for choice of floor finish (Figure 1.4).

Principles of modern building noted that the likelihood of problems from local distress in a floor was low with a floor such as one of solid reinforced concrete which is capable of effective lateral distribution of loads, whereas the effect of concentrated loads might be more serious where the floor is built up from a series of separate adjacent beams. The extent of load sharing between beams depends on the character and extent of the connections between them formed, for example, by grouting or by a surface layer such as a screed or other decking. In a timber floor the load is shared between joists by the action of the floorboarding. In domestic scale buildings, no example of serious structural failure was recorded in the Quality in Housing database which was not accompanied by other factors which directly weakened the floor, such as rot.

In respect of excessive loading and consequent potential displacement of floor beams and girders, perhaps those most at risk are bresummers (or breastsummers or summer beams), where the loading, characteristically, is asymmetric. It should also not be forgotten that loads due to rehabilitation and replacement work must be taken into account. BRE investigators have seen a number of cases where floors have been overloaded by materials temporarily stored before use.

BRE Good Building Guide GBG 10[22] shows how to provide temporary support for floors when replacing lintels or creating or enlarging openings in external walls. Before starting work a check must be made on the size of load which will bear on the temporary supports which will be carried on the floor (Figure 1.5).

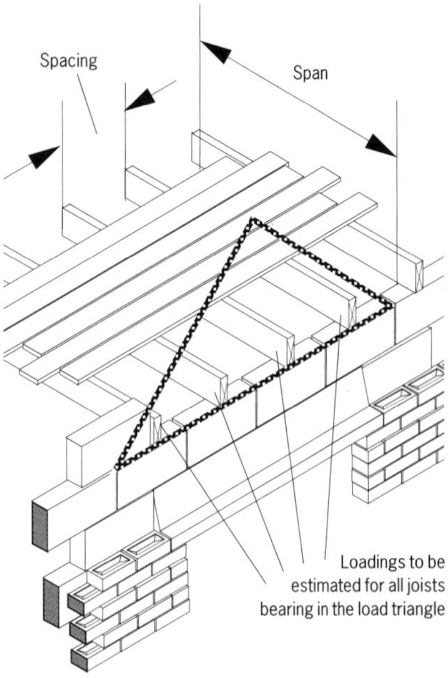

Figure 1.5
Estimating loads from timber suspended floors when needling for lintel replacement

Lateral support and restraint

Lateral support is often required to be provided by floors to walls, and such requirements are set out for new buildings; for example in Table 11 of Approved Document A of the Building Regulations 1991. All external and compartment or separating walls greater than 3 m long will require lateral support by every floor forming a junction with the supported wall; and all internal loadbearing walls of whatever length will require support from the floor at the top of each storey.

Straps may not be needed if the floor has an adequate bearing on the wall – for example 90 mm in the cases both of the bearing of timber joists and of the bearing of concrete joists.

BRE site studies have demonstrated that these requirements have not always been complied with in practice, and surveyors should be aware that a proportion of relatively new construction will be deficient in this respect. Information from Housing Association Property Mutual indicates that lateral restraint strapping was not provided in 1 in 5 cases where such strapping should have been provided in accordance with BS 8103:Part 1[23] and in reinforced concrete suspended ground floors where, for example, the floor was more than 1 m above ground level (Figure 1.6).

Older buildings with much thicker loadbearing walls than have become common in recent years may not need special provision for lateral restraint. Indeed, where these buildings have adequately withstood the test of time, and no significant alterations are proposed, there is little point in providing strapping.

Material change of building use

It is important to carry out an appraisal of the structural design of floors where there is a proposed change of use of the building. Indeed, compliance with the building regulations for England and Wales is required in particular circumstances.

Load testing

Load testing of a structure involves applying test loads to an existing structure to determine whether it is satisfactory (BRE Digest 402[24]). Loads tests may be needed when there is a change of use of the structure or when deterioration of the structure is suspected, though testing is also used when it is apparent that original specifications have not been complied with. Load testing may even be the only way of assessing the serviceability of some existing structures where materials testing and calculations provide insufficient evidence.

When load testing is applied to floors which consist of separate components, and which are or are not tied or cemented together, it may be necessary to isolate sections so that load sharing between components is eliminated and it is certain that the full intended load can be carried. Alternatively, the whole floor may be loaded, as is done with integrally cast floors.

Since it is also necessary to eliminate the effects of variables such as temperature in recording resulting deflections and rotation, instrumentation is extensive.

Load testing is expensive and is used only infrequently in structural appraisal.

Impact resistance

The importance of impact resistance, both for the floor as a whole as well as the flooring, depends on particular circumstances; but it is generally the case that the solidity of the deck or the bedding substrate within the floor will be one of the most important factors in the resistance of the finish to indentation.

Impact resistance is not usually the most critical factor in the performance of floorings, and it is probably unrealistic to expect that under accidental impacts the flooring will not sustain some damage, though the floor itself should not be in danger of collapse.

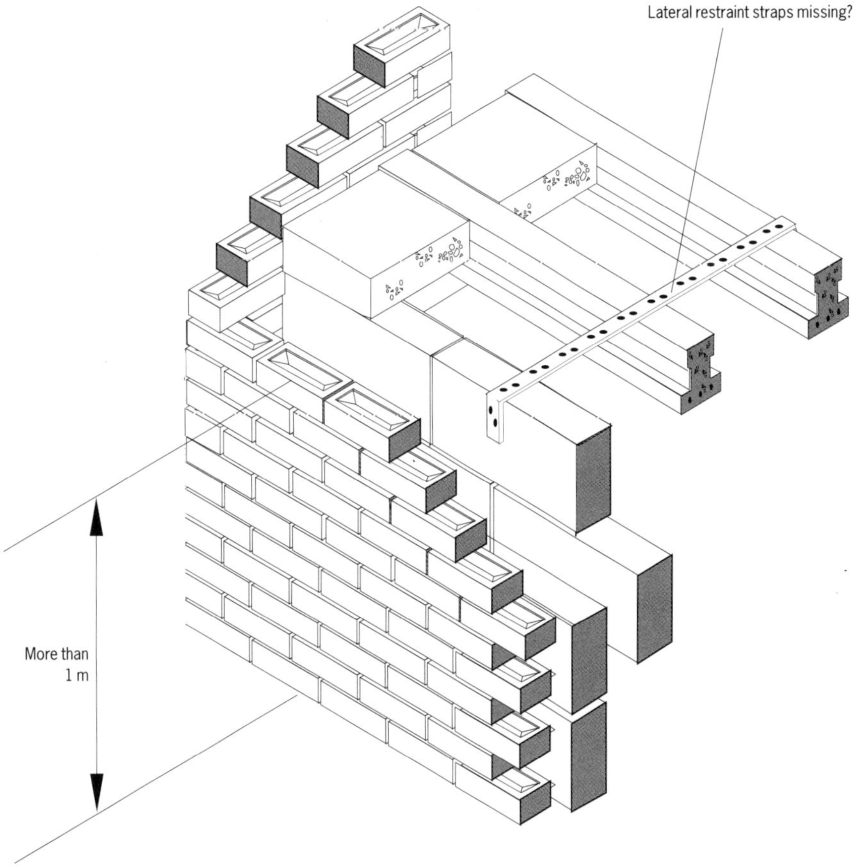

Figure 1.6
Strapping should be provided in reinforced concrete suspended ground floors where the floor is more than 1 m above ground level

Indentation resistance and recovery

It may not matter that a floor finish indents under load, provided that it recovers when the load is removed; materials vary greatly in their recovery after loading. (The properties of individual materials are discussed in Chapters 5–8.) Surface texture and pattern may help to disguise indentations which do not fully recover. It is fortunate that fashions in footwear have changed so as to reduce the common incidence of indentation by stiletto heels.

In some situations, indentations may collect aggressive fluids and become centres of erosion, but what is tolerable in any particular situation is a matter for individual decision.

Resistance to indentation (poinçonnement) has been considered by the European Union of Agrément (UEAtc) in relation to thin floorings used in domestic scale buildings (UEAtc Method of Assessment and Test No 2[(25)]). Tests are designed to provide assessment of the effects of dynamic and static furniture loads, falling objects, and the action of sharply pointed footwear. The Method of Assessment and Test is examined further in Chapter 1.9.

Chapter 1.2 Dimensional stability

This chapter covers movements, vibrations and deflections. Particular mention is made of thermal and moisture movements, and of dynamic movements from foot and small wheeled traffic which affect the comfort of building users.

Movements

Expansion and contraction of any part of the building fabric subjected to variations of moisture content or temperature will have the potential to cause problems if not accommodated in the design. As a general rule, all common building materials will be subject to thermal expansion and contraction. So far as materials used in floors and flooring are concerned, it is the larger components which need most consideration, especially where the building is only intermittently heated. Fortunately, though, movements will be much smaller than in roofs, for instance, because the internal environment will be that much less susceptible to external factors such as weather.

Moisture movement is mainly a property of porous materials; thus concretes have reversible movements in the range 0.02–0.2%, and softwoods sawn tangentially across the grain in the range 0.6–2.6%, and sawn radially in the range 0.45–2.0%. Concretes also have irreversible drying and carbonation shrinkage in the range 0.02–0.1%. In most real situations there is some restraint, though rarely complete restraint, afforded to materials undergoing such movements. Values for both moisture and thermal movements are given for particular materials in appropriate later chapters.

Most substrates under floorings are affected to a greater or lesser degree, and the effects of movement can be sufficient to cause rippling in floorings if sufficient precautions are taken to prevent it.

Where two differing materials are joined, differential movement can occur, which usually exacerbates the problem. The best practice is to try to accommodate movements at the smallest and most elemental level, provided this does not prejudice the structural function. Thus, provided floor components are simply shaped, not too thick, not too large, have movement tolerant fixings and have joints at their periphery which accommodate movement, the movement strains and the corresponding loads (stresses) are generally small and can be borne by the element or deck. With large structures, if one or more of these requirements is not met, stresses may accumulate over large areas and damage may result. In small structures such as detached domestic houses, it is often possible to omit explicit movement design and depend on restraint from other elements to accommodate movement loads, though floors in larger span buildings with higher loadings are another matter entirely. In timber floors the relatively high

The mechanism and causes of rippling in flooring

Rippling occurs over discontinuities in substrates. These discontinuities are normally either:

- daywork joints
- contraction joints
- shrinkage cracks
- joints between timber boards or sheets

The substrate over which rippling most often occurs is the unbonded sand and cement screed laid directly over a DPM, and where the screed has had an excessively long drying time: invariably six months-plus. The problem can also be found over a concrete or timber substrate.

The mechanism depends on the discontinuities or gaps in the substrate being at their maximum width (either through low temperature or drying shrinkage) at the time of laying flooring. Ambient conditions then change, the substrate expands, the gaps close, and the flooring can do no other than form a ripple over the discontinuities.

The cause of change in a substrate that leads to rippling may be one or more of the following:

- expansion of the substrate due to moisture, either introduced from an underlayment or from a water based adhesive
- moisture redistribution within a thick slab, where the top layer is very much drier than the bottom at the time of laying the flooring, and subsequently the moisture rises to cause expansion in the upper layer
- laying flooring at low temperatures. When the building is then brought up to its running temperature, the substrate expands and closes the gaps
- laying flooring on a timber substrate when the timber is at a low moisture content. Subsequent adjustment to ambient conditions causes the timber to swell and the joints to close

The floorings mainly affected are:

- flexible and semi-flexible PVC
- linoleum
- some relatively thin resins

moisture movement across the grain can lead to appreciable shrinkage in the section, causing gaps to open under skirtings.

Progressive movement may conceivably occur at the edges of floors adjacent to external walls, mainly as a result of ratcheting. Where movement in a material resulting from contraction, say drying shrinkage, causes cracking, debris can fill the crack. Expansion cannot then close the crack because of the debris, in turn causing the material, and perhaps adjacent materials or elements, to move. Successive contraction and expansion movements, with more debris accumulating in the cracks, results in ratcheting. Design information on structural movement is given in BRE Digests 227–229[26].

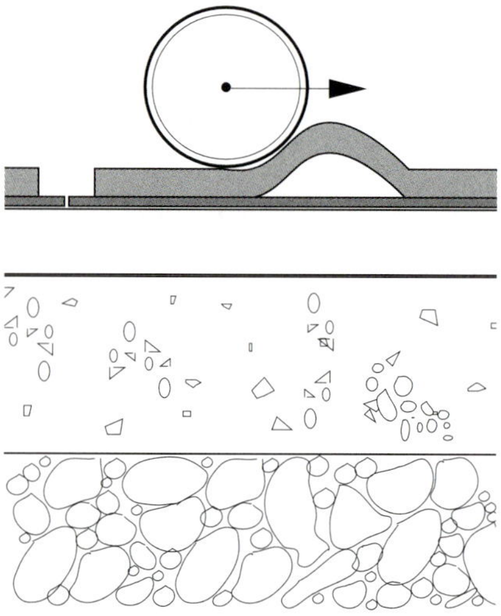

Figure 1.7
The 'bow wave' effect

The 'bow wave' effect

With certain thinner resilient floorings, the effect of heavy wheeled loads on the material leads to a tendency for the material to be squashed and to rise slightly in front of the wheel. This causes a tensile force tending to lift the material from the adhesive. Where the tensile force exceeds that of the adhesion or bond strength, detachment occurs, leaving the material subject to further deterioration (Figure 1.7).

The adhesive may retain some tackiness so that the flooring is restuck by the passage of the wheels, but it will eventually fail again.

Figure 1.8
Vibration is a common problem in converted buildings

Vibrations

Dynamic movements and vibrations induced by foot traffic need to be taken into account, particularly whenever change of use is contemplated (Figure 1.8).

Long span lightweight floors are especially susceptible to vibrations which can be detected by building occupants. Although vibrations produced by people walking on the floor can prove annoying to other users, these do not create most concern. It is when floors are used for dancing or organised keep fit exercises – with people moving in unison, giving rise to resonance in the floor, and which raise potential safety rather than serviceability issues – that the greatest problems may be encountered. The phenomenon has been generally appreciated for many years – witness the military instruction to break step when marching over bridges.

Current design guidance for newly built special purpose floors is not clear, and guidance for the examination of existing floors when use changes is certainly no better. Static analysis alone is patently insufficient. A simple design criterion which has been used in the past is that the fundamental frequency of a dance floor should be two or three times the highest expected frequency of the load[27, 28].

Even a well designed and well constructed domestic floor in timber moves a millimetre or two when walked on, and this movement is perceptible to anyone standing on that floor.

Movement can be minimised by building in bearings tightly (encastred bearings), but remedial work of this nature is largely impracticable.

Floor designs can be analysed using various different methods to see whether they would have acceptable vibration characteristics in service. Floors can also be tested to assess human reaction to vibrations induced in the floor. The addition of a suitable floating floor

> **Case study**
>
> **Noise and vibration from an adjoining railway affecting housing**
>
> The owner of housing adjoining a railway line sought advice about excessive noise and vibration from passing trains. Recordings were made at one particular property of the effects of train movements using microphones both inside and outside the house and a vibration pick-up on the thermoplastic tiles of the lounge floor. These recordings were analysed in the laboratory to obtain noise and vibration measurements. The results showed that it would be possible to introduce palliative measures against the noise, by sealing windows and providing sound attenuated ventilators, but that measures to reduce the effects of vibration in the floors would be expensive and probably not worthwhile.

above has little effect on the measured properties but produces a considerable difference in human perception to footfalls.

This is a specialised field but some further information is available in BRE Information Paper IP 17/83[29].

Movement joints

Floors need to be provided with movement joints which reflect the positions of movement joints throughout the whole structure, and such joints should continue through the floor finish, and normally also provide continuity in other aspects of the performance of the floor such as its fire resistance. Problems which arise in this connection are dealt with in the chapters covering the different types of floor structure and finishes.

Deflections

The amount of deflection which can be tolerated in a old floor may depend more on the judgement of the occupants than on the actual construction of the floor and the effects of deflections on the integrity of finishes such as plaster ceilings and partitions carried on the floor. For many years there was a rule of thumb that deflection under full load should not exceed 0.003 of the span (*Principles of modern building*[6]). As will later be seen (Chapter 4.6), duct covers which are insufficiently robust can give rise to deflections which lead to spalling of tile floorings.

Jumping on old suspended timber floors to test the deflections should be done with circumspection.

Chapter 1.3 Thermal properties

This book is not the place to discuss the various requirements for thermal insulation in terms of thermal transmittance (U values). Indeed, the economic and other criteria which may be used to meet the values given in the national building regulations will vary according to building type, fuel used, its relative cost, and a host of other factors. What is important is where that insulation goes within the floor, the fact that it is laid consistently, filling every space and avoiding cold bridges, and that it is located on the correct side of any item of construction which functions as a vapour control layer.

The rate of heat loss through a floor situated next to the ground varies with its size and shape. It is possible, if the floor is large enough, to achieve a target U value of 0.35 W/m^2 °C without the use of any additional thermal insulation, but only in floors over 440 m^2. For most housing, therefore, as for other small buildings, it will be necessary to add thermal insulation to meet this target value. Figure 1.9 shows the appropriate relationship of width to length for floors which satisfy this requirement.

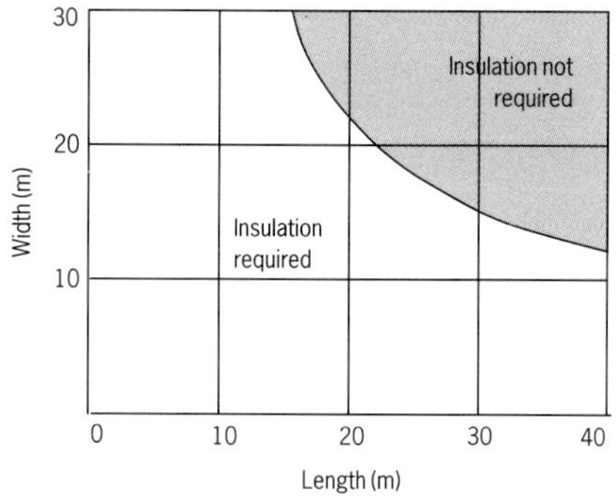

Figure 1.9
Floor width-to-length relationships achieving 0.35 W/m^2 °C without the need for additional thermal insulation

For floors where thermal insulation must be added, the thickness will depend upon the thermal conductivity value of the material, and the shape factor of the floor. The shape factor is a simple ratio of heated perimeter length divided by heated perimeter area, or P/A.

The thermal insulation value of materials reduces with increased moisture content. It follows that materials which do not absorb water are needed where prolonged wetting, for example through dampness rising from below from high water tables, is inevitable. Some thermal insulation materials are more vulnerable than others to this reduction in efficiency when wet. In most buildings, insulants are kept at acceptable moisture contents by protecting them from rain during construction when the floor is open to the weather, and designing to avoid the buildup of condensation when the building is in use.

The insulation value of the floor can be degraded by thermal bridges where high thermal transmission materials penetrate layers of low thermal transmission material, such as may happen at thresholds. Thermal losses due to cold bridges are often ignored in calculations, especially where thin sections are involved, but these and other components and materials such as concrete floor beams become more important as thermal insulation standards increase. Some cold bridges also have serious implications because they produce inside surface temperatures below the dewpoint of the air, leading to selective condensation on parts of the flooring (Figure 1.11). Suitable designs can overcome or reduce cold bridging to acceptable levels.

In suspended ground floors, air movement into and within the void and especially through layers of low density insulation material, can reduce the thermal efficiency of the floor considerably. Sealing at joints and around areas where services

Calculation of a typical perimeter-to-area ratio

Calculation of a typical perimeter-to-area ratio

Perimeter P = 5 m + 3 m + 3 m + 10 m + 8 m (the unheated porch is ignored, together with the separating wall, since it is assumed that the adjoining house is heated) = 29 m

Area A = 56 m^2 + 9 m^2 = 65 m^2

Ratio P/A = 0.45

From Table 1.1, it can be seen then that the appropriate U value will be met by, for example, extruded polystyrene of 50 mm thickness.

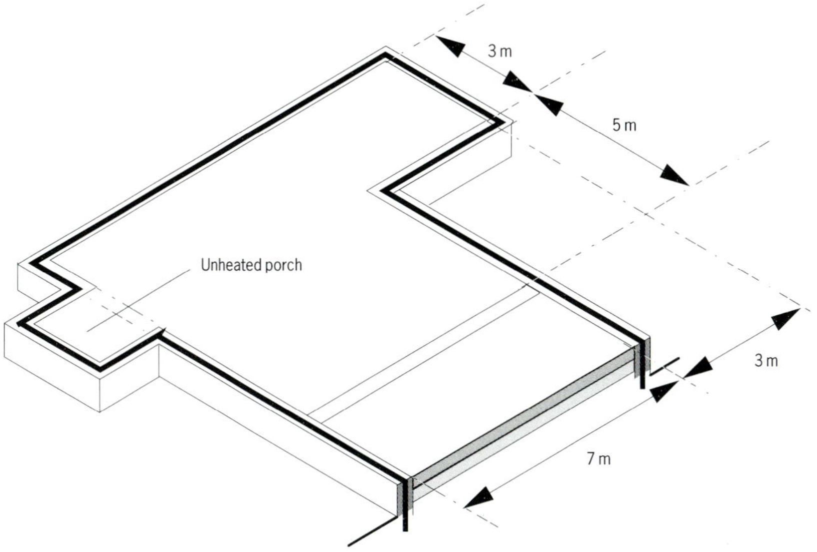

Figure 1.10
Floor plan of a semi-detached house used in the above calculation

Table 1.1
Required insulation thicknesses

Insulation material and thermal conductivity (λ)	Given the shape factor 0.45, appropriate thicknesses are as follows: required insulation thickness (mm) P/A = 0.35	P/A = 0.4	P/A = 0.6
Polyurethane (λ = 0.023)	25	25	40
Extruded polystyrene (λ = 0.027)	35	35	50
Mineral fibre slab (λ = 0.033)	35	40	50
Expanded polystyrene (λ = 0.037)	40	40	60

layer is intended to provide support for a screed or flooring, it will need to have certain minimum structural and other properties such as:

- resistance to loads
- dimensional stability
- chemical compatibility with the layers above and below

The position of thermal insulation will determine the thermal properties of a ground or basement floor. Where the insulation is placed as a sandwich between the base and screed in a solid floor, a quick warm up is achieved as only the screed is heated. However, once the heating is switched off the temperature will fall rapidly as the screed has little heat capacity. If the insulation is placed below the slab then, because of the large heat capacity of the concrete, the floor will be slow to respond to the heating being switched on but slow to cool when it is switched off.

Because most thermal insulating materials used in floors have some resistance to moisture vapour, but considerably less than that of a DPM, it is necessary with some constructions to provide additional moisture barriers to control constructional water. These are often referred to, rather loosely, as vapour control layers, and are nearly always of polyethylene sheet. This topic is dealt with in Chapter 1.4.

High material and labour costs are penetrate the insulation is important. Undesigned air movement within the floor void may also carry water vapour to areas where condensation can cause problems. U values, however, can be approximated (BRE Information Paper IP 3/90[30]).

For illustrative purposes, in most of the diagrams which follow, a U value of 0.35 W/m^2 °C is assumed. Thermal insulating material which is susceptible to moisture intake from, for example, condensation, rising damp or rainwater ingress, will offer only reduced values of thermal insulation.

Where the thermal insulation

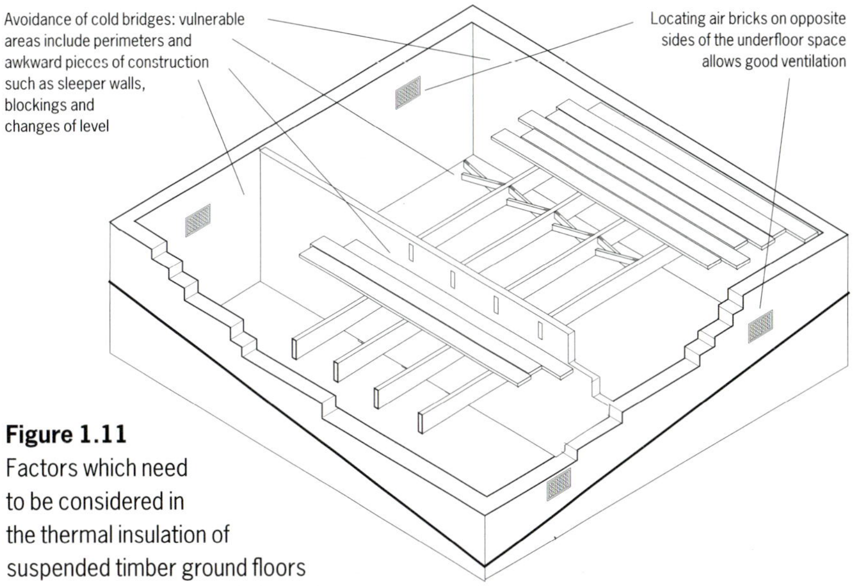

Figure 1.11
Factors which need to be considered in the thermal insulation of suspended timber ground floors of housing undergoing rehabilitation

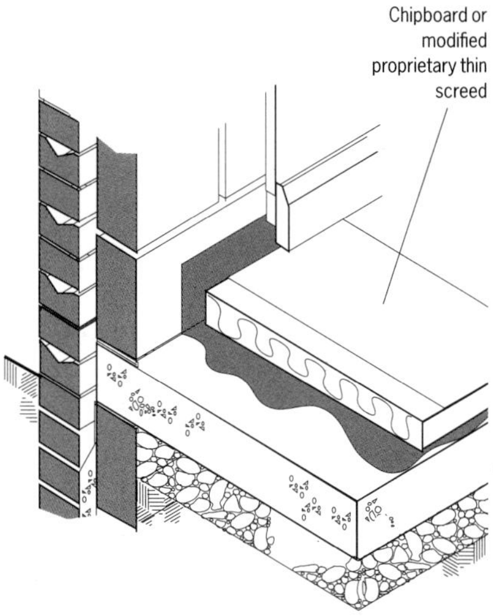

Figure 1.12
Replacing an existing unbonded 50 mm screed with a layer of thermal insulation under a panel product

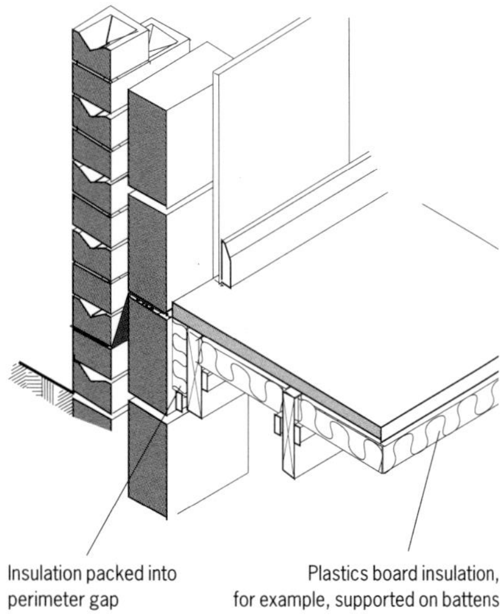

Figure 1.13
Where access to the underside is straightforward, insulating a suspended floor is well worthwhile. The preferred method is to use insulation boards suspended on battens fixed alongside the joists

often involved in retrofitting floor insulation measures of all types. It can also be technically difficult with solid floors because of the thickness of insulation required; floor level changes are not easily accommodated (Figure 1.12). As a result, fitting these measures in isolation usually proves not to be economically worthwhile. However, assuming some level of remedial work is necessary and that there will be an annual fuel price rise of 3% in real terms over the lifetime of the measure, insulation measures may become cost effective. Under these assumptions, insulating suspended timber floors with fibre quilt (Figure 1.13) and solid floors with EPS panels can increase cost effectiveness. Because of the pattern of heat loss from floors, the savings are particularly sensitive to changes in floor area, house type and shape. The detached house benefits the most from all types of floor insulation.

Further guidance is given in *Thermal insulation: avoiding risks*[31].

For construction which involves completely replacing the existing floor, the various building regulations apply[14, 32, 33].

Where the ground floor of a detached or semi-detached house is being completely replaced, or for that matter new floors are being installed in extensions, it is cost effective to insulate with 25 mm insulation either below a solid floor or above a beam and block floor. This can be proved by calculation and verified by test.

Using a highly insulated beam and block construction, calculations show it is possible to achieve a U value of less than 0.2 W/m^2 °C for a terraced, semi-detached or detached house cost effectively. For many uninsulated terraced houses the U value of the row of terraces will be less than 0.45 W/m^2 °C and insulation will not be needed. For suspended timber floors it is possible to insulate all three types of house cost effectively with up to 150 mm insulation. Again this could achieve a U value below 0.2 W/m^2 °C.

The effect of adding floor insulation has actually been tested by BRE in a matched pair of houses both before and after the thermal insulation standards in the Building Regulations 1991 were uprated. Measurements before and after the uprating indicated that in an average heating season the floor insulation should give a saving of 4 GJ, equivalent to about 0.1 GJ m^2. It is therefore well worthwhile adding thermal insulation to a floor when other improvements or replacements are called for.

Basement floors

The heat flow pattern in basements is rather complex and time dependent. The U values can be calculated using the steady-state component averaged over the basement, providing an approximation of the heat losses

Table 1.2
U values for basement floors

Perimeter-to-area ratio (m^{-1})	Basement depth H (m) 0.5	1.0	1.5	2.0	2.5
70.1	0.21	0.19	0.18	0.17	0.17
0.2	0.34	0.32	0.29	0.28	0.26
10.3	0.45	0.41	0.38	0.35	0.33
10.4	0.55	0.50	0.45	0.42	0.39
0.5	0.63	0.56	0.51	0.47	0.43
0.6	0.70	0.62	0.56	0.51	0.47
0.7	0.76	0.67	0.60	0.54	0.49
0.8	0.82	0.71	0.63	0.57	0.52
0.9	0.87	0.75	0.67	0.59	0.54
1.0	0.91	0.79	0.69	0.61	0.55

which is adequate for most purposes (BRE Information Paper IP 14/94[34]). The deeper the floor is in the ground, the better the U value (Figure 1.14).

Alternatively, the table on the opposite page, taken from IP 14/94, gives U values of uninsulated basement floors in terms of the perimeter-to-area ratio (P/A) and the basement depth H. Linear interpolation is appropriate.

Edge insulation in existing floors

It may not be obvious to the surveyor inspecting recently built dwellings that they may well have thermal insulation set in the plane of the wall, either within the cavity (Figure 1.15) or inside the inner leaf (Figure 1.16). Later chapters contain information about particular floor types.

Exposed soffit floors and balconies

If a floor has a soffit exposed to the outside, or if a reinforced concrete floor continues through the external wall to form a balcony in a building built before, say, the mid or late 1960s, it is likely that thermal insulation will be inadequate by current standards (Figure 1.17). Indeed, such balconies function, unintentionally, as cooling fins on the outer surfaces of buildings.

Consideration needs to be given to applying suitably robust thermal insulation to the soffit of the balcony; also to continuing the insulation into the building (Figure 1.18 on page 24). The material should have a thermal resistance of not less than 0.6 m^2K/W. Flying

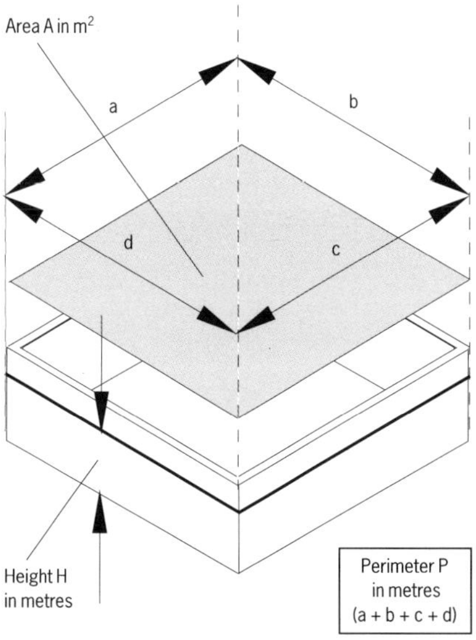

Figure 1.14
Criteria for calculating the perimeter-to-area ratio and depth (H) of a basement for establishing the U value

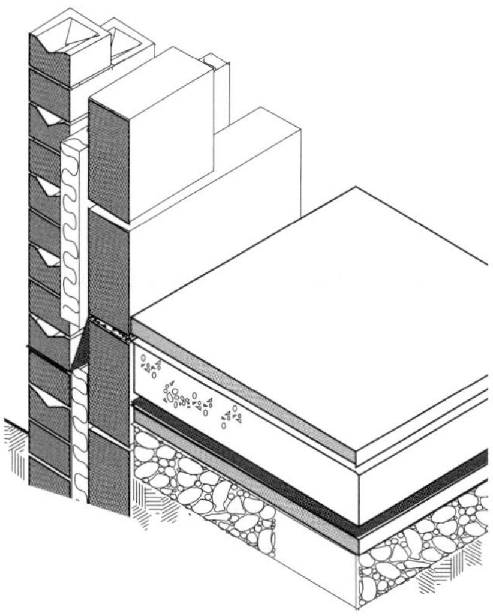

Figure 1.15
Thermal insulation placed within the cavity should cover the edge of the floor

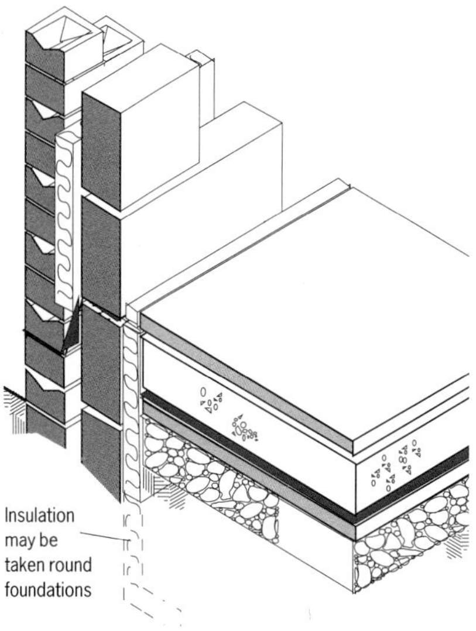

Figure 1.16
Thermal insulation placed inside the inner leaf of the external wall should cover the edge of the floor

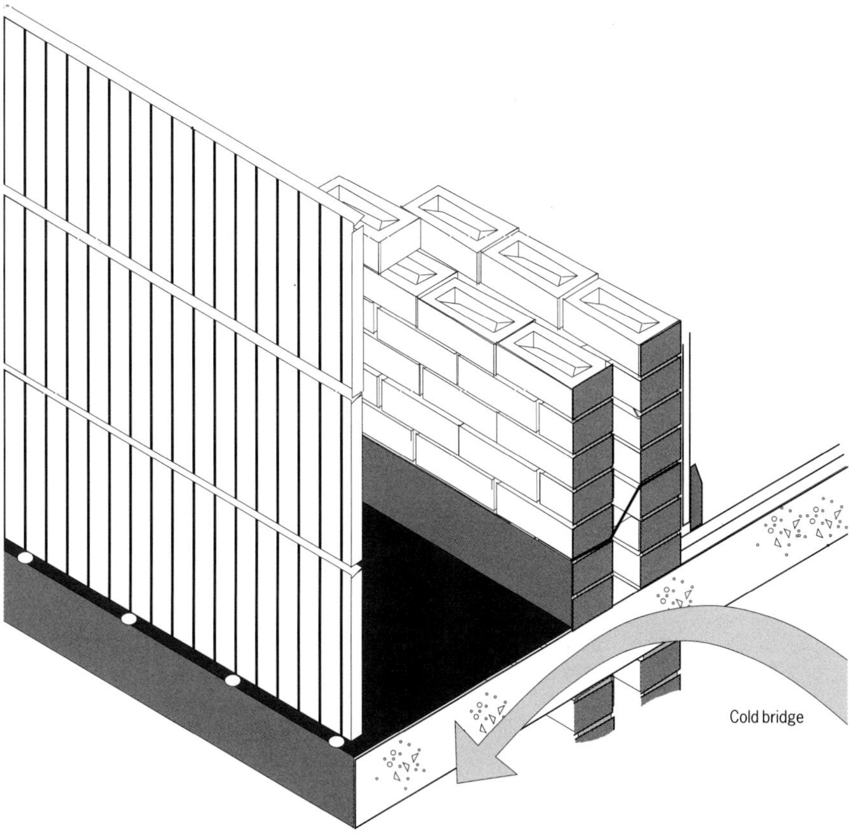

Figure 1.17
Heat loss through a balcony extension can generate condensation on the cold surfaces of the slab inside the building

floors (floors the undersides of which are directly open to the external air) should be insulated over their entire underside surface, too, to the same standard. *Thermal insulation: avoiding risks* gives more information.

Predictions of the heat losses at intersections of floors with walls, particularly in the corners of internal spaces, are difficult to visualise. Finite element analysis may be useful in these circumstances.

Service entry points

When renewing water mains entering buildings under all kinds of floors, but especially suspended floors, it is very important to ensure that protection against frost is adequate (Figure 1.19).

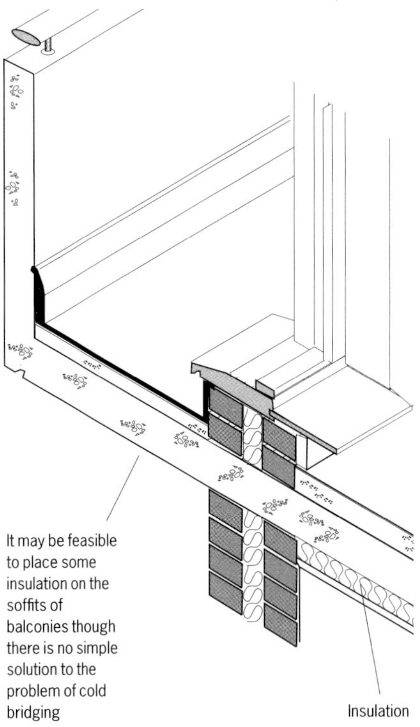

Figure 1.18
Thermal insulation can be placed on the underside of a suspended floor within a building where it adjoins the external wall

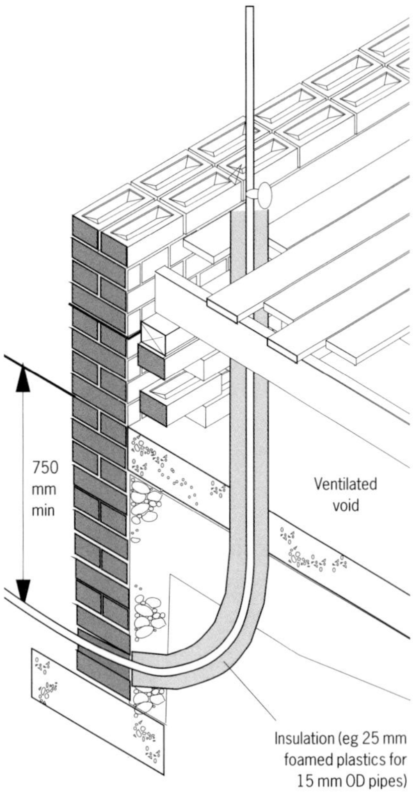

Figure 1.19
A water service entry pipe within a suspended timber ground floor void must be protected from frost

Chapter 1.4 Control of dampness and condensation, and waterproofness

Not all floors are subject to dampness problems, but those in contact with the ground in basements or ground floors will clearly be most at risk and therefore measurements should be taken (Figure 1.20). Moreover the risk should not be discounted altogether in the case of suspended floors, particularly at bearings on external walls. Information from Housing Association Property Mutual indicates that about 1 in 10 of newly built dwellings inspected by them during the early 1990s were not adequately protected against rising damp.

The form of any failure depends on the flooring material and the adhesive. Flexible sheet materials like PVC, linoleum and rubber commonly curl or blister, while tiles made from similar materials curl or tent at their edges. The basic trouble in most cases is some expansion of the flooring material accompanied by loss of adhesion between it and the base. Often any moisture expansion is aggravated by stretching of the flooring material by traffic after adhesion has been lost. Many adhesives soften in the presence of moisture. This adverse effect on flooring and adhesive is often made worse by the fact that the moisture has a high pH (highly alkaline) because it contains alkalis derived from the cement in the concrete base or screed to which it is fixed. Timber block and strip floors fail by disruption brought about by expansion of the wood following moisture take up. Carpets and other textile floorings may become debonded and ruck because the adhesive has softened. In addition there is the possibility that some fibres or backings will rot.

Figure 1.20
Hygrometer box, with thermally insulated lid removed, sealed to a screed on which rubber flooring was bubbling. The reading is 90% RH

There are four main possible sources of moisture in floors to consider:

- excess constructional water
- ingress of water from the outside
- condensation
- water spillages and leaks

Each will now be dealt with in turn.

Excess constructional water

Water is added to a screed or concrete mix to make it workable, and is more than that which is required for hydration of the cement. The water for workability usually amounts to between half and two thirds of the water used for mixing. This excess must be allowed to dry out before fixing impervious floorings such as PVC sheet, or moisture sensitive floorings such as wood or textile, as both of these categories can be affected. If moisture sensitive flooring is used, an extra surface DPM will be needed (Figure 1.21).

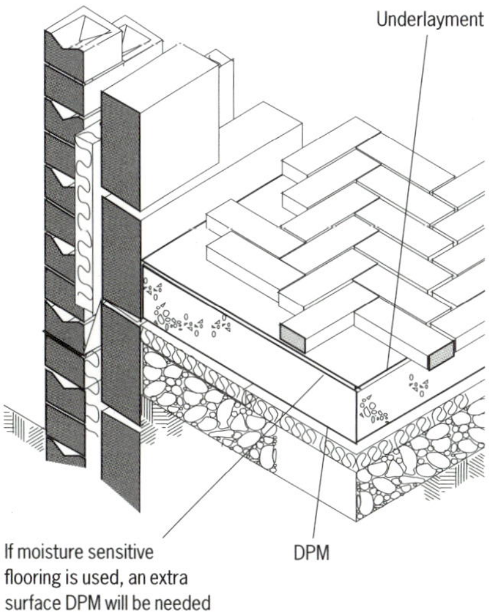

Figure 1.21
Where construction times are too short to allow adequate drying time, precautions may need to be taken to avoid construction water affecting any moisture sensitive flooring

Table 1.3

Drying times for screeds and bases

This table is given for guidance. The actual drying time of a screed or slab depends on many factors. Before laying any moisture sensitive floors such as PVC, linoleum, wood blocks or carpets, the moisture conditions should be checked with a hygrometer. (BS 8203[35], and BRE Digests 163[36] and 364[37])

Construction	Estimated drying times
Screed, 50 mm	4–6 weeks
Screed, 75 mm	6–8 weeks
Concrete, 100 mm	4–6 months
Concrete, 150 mm (or screed, 50 mm, plus concrete, 100 mm, with no DPM between)	6–12 months
Concrete, 200 mm	8–20 months

Water moves through concrete partly by capillary action but mainly by diffusion. (Water moving through concrete is a very slow process.) Unless the atmospheric humidity is high, that is to say in excess of 90% relative humidity (RH), water will evaporate from the surface of a base as fast as it reaches it. Moisture is never visible once the initial surface water has dried off. The result is that the surface may look dry even though there is still a considerable amount of water below the surface.

The rate of drying, and therefore the time to dry a concrete base or screed before floorings can safely be fixed depends on a number of factors. These include the mix proportions, amount of mixing water added, the temperature of the floor, the relative humidity of the air and the thickness of the base or screed. A rule of thumb method often quoted for forward planning purposes is to allow a day for each millimetre of thickness for screeds. This has worked well for thicknesses up to 75 mm, although many sand and cement screeds laid semi-dry will be sufficiently dry well within the predicted times. Thus a sand and cement screed 50 mm thick laid on a DPM usually dries within 4 to 6 weeks. Proprietary screed systems are available which dry considerably more quickly than conventional screeds.

The rule of thumb method of predicting drying times does not apply to thick concrete slabs. Commonly, slabs 150 mm thick with a DPM immediately below, take between 6 and 12 months to dry, but there have been cases where they have not been dry even after 18 months. The mix design of the concrete appears to be a big factor, particularly the cement content and the cement:water ratio, but this is not an area which has been well researched.

Drying cannot be assumed to start until the shell of the building is watertight; indeed, rainwater falling onto an uncovered slab can add more water than was originally in the concrete (Figure 1.22).

Where screeds are bonded directly to the concrete bases the problem of moisture is compounded. Often some drying of the slab takes place before the screed is laid. However, the slab is intentionally re-wetted at the time of laying the screed to reduce suction. With this type of construction, the total thickness of screed and base above the DPM must be taken into account and, for forward planning, the start of drying taken from when the slab was laid.

Where it is impracticable to allow sufficient time for the concrete base to dry out, a DPM or vapour control layer must be laid between the wet construction and the sensitive flooring (Figure 1.23). Conventionally this is achieved by providing a DPM on top of the slab and covering it with a screed which, because it is unbonded, should be a minimum of 50 mm thick. The application of an epoxy bonding agent (which also acts as a DPM) to the surface of the base enables bonded screeds less than 50 mm to be laid.

Figure 1.22

Rainwater can find its way onto newly laid floors, particularly if the building is not weatherproof. Here there is a veritable lake in which the columns and the soffit of the ceiling are reflected

Proprietary surface DPMs based on epoxy resins can be applied to the surfaces of screeds or concrete bases, or, where the old substrate is not in good condition, over a moisture tolerant skim coat of underlayment. They are most often used where specifiers have misjudged how long a construction will take to dry and there is a need to lay the flooring on a base which is still wet. These DPMs are also very useful where the use of a building is being changed and there is no DPM in the existing floor. For both of these applications, solventless epoxy resins have a good track record.

As already emphasised, the problem of providing adequate drying time for thick slabs has been known for many years. BRE Digest 54[38], first published in 1965, states that a 150 mm slab would take at least six months to dry sufficiently for a moisture sensitive finish to be laid, and BS 8203 (Clause 11.3), published in 1987, states that concrete 150 mm thick may require as much as one year to dry from one face only. There is some evidence that these lessons still have not been understood by the industry.

Although there is little experimental evidence one way or the other, BRE experience is that power floated slabs, having a dense and therefore relatively impervious surface, dry out more slowly than trowelled slabs. Hygrometer readings from instruments left in position for at least 72 hours should provide a good indication of moisture conditions, even of power floated slabs.

Water ingress from the outside

Water which can potentially move from the ground through the base hardcore and concrete, or which can be conducted from wet external leaves of walls to the edges of sensitive floor construction, must be prevented from doing so by effective dampproofing (Figure 1.24). This is especially crucial in basement floors with high water tables where tanking sufficient to withstand the hydrostatic pressure is required. In the case of a basement it may be necessary to ensure that sufficient mass is available in the building above to prevent flotation of the basement.

If the surface of the screed or base is dry when flooring is laid, moisture from below the foundations can still rise after many months and failure of the flooring occur a year or two later.

Condensation

Condensation will occur when the surface temperature of the floor is below the dewpoint temperature of the atmosphere for a sustained period of time. The dewpoint will vary according to the air temperature and the relative humidity. Condensation occurs when the relative humidity of the air in direct contact with the cold surface rises to 100%.

The two most common situations in which condensation occurs on a floor are:

- where the floor is adjacent to an exterior perimeter wall and there is a loss of heat from the floor to the outside via a cold bridge (Figure 1.25 on page 28)
- where the floor has a high thermal capacity and the floor temperature is unable to follow rapid changes to the air temperature which often falls below the dewpoint. This phenomenon particularly affects floors in warehouses in weather conditions where a cold spell is followed by a warm front.

It is not normally necessary to provide a vapour control layer in floors to prevent moisture ingress downwards, though it may be required for special reasons; for example to prevent water vapour entering from beneath the construction to affect sensitive materials.

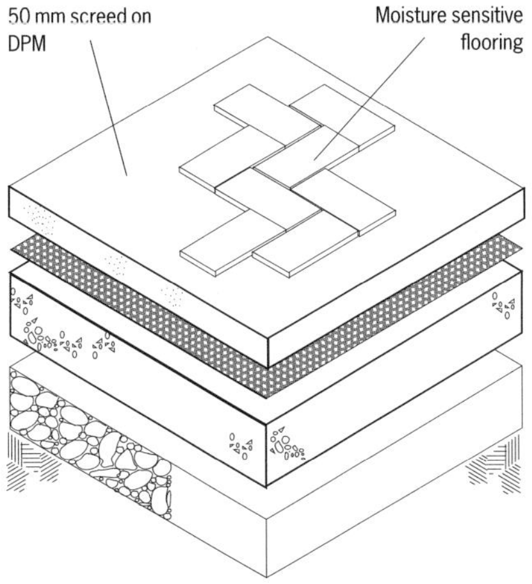

Figure 1.23
If the DPM is placed between base and screed, only the screed needs to be dry

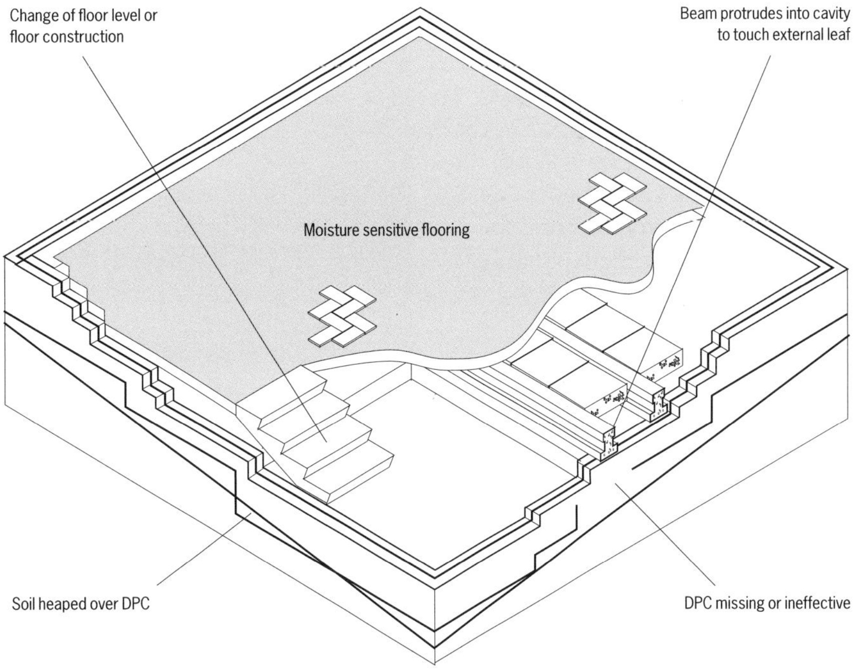

Figure 1.24
Routes for moisture from the exterior to sensitive floor and flooring materials, even where the DPM is present and effective

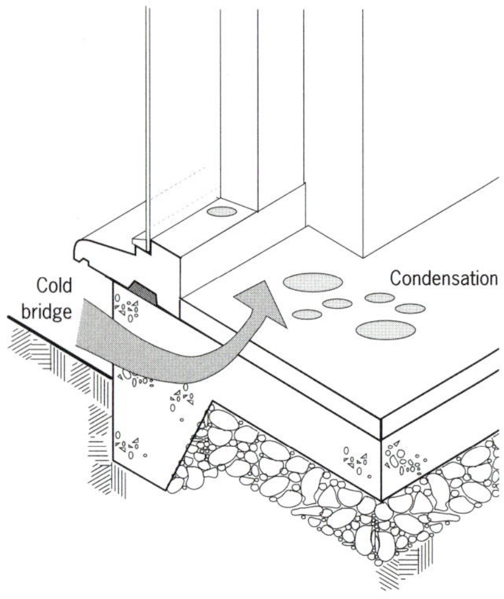

Figure 1.25
Condensation caused by cold bridging at the exposed edge of a floor slab

Of the various terms that are used to describe a barrier to water vapour, 'vapour control layer' is preferred to either 'vapour check' or 'vapour barrier' to emphasise that the function of the layer is to control the amount of water vapour entering construction. As the achieved vapour resistance will depend at least as much on workmanship as on design and the integrity of the materials used, it is not realistic to specify a minimum vapour resistance to be achieved for the layer as a whole, though for the material to qualify as a vapour control layer it should have a vapour resistance greater than 200 MNs/gm.

Plastics films are the most usual materials for forming a vapour control layer in a floor construction. Joints in a flexible sheet vapour control layer should be kept to a minimum. Where they occur, they should either be overlapped by a minimum of 100 mm and taped, or sealed with an appropriate sealant, and should be made over a solid backing. Tears and splits should always be repaired with jointing or sealing as above. Penetrations by services should be kept to a minimum and carefully sealed at interfaces. Draughts of moisture laden air through gaps in vapour control layers are more significant than normal still air diffusion through materials; it is therefore of greater importance to avoid holes in the vapour control layer than to take elaborate precautions for sealing laps in the layer.

Water spillages and leaks

Spillages and leaks have frequently been found to be the cause of dampness in investigations carried out by BRE Advisory Service (Figure 1.26). Too lavish a use of water for cleaning purposes has also been known to cause surface breakdown of sensitive floorings, and, in some cases, to accumulate on DPMs. The water, often carrying residues of corrosive products, will pass through joints between impervious coverings, or, at the edges of the material, under skirtings or external door thresholds.

Materials for DPMs

Although concrete bases of good quality laid directly on the ground can be relatively impervious to the passage of 'liquid' water, they cannot be expected to stop all moisture rising from the ground. It is therefore necessary to provide some form of protection from damp. The moisture

Figure 1.26
Deterioration of a chipboard deck laid on expanded polystyrene. An unsealed joint in the polyethylene vapour control layer has allowed water from a plumbing leak to rise to the surface

barrier usually consists of a membrane either laid under the slab or sandwiched between the slab and a screed. Undoubtedly the ideal membrane is completely impervious to moisture, either as water or water vapour. Suitable materials have existed (and still do exist), but they have tended to be difficult to handle because they had to be applied in the fluid state and were expensive, such as hot applied bitumen. Other materials like bitumen solutions, bitumen/rubber emulsions and coal tar/rubber emulsions have been used for many years; and from around 1960, polyethylene sheeting has also been used. All these materials resist the passage of liquid water, but are not impervious to water vapour. Some permeability of the membrane can be accepted provided it is less than the flooring material it is protecting. The degree of protection required depends on a number of factors which include the moisture sensitivity of the flooring, the adhesive used to fix it, and the site conditions. Usually little is known in advance about site conditions, but the properties of floor finishes, in terms of their resistance to rising ground moisture, are well known and are summarised in Table 1.4 (page 30).

The range of membrane materials which have in the past been found suitable for providing the necessary protection is shown in Table 1.5 (page 31).

Materials which can provide effective sandwiched or buried DPMs, if properly laid, include:

- hot applied pitch
- hot applied bitumen
- polyethylene sheet
- polyethylene sheet backed with bitumen
- bitumen sheet
- bitumen/rubber emulsions
- cold applied bitumen
- cold coal tar emulsions
- epoxy resin (also called epoxy bonders)

Materials which can provide effective surface DPMs, if properly laid, include:

- epoxy resins
- pitch/epoxy resin mixtures
- polyurethanes
- mastic asphalt, not less than 12 mm thick
- pitchmastic

Concrete bases containing proprietary 'waterproofers' are not an acceptable substitute for a properly laid DPM.

See Chapter 1.5 for comments on the use of materials for DPMs containing volatile organic compounds (VOCs).

Waterproofness of floorings

In certain circumstances, floors need to be completely waterproof, especially where the materials from which they are made are at risk of deterioration. Such circumstances arise, for example, in the bathrooms and kitchens of dwellings, or in manufacturing areas where water for cleaning is used on a large scale. There is no criterion which can be applied generally, and each case must be treated on its merits. Indications of the characteristics of the various floorings are given in Chapters 5–8.

Table 1.4
Effects of rising moisture on floor finishes

Group		Material	Properties
A	Finish and dampproof membrane combined	Pitchmastic flooring Mastic asphalt flooring	Resist rising damp without dimensional or material failure
B	Finishes that can be used without extra protection against damp	Concrete Terrazzo Clay tiles	Transmit rising damp without dimensional, material or adhesion failure
		Cement/polymer latex Cement/rubber latex Cement/bitumen Composition blocks (laid in cement mortar)	Transmit rising damp slowly without dimensional or material failure and usually without adhesion failure
		Wood blocks (dipped and laid in hot pitch or bitumen)	Transmit rising damp slowly without material failure and usually without dimensional or adhesion failure. Only in exceptional conditions of site dampness is there risk of dimensional instability
C	Finishes that are not necessarily troublefree but are often laid without protection against damp	Thermoplastic flooring tiles PVC (vinyl) asbestos tiles Acrylic resin emulsion/cement Epoxy resin flooring	Under severe conditions, dimensional and adhesion failure may occur. Thermoplastic flooring tiles may be attacked by dissolved salts
D	Reliable protection against damp needed	Magnesium oxychloride	Softens and disintegrates in wet conditions
		PVA emulsion/cement	Dimensionally sensitive to moisture. Softens in wet conditions
		Polyester resin flooring Polyurethane resin flooring Rubber Flexible PVC flooring Linoleum Cork carpet Cork tile	Lose adhesion and may expand under damp conditions
		Textile flooring	Dimensional and material failure, and usually adhesion failure, occur in moist conditions
		Wood block laid in cold adhesives Wood strip and board flooring Chipboard	Acutely sensitive to moisture with dimensional or material failure

Table 1.5
Materials for membranes

Material	Standard or grade	Position	Comment
Hot applied			
Mastic asphalt	Mastic asphalt for flooring (natural rock asphalt aggregate, formerly to BS 1410, now withdrawn). Mastic asphalt for flooring (limestone aggregate,formerly to BS 1076 and BS 1451, both now withdrawn). BS 6925 [39] Type F now applies	Surface	Used as a floor finish. If used as an underlay to a floor finish, the thickness should have been not less than 12 mm. A compressible underlay was not recommended but glass fibre may have been used
	Mastic asphalt (limestone and natural rock aggregates, formerly to BS 1097 and BS 1418, both now withdrawn). BS 6925 Types T and R now apply	Sandwich	When loaded, can withstand hydrostatic pressure
Pitchmastic	Pitchmastic flooring (formerly to BS 1450 and BS 3672, both now withdrawn)	Surface	Normally used as a floor finish but may have been a surface membrane to protect other finishes. Its indentation characteristics make it less suitable than mastic asphalt. The material is no longer available
Pitch	Coal tar pitches, Grade R and B40 (formerly to BS 1310, now withdrawn)	Sandwich	Should have been laid on a primed surface to give an average thickness of 3 mm (3 kg/m^2). The material is no longer available
Bitumen	Should have a softening point of 50–55 °C. This corresponds to a penetration number of 40–50 at 25 °C	Sandwich	Should have been laid on a primed surface to give an average thickness of 3 mm (3 kg/m^2)
Cold applied			
Bitumen solutions, coal tar pitch/rubber emulsion, or bitumen/rubber emulsion	Not defined by any BS specification or code	Sandwich	BS CP 102 [40] recommends 0.6 mm min thickness but this is for broad guidance only. The solids content will usually have been adjusted to give adequate coverage by two or three coats. The material should not have been thinned by dilution or spread in thinner coats than recommended by the manufacturer
Pitch/epoxy resin Solventless epoxy resin	Proprietary only	Surface Sandwich	Although applied in thin layers, the material is strong enough and its adhesion to concrete is usually sufficient to have made it a satisfactory base for a variety of floor finishes. It cannot tolerate any cracking in the surface to which it is applied
Sheet material			
Polyethylene film	125 μm (500 gauge) thick*	Sandwich	Joints must have been properly sealed; the welting method normally used appears satisfactory. Where there was risk of damage by subsequent screed laying operations, material about twice as thick, 250 μm (1000 gauge), may have been used
Composite	Polyethylene and bitumen, self-adhesive; thickness in excess of 1.5 mm	Sandwich Below concrete	Adhesion simplified joint treatment and reduced risk of tearing
Bitumen sheet	Bitumen sheet to BS 743 [41]	Sandwich Below concrete	Joints should have been properly sealed

* Although polyethylene sheet 125 μm (500 gauge) thick is sufficiently impervious to be used as a DPM, it is not very robust and is easily damaged. Paragraph 3.5 of the Building Regulations 1991 Approved Document C [42], recommends that polyethylene should be at least 300 μm (1200 gauge) thick or 250 μm (1000 gauge) if in accordance with appropriate BBA certificate or to the Packaging and Industrial Films Association (PIFA) standard

Chapter 1.5 Comfort and safety

Comfort and safety are two functions which may be required of flooring. They are less amenable to quantitative measurement and specification than other functions, though they are not of less importance (Figure 1.27). Indeed, aspects of safety should be of major concern to all specifiers of flooring, whether new or replacement.

Warmth to touch

Although a floor can be made to play a relatively small part in the heat losses from a building by suitable choice of thermal insulation, it is often the rate at which bare feet or thinly shod feet lose heat to the floor which can be of more immediate importance to the building user. The thermal characteristics of the flooring can be of crucial importance in buildings such as children's nurseries (Figure 1.28). On the other hand, comfort can be very subjective. One case known to BRE involved complaints about the coldness of a concrete floor which ceased when the floor was painted red.

A suggested figure of 45 kJ/m^2 per minute, representing the maximum rate of heat loss where warmth to the touch is very important, has been discussed at international level. The actual rate can be calculated from the thermal conductivity of the various layers of known thickness of floors and flooring. If adopted, it would rule out many thin floorings laid on dense substrates.

If a floor can be found to be too cold, it is reasonable to assume that it could become too hot. Excessive heat in a floor might affect people and might also lead to degradation of the materials forming the flooring. In no case should the temperature reached by the flooring exceed 27 °C; some materials are affected at lower temperatures. These effects are dealt with in Chapters 5–8.

Figure 1.28
While two different materials can have the same temperature, one can seem to be warmer (or colder) than the other. It is a phenomenon that is explained by the different rates of heat loss by the human body into the materials. For example the carpet in this nursery may seem warm to the child and its carer even though it will have the same temperature as the concrete slab beneath it

Figure 1.27
A wood strip floor in a sports hall: comfort and safety are particularly important requirements for the flooring in this type of situation

The European Union of Agrément has suggested a formula for considering the appropriate thermal comfort requirements of a flooring material:

$$\sqrt{\frac{k}{pc}}$$

where k is the thermal conductivity (W/m °C), p the bulk density (kg/m^3) and c the specific heat (J/kg °C) of the material (UEAtc Method of Assessment and Test No 2[(25)]).

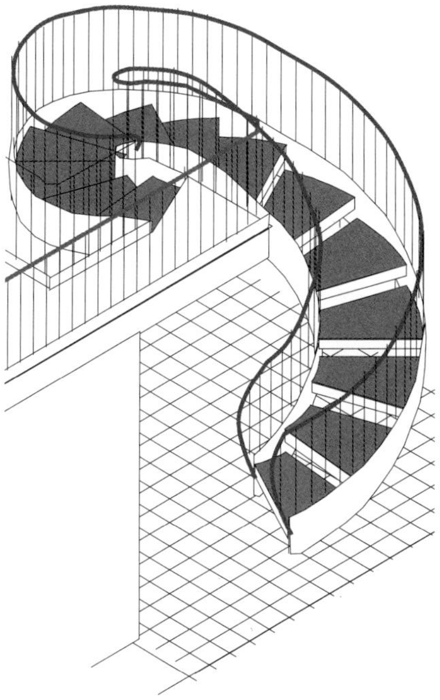

Figure 1.29
The design of helical stairs is covered in BS 5395:Part 2

A test method is available to determine the quantity of heat lost in W(Jm^2) at the end of one minute and at the end of ten minutes.

Resilience

Many floor finishes have a characteristic feel which is rather difficult to define and, hence, equally difficult to select objectively when offered as a choice. Subjective views often predominate, as in the case of some people such as sales staff preferring a hard floor to a deep pile carpet which makes them feel tired. There may be problems with textile coverings; for example certain types of open weave barrier matting without metal armouring can present wheelchair users with problems.

Accidents and safety

Accidents

The contribution of the floor finish to accidents is an important issue. This is true not only for industrial locations, where the risks may be more obvious, but also in the home.

Current UK statistics for accidents on level surfaces within homes (which can be interpreted for all intents and purposes as floors) show that, on average, there are 127 fatalities and tens of thousands of non-fatal accidents per annum resulting in injury out of a total of just over 250,000 accidents. Falls on and from stairs are also estimated in the UK at around 250,000, and falls at other changes of level (eg single steps and thresholds) at around 200,000 (*Building regulation and safety*[43]). (Figures for fatalities on stairs and at other changes of level are not available.)

Comparable figures for other building types are not published separately, but those for leisure activities in buildings in the UK other than the home include estimates of accidents on the same level at over half a million per annum. It therefore places responsibility on all concerned with the specification and maintenance of floors, flooring and stairs to be vigilant on matters of safety.

Safety

Responsibility for the safety of people engaged in building operations is not only confined to the contractor the specifier also shares the responsibility in the sense that they must consider safety aspects of installing the items they are specifying, consider the risks, and decide whether to select alternatives. The specifier is also obliged under current legislation to provide information about health and safety issues for all items specified.

Compliance with all relevant items of safety legislation is outside the terms of reference of this book, but reference may be made to the various health and safety regulations[44–47] and to the *Construction Safety Handbook*[48].

When considering the replacement of floorings in buildings, it is important to consider the slipperiness characteristics of new materials and the lighting needs for all potential users – especially physically and visually handicapped people.

Where features for stairs need to be checked, for example for conformance with the various national building regulations, the relevant detailed criteria will be found in the British Standards codes of practice for the design of straight, helical and spiral stairs (Figure 1.29), and industrial type stairs, permanent ladders and walkways in BS 5395: Parts 1–3[49].

Nosings are the most crucial parts of stairs from the point of view of safety. Prefabricated nosings are available in a very wide range of profiles and finishes to suit the materials forming the main part of the tread surface. The aim is to provide the right compromise between slip resistance and tripping hazard for both upward and downward travel.

Most nosings consist of an extruded aluminium section into which is inserted a strip of PVC to which a proportion of carborundum has been added. These nosings need to be firmly screwed to the substrate of the tread and monitored afterwards to make sure they do not loosen. An alternative, less satisfactory, solution for thinner finishes is offered by plastics or rubber mouldings which are bonded to the substrate.

One further risk area which needs to be considered is the sizes of gaps in flooring grilles. There is some difference of opinion on this matter. It can be argued that warnings are sufficient for those floors surrounding industrial processes where the public are not admitted. In public areas, particularly at entrances to buildings, consideration should be given to limiting the mesh clear gap size to 8 mm. This should allow snow, rainwater and most stones picked up on moulded rubber shoe soles to fall through, while at the same time preventing narrow heels and umbrella ferrules from penetrating the gap.

Slip resistance

Slip resistance is a significant component in the avoidance of accidents, but it is also one of the most complex attributes of floorings in buildings to assess. The more important factors are described in the feature panel on page 34.

The factors affecting slipperiness

Everyone knows the unpleasant sensation that is felt when the foot fails to grip an unexpectedly slippery surface. Although a recovery of balance usually prevents anything more serious than an uncomfortable jolt, the slip may result in a fall with serious consequences, particularly for the elderly.

People slip because at that instant the frictional force between the shoe and the floor is too small to resist the horizontal component of the force applied to the floor. When a person, with his or her leg inclined at an angle to the vertical, places the foot on the floor, a force is applied to the floor which may be resolved into a horizontal component (H) and a vertical component (V). To walk without slipping, the horizontal component must always be less than the frictional force (F) between the shoe and the floor, in other words for safety:

$F > H$

where $F = \mu V$ and μ is the coefficient of friction, which depends on the nature of the shoe and floor materials, and any third body interference such as water, dust, polish etc.

It follows that, for safety:

$\mu V > H$

or $\mu > H/V$

Conversely, if H/V exceeds the coefficient of friction, a slip will occur, but whether that leads to a fall depends on many other factors.

Work at the Building Research Station (BRS) in the late 1950s – published as research findings in 1961 (BRS National Buildings Study Research Paper No 32[50]) – to measure the forces applied by the foot in walking, was primarily aimed at providing data for designing an abrasion machine. However, part of the data could usefully be used to measure the H/V ratios that were produced in walking both in a straight line and round corners on the level. Analysis of the data showed that the average H/V produced was between 0.16 and 0.22 depending on the direction of movement and sex of the person. Only one person in a million was likely to exceed a value of 0.4. Since that time the concept that the coefficient of friction between footwear and floor should not be less than 0.4 has been widely used.

More recently, some of the original experimental data has been reassessed (*A brief review of the historical contribution made by BRE to slip research. Slipping – towards safer flooring*[51]): it indicates the level of risk of people exceeding other H/V ratios. The values are reproduced in Table 1.6.

Because of the constraints of the experiment, any conclusions drawn from the table have limitations and should be subjected to the following qualifications.

- Only 124 sets of data were available for analysis (87 men participants, 37 women).
- All participants were fit and able and between the ages of 18 and 60 years
- The experiment was conducted at 100 paces to the minute, and other than turning a corner, there were no sudden changes of direction or pace, or stopping or starting.
- Any person who walks in a manner that he or she regularly exceeds an H/V ratio of, say, 0.35 will often slip. That person will soon learn from painful experience to adjust their gait to lower the H/V ratio.

  (The risk analysis assumes that by statistical means the chance of a person exceeding a certain high H/V ratio can be extended from a small sample.)

Table 1.6

Coefficients of friction between foot and floor

Risk 1 in:	Straight walking	Turning: left foot	Turning: right foot
1 000 000	0.36	0.40	0.36
100 000	0.34	0.38	0.34
10 000	0.29	0.34	0.33
200	0.27	0.31	0.32
20	0.24	0.27	0.29

Measurement of slipperiness

Many instruments have been designed for measuring the slipperiness of flooring (about 80 at the last count). Many operate quite arbitrarily as little thought has been given to the loads and speeds involved when people slip. Some instruments consistently give values of coefficient of friction higher for wet floors than dry, which is completely at variance with common experience.

Since 1960, whenever BRS or BRE has been asked to make an assessment of the slip resistance of any flooring, the Transport Research Laboratory (TRL) pendulum tester has been used (Figure 1.30 and RRL Road Note 27[52]). This apparatus has a pendulum which traverses 125–127 mm across the floor being tested. It measures the frictional resistance between a slider mounted on the end of the pendulum and the floor surface.

Much work was done by the Greater London Council (GLC) Scientific Branch in the late 1960s and early 1970s using this instrument at accident sites. It published the following assessed values of resistance to slipperiness and related categories of safety (GLC Bulletin No 43[53]):

- less than 19 dangerous
- 20–39 marginal
- 40–74 satisfactory
- above 75 excellent

These values of slip resistance are read directly from the pendulum scale. They are called slipperiness coefficients or slip resistance values (SRVs) and are about 100 times greater than the coefficients of friction. The values have been used widely in assessing the slipperiness of floorings.

In the late 1980s there was some criticism that the rubber slider of the TRL pendulum did not provide sufficient discrimination between wet and dry flooring surfaces. A much harder rubber was developed by the Rubber and Plastics Research Association (RAPRA) in an attempt to overcome this criticism. Unfortunately, this rubber, known as 4S, had poor abrasion resistance and it has been shown that the rubber's texture can be affected by the last flooring it was used to test†. Particularly, it could affect the results of the next flooring tested in the wet.

The TRL pendulum appears in a number of British Standards, including BS 8204:Part 3[54], Part 4[55] and Part 5[56], and BS 7044:Part 2: Section 2.2[57].

It became apparent that not all operators were following the same procedure when using the TRL pendulum. To overcome this shortcoming, the UK Slip Resistance Group, of which BRE is a member, has issued guidance (*The measurement of floor slip resistance*[58]) so that all operators use the machine in the same way to assess floorings.

Since 1989, work at the Health and Safety Laboratory, Sheffield, has shown that in wet conditions the roughness or surface texture of flooring is of crucial importance to slip resistance. Irregularities on the floor surface can break through a water film to establish contact between shoe heel or sole and the flooring. On very smooth surfaces, complete hydrodynamic lubrication is possible. It is not certain what type of roughness needs to be measured, but research has shown that peak-to-trough roughness (Rtm) gives a useful guide to the likely slip resistance in wet conditions. It is generally considered that an Rtm roughness value of at least 10 μm is required for all floorings. The UK Slip Resistance Group recommends that roughness measurements are made at the same time as the TRL pendulum measurements are taken.

Figure 1.30
The TRL pendulum tester. BRE has adapted it for measuring the slip resistance of flooring

† At the time of writing, this matter had not been resolved.

Materials

A flooring material cannot have a unique slip resistance (or coefficient of friction) which applies in all circumstances. It has to relate to some heel or sole material under specific conditions including whether the floor surface is wet or dry, wax polished etc. For example some types of microcellular polyurethane rubber solings consistently give higher slip resistance values than most other solings, and PVC solings tend to give lower values. Also, that wet floorings are more slippery than dry ones, and that oily floors are even worse than wet ones.

Most floorings when dry have adequate slip resistance, but many smooth floorings can become dangerously slippery when wet. The wet slip resistance of all floorings is characterised by its roughness or texture: the smoother the flooring, the more slippery it will be in the wet.

The choice of a material for a floor must take into account not only its properties when new but also its probable behaviour when worn. Taking concrete as an example, it may start relatively smooth, and therefore not be very good in the wet. In use, it may wear to provide a texture which will improve the slip resistance, particularly in the wet. On the other hand concrete finished with a fine texture may wear smooth in use and become dangerously slippery when wet. The slip resistance of concrete can be improved and maintained by introducing carborundum or aluminium oxide granules into the surface layer. Similarly, ceramic tiles can be finished with a specially roughened surface, either by introducing hard materials into the clay before firing or by moulding a pattern on the surface. Special grades of PVC flooring are available which incorporate particles of carborundum or aluminium oxide. Resin floorings commonly have hard materials sprinkled onto the surface before they set to improve wet slip resistance. In all these cases the intention is to provide the flooring with some texture to improve and maintain slip resistance in the wet.

Old floors may have become slippery in use because they have worn smooth. Some, like concrete, can be improved by shotblasting the surface to provide a texture. Others can be improved by coating with epoxy resin into which hard aggregate, like carborundum or crushed flint, is sprinkled; a second coat of resin is usually then applied to partially close in the surface. Most treatments to existing floors will alter their appearance.

However, adequate performance with respect to slipperiness is also bound up with maintenance, an aspect which it is impossible for the specifier to control. Both dirt accumulation and polish can make a floor more slippery. Any floor can be made dangerously slippery when wet by polishing, and, unfortunately, a high degree of polish is associated with cleanliness in the minds of many people involved in the cleaning process.

Certain synthetic sealers are specially formulated to provide slip resistance for floors. Polishes made from soft waxes tend to make hard floors more slippery. A well known hazard occurs when timber floors are maintained with wax polish; the polish can be tracked onto adjacent hard floors, like ceramic or terrazzo tiles, which become dangerously slippery, even in the dry.

Generally the smoother and less porous a floor surface, the easier it is to clean. However, for safety in the wet some texturing of the floor surface is required. If the texture is made too coarse then it becomes very difficult to clean. The ideal is to provide a floor which has sufficient texture to provide slip resistance in the wet but still allows it to be easily cleaned. The degree of roughness required for slip resistance in the wet is not as coarse as is often thought. Peak-to-trough roughness (Rtm) values of between 10 μm and 20 μm are usually adequate for areas which are frequently or permanently wet like industrial kitchens, dairies and breweries. Where liquids with higher viscosity than water (eg mineral oil) are frequently spilt, higher roughness

Table 1.7
Coefficients of friction between floorings and smooth rubber soles (broad range)

Material	Dry*	Wet+
Carpet	> 0.7	0.4–0.6
Ceramic tile, smooth	> 0.6	0.2–0.3
Ceramic tile with carborundum in finish	> 0.6	0.3–0.5
Concrete, smooth	> 0.5	0.2–0.3
Concrete with carborundum in finish	> 0.6	0.3–0.5
Cork	> 0.7	0.3–0.5
Linoleum, smooth	> 0.7	0.2–0.35
Mastic asphalt, smooth	> 0.7	0.2–0.3
PVC, smooth	> 0.7	0.2–0.3
PVC with carborundum or aluminium oxide in finish	> 0.7	0.3–0.45
Resin, smooth	> 0.7	0.2–0.3
Resin with carborundum in finish	> 0.7	0.3–0.5
Rubber, smooth	> 0.7	0.2–0.3
Stone	> 0.5	0.2–0.45
Terrazzo, smooth	> 0.5	0.2–0.3
Terrazzo with carborundum in finish	> 0.6	0.3–0.5
Thermoplastic or vinyl asbestos tiles	> 0.7	0.2–0.35
Wood	> 0.5	0.2–0.4

* Clean, no polish (contamination can greatly affect values)

\+ On very smooth floorings of all types, the coefficient of friction can be as low as 0.1. Coefficient of friction in the wet is greatly dependent on surface texture of flooring

Figure 1.31
Stair nosings are generally the most critical factor in safety on staircases. Most nosings are compromises between safety and propensity to tripping

values may be required. A floor which has adequate slip resistance will not remain so if contaminants are allowed to build up such that the texture is masked. These conditions are as unnecessary as they are undesirable. For large areas, machines for sweeping, washing, scrubbing and drying are readily available.

In some industries like food processing, it may be necessary to clean the floors during shifts to keep them safe and hygienic.

Because the coefficient of friction of any flooring is a function of many factors, like the nature and texture of the surface and the shoe material with which it is in contact, it is not possible to allocate a precise value. However, Table 1.7 on page 35 indicates the broad range of coefficients which can be expected when smooth rubber solings are in contact with various types of flooring.

Bare foot slip resistance in wet areas like swimming pool surrounds and showers is complex and beyond the scope of this book.

Although the coefficient of friction of not less than 0.4 for safety is well established for walking on the level, higher values will be required for slopes and ramps. Hand rails provide additional security for the user.

Smooth floorings which are continuously dry and safe may become temporarily unsafe during cleaning operations. For good housekeeping, warning signs should be erected.

Certain synthetic sealers are specially formulated to provide slip resistance for floors.

Steps and stairs

Staircases provide one of the most hazardous situations in a building (Figure 1.31), as already mentioned in the section on accidents. So far as the horizontal surfaces (ie treads) are concerned, the following points will perhaps need to be monitored and reported on in any consideration of maintenance and refurbishment.

Changes of level and of material need to be made obvious to users, and this generally means good lighting conditions, a contrast in floorings either side of a change of level, or a strip of contrasting appearance such as a nosing (Figures 1.32 and 1.33). At a change of level, flooring such as a textile which exhibits a strong or dazzling pattern, and which can camouflage the actual situation of the step or nosing to the

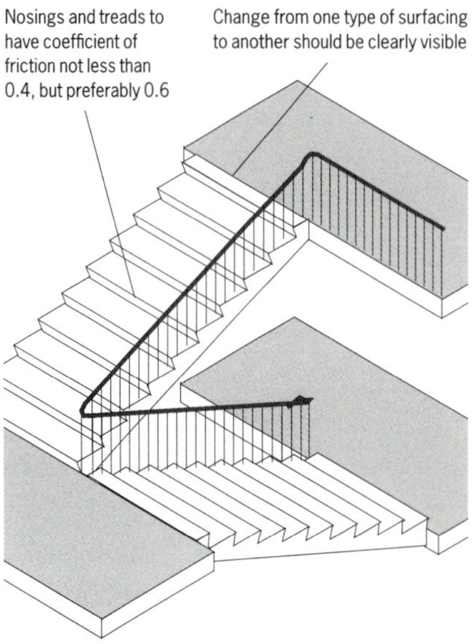

Figure 1.32
Staircase and landing nosings and treads need to be slip resistant and easily visible

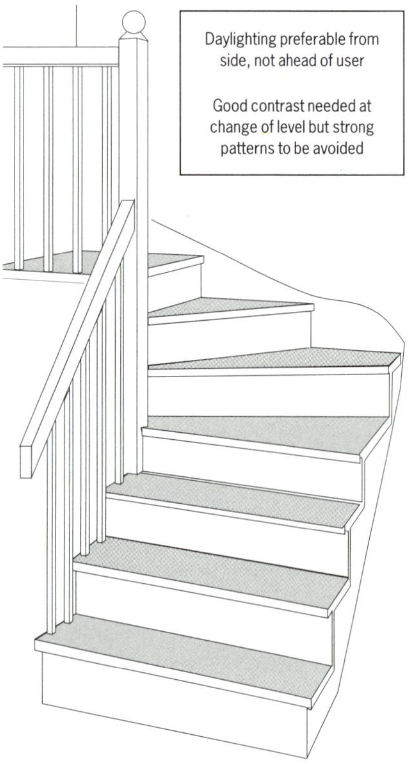

Figure 1.33
Lighting conditions are crucial to safety on staircases, especially those with winders

user, should be avoided. This phenomenon of indiscernible change of level will be easily understood by anyone who has had to walk down a stationary escalator which has shiny steel ribbed treads where one tread blends with that beneath it.

If a single step is isolated from any other change of level or occurs in an unexpected place, it is likely to cause a fall (Figure 1.34). Where a change of level is expected, at thresholds or entrance to buildings, it is not dangerous for the able bodied. A single step which already exists can be made safer by a definite change of colour and good lighting. An adjacent handrail may be provided to indicate a change in level.

It is probably better, where circumstances permit, to replace the step with a ramp. Ramps or slopes should not be steeper than 1 in 12, and preferably be shallower than 1 in 15.

With the increase of factory made building components, thresholds to internal doors are likely to become more common. If fitted carpet is used each side of a thin section it is not dangerous, but if a thick section is used it should be tapered off. A threshold associated with a change of level is not so dangerous since it should be obvious.

Figure 1.35
There are obvious hazards at a tiled surround to a swimming pool. A contrasting edge helps to improve safety

Figure 1.34
Changes of level in unexpected places could lead to people tripping or falling

Clause 5.25 of the Building Regulations 1991 Approved Document B: Fire Safety[59] recommends that the floorings of all escape routes should be chosen to minimise their slipperiness when wet.

Protection against radon and methane

Protection against radon and methane from the ground is a relatively new development in buildings. Although the problems posed by the two gases are different, where they occur together in the same property they can sometimes be dealt with simultaneously. However for existing buildings any radon reduction measures proposed, where there may be a methane problem as well, need to be designed to take this into account; for example fans will need to be protected against sparking.

Radon

Radon is a colourless, odourless gas which comes from the radioactive decay of uranium and radium. Uranium and radium (which itself is a decay product of uranium) are to be found in small quantities in all soils and rocks, although the amounts vary from place to place. They are particularly prevalent in areas of the country where granite predominates.

Radon in the soil and rocks mixes with the air in the ground and rises to the surface where it is quickly diluted by the atmosphere. However, radon which enters enclosed spaces within or underneath buildings can reach relatively high concentrations in some circumstances.

When radon decays it forms tiny radioactive particles which may be breathed into the lungs. Radiation from these particles can cause lung cancer which may take many years to develop. In addition, the combination of smoking and exposure to radon are known to increase greatly the risk of lung cancer.

Most buildings in the UK do not have significant radon levels. The problem of radon exists mainly in the SW of England, the north Midlands, parts of the Pennines, the Grampians and Highlands of Scotland, and parts of Northern Ireland.

Concentration levels of radon gas within buildings are determined often by the radon entry rate. This in turn is determined principally by building characteristics (especially floor construction) and the pressure differences between soil, gas and air within buildings. The phenomenon affects ground and basement floors. With suspended ground floors that are well ventilated, the risk of radon percolating into the dwelling may prove to be acceptably low, though the risk is higher where airbricks are absent, blocked or incorrectly placed occur (Figure 1.36), or where there are large gaps around service entry points (Figure 1.37).

For new-build, the advice given in *Radon: guidance on protective measures for new dwellings*[60] and *Construction of new buildings on gas-contaminated land*[61] should have been followed. Whether protection measures are required against radon depends on the geographical location of the building.

For existing buildings in those areas of the country known to produce high levels of radon, or where it is suspected that the level might be high, the first action should be to undertake a radiological survey of the buildings. This can be carried out by placing detectors, which can be obtained from the National Radiological Protection Board (NRPB) and other commercial suppliers, in suitable places.

Remedial measures will depend on the radon level measured. The Government has set a recommended action level of 200 Bq/m^3 and it is recommended that for buildings with levels above this value, action should be taken to reduce them.

The main methods of remedy in existing dwellings and protection of new dwellings are:

- radon sumps
- positive pressurisation
- subfloor ventilation
- sealing and ventilation of dwellings

Whereas modern in situ floors above good permeable fill are fairly easy to deal with by means of a radon sump, old concrete floors, often poorly constructed on beaten earth, are much more difficult. The experience with suspended timber floors has been more variable. Natural ventilation, mechanical supply ventilation and mechanical extract ventilation are three techniques that have been used successfully as radon remedial measures (Figure 1.38). Preliminary results of ongoing research suggest that supply ventilation is more effective than extract ventilation, particularly where the radon levels are very high.

A sump works by changing the pressure field in the soil below a building so that air flows from the building down into the soil, preventing radon entry. It may fail if the pressure is not reversed across all of the floor. The heating cost of a sump system due to extra ventilation must also be taken into account. Without adequate information about the soil below a building it is not possible to predict with accuracy the way a sump would operate. For radon, the sump usually operates in suction. Only where the soil is very permeable, or methane is also present, would a pressurised sump be recommended.

Where it is already known that the building undergoing rehabilitation has a radon problem, simple remedial measures can be included with the other works

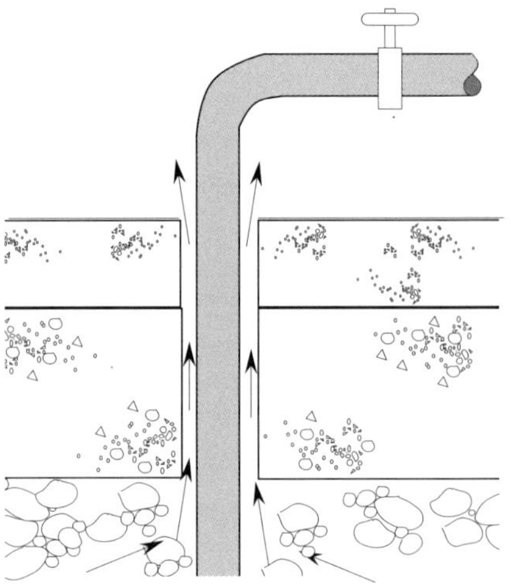

Figure 1.37
Radon may enter habitable spaces at large gaps around service entry points

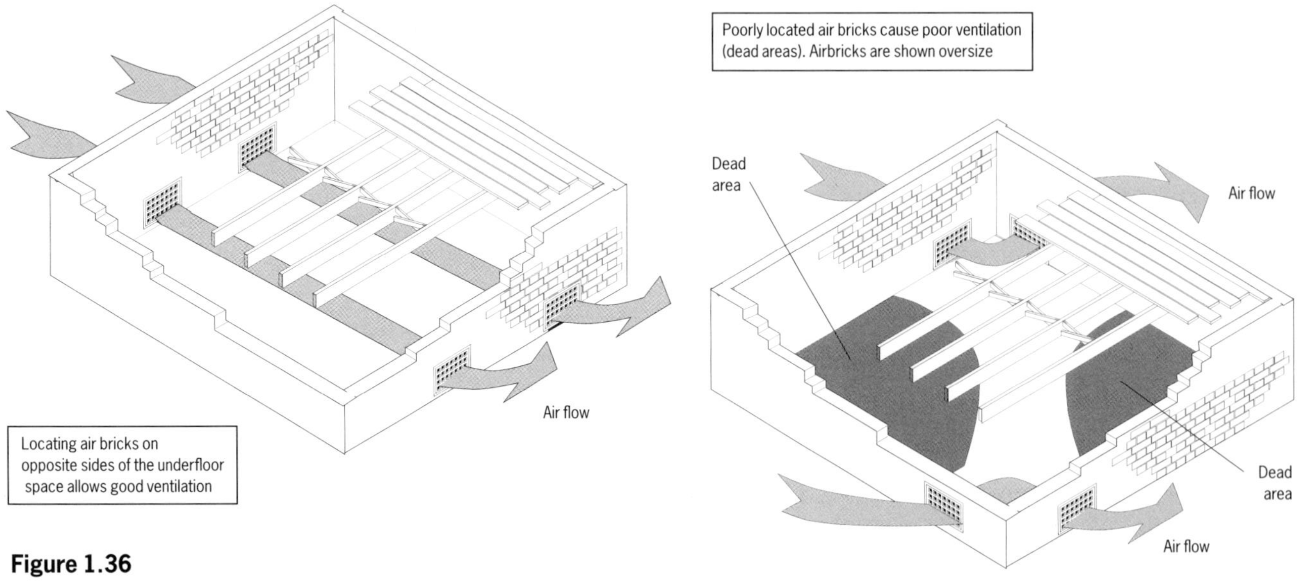

Figure 1.36
Effective and ineffective ventilation routes in traditional suspended timber floors

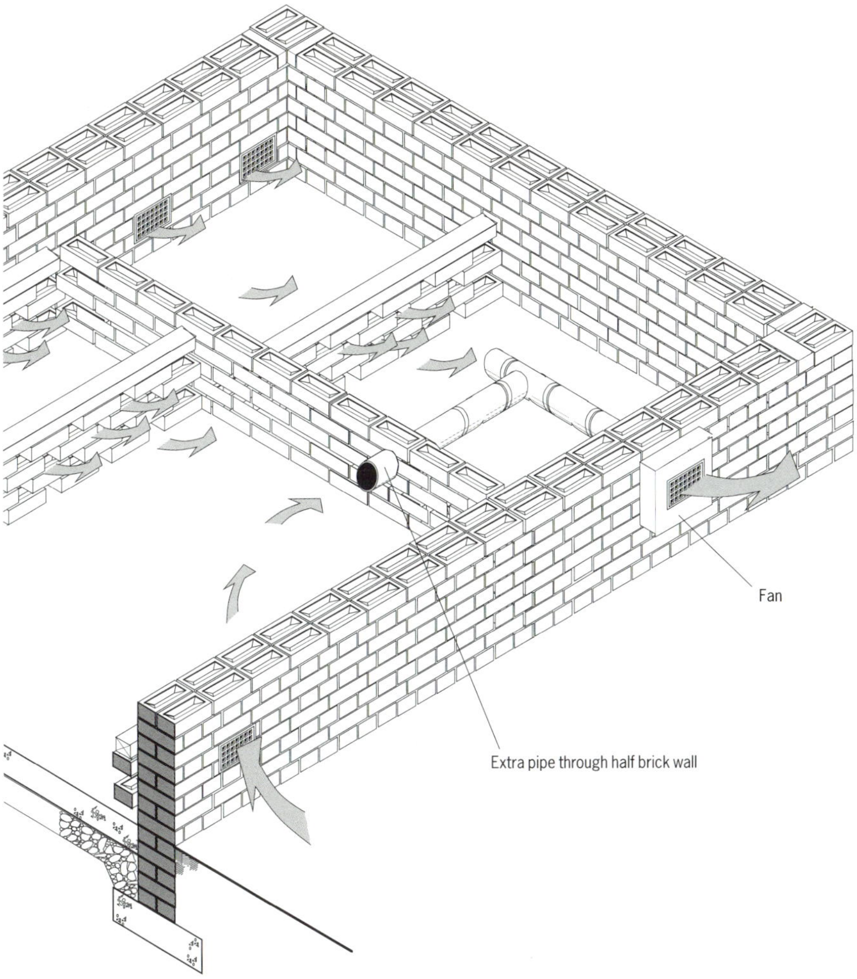

Figure 1.38
A fan inserted into the external wall, manifolded to ensure that all internal underfloor voids are depressurised, is one way of removing radon from the voids and thereby reducing its entry into a dwelling. However, in many properties there will be sufficient missing perpend joints, leading to relatively high air leakage rates through walls, to obviate the need for installing a manifold

using track etch detectors. Some of the dwellings were protected while others were not. Four factors were considered for their likelihood to affect the annual average indoor radon level:

- the radon level of the area
- the type of floor construction
- the presence or absence of a protective membrane in the floor
- whether the dwelling was detached or attached to other dwellings

The radon level of the area is determined by the proportion of existing houses in the area above the action level. The type of construction examined was generally in situ concrete floor or beam and block floor. Houses were situated on a total of 33 sites in Devon and Cornwall offering a variety of both terrain and housing types. Statistical analysis showed that the effect of the floor membranes in reducing radon levels was significant. The effect of the floor construction was, however, less significant. Neither the radon levels of the areas in which the dwellings were situated, nor whether the dwellings were detached or attached, had statistically significant effects on the radon levels.

(Figure 1.39). In cases where the building has not been or cannot be monitored for radon prior to carrying out works, it might be appropriate to incorporate precautionary measures. Such measures can help to reduce indoor radon levels, and make it easier to remedy any future problem. This is of particular relevance when converting redundant farm or other outbuildings into living accommodation, or carrying out major works such as floor replacement in older properties.

Field trials have taken place to verify the contribution of various radon prevention measures implemented under UK building regulations. Measurements were carried out in a total of 423 dwellings

Figure 1.39
Radon extraction equipment has been partly installed under the suspended ground floor of this dwelling. The sump is in the site concrete, beneath the vertical pipe

Using the BREVENT 'radon parameter', the levels of radon in a house with an underfloor duct have been compared to an identical house without the duct. It was found that the performance of the duct depends very much on the weather conditions. For house characteristics that gave a high radon parameter the duct did have a beneficial effect; however, with characteristics that gave a low radon parameter the added effect of the duct was negligible. The results indicate that, for a house with a radon problem the reduction in indoor radon levels that can be achieved by sealing the gaps in the flooring and increasing airbrick area is comparable to and more reliable than an underfloor duct.

Reducing radon levels in houses may involve depressurising the soil or fill under the houses using ducts and fans. The fans are often mounted in the roof spaces. The noise generated by these fans, and perhaps also by fans installed in underfloor sumps, is potentially disturbing to occupants. However, with good design these difficulties can be usually overcome (BRE Good Building Guide GBG 26[(62)]).

Remedial measures are available for concrete floors laid directly on the ground but some of the principles could be also used with suspended concrete floors and, to a lesser extent, with floors consisting of large stone slabs.

Methane etc

Incidents involving landfill gas in buildings have increased since the 1970s as shortages of suitable building land became more acute and landfill sites were brought into use. The principal components of landfill gas are methane (which is flammable) and carbon dioxide (which is toxic), so if either enters a building it can pose a risk to both health and safety.

Landfill gas can enter buildings through gaps around service pipes, cracks in walls below ground and in floor slabs, construction joints and wall cavities. It can also accumulate in voids created by settlement beneath floor slabs, in drains and soakaways, and in confined spaces within buildings such as cupboards and subfloor voids.

Preventive measures include installing low permeability ('impermeable') gas barriers and high permeability layers from which gas can be extracted in a controlled manner[(61)]. A simple way of protecting a building from gas in the ground is to have a gas proof membrane in conjunction with a reinforced concrete floor slab above a granular venting layer. This layer should have a minimum thickness of 200 mm. The membrane can either be sandwiched between the floor slab and the granular layer which would then need to be blinded, or it can be positioned above the floor slab and protected with a floor screed. Service penetrations through the membranes should be kept to a minimum. The granular material should be highly permeable (no fines) and not heavily compacted. The gas can then be vented to the atmosphere by means of a granular trench around the perimeter of the house. Regularly spaced pipes (every metre or so) can provide a pathway through foundation walls into the trench.

Instead of a granular venting layer, a passively vented subfloor void can be used to protect buildings. Precast concrete flooring with a gas proof membrane and protective floor screed would be suitable. The recommended minimum area of ventilation for a subfloor void is 1500 mm^2 per metre run of wall or 500 mm^2 per square metre of floor area, whichever gives the greater area of opening. The void should be well cross-ventilated (airbricks located on opposite walls) and there should be plenty of air gaps in sleeper walls to ensure that there will be no pockets of still air where gas could accumulate. Air should be allowed to flow freely through the airbricks and they should not be blocked by debris (leaves etc) or covered up by further construction around the building.

Suspended timber floors will be found to be unsatisfactory on landfill sites where there is a risk of gas contamination. Attempts at remedial work on such floors should take into account the risk of inducing rot where gas proof membranes are placed over the structural floor.

Gas detection equipment can be used in commercial and industrial buildings. Monitoring can be either periodic, carried out using portable equipment, or continuous, using permanently installed gas detectors. (It is recommended that buildings be evacuated if the indoor methane and carbon dioxide concentrations exceed 1% by volume and 1.5% by volume respectively.)

Aldehydes and volatile organic compounds

Some binders used in the manufacture of wood based products and boards, and adhesives for timber laminates used in floorings, contain aldehydes, formaldehyde in particular. The vapour is an irritant and, in sufficient concentrations in the internal atmospheres of buildings, can cause personal discomfort. In old buildings which were less well sealed than their current counterparts, the concentrations tend to be low. The specifications of the boards were amended when the problem became apparent and new boards conforming to the latest specifications should not present these problems.

Some products emit other volatile organic compounds (VOCs) which, in sufficient concentrations, may be injurious to health. They include solvent naphtha and related compounds from DPM materials, and plasticiser from degraded PVC floor coverings. To enable specifiers to select suitable products, information is available about the types and amounts of VOCs emitted. Tests are carried out according to a guideline drawn up by a working group of the European Communities (European Concerted Action: Indoor Air Quality and its Impact on Man Report No 11[(63)]).

So far as DPMs are concerned, the main sources of VOCs are those based on coal tar which may contain naphthalene and related products. Bitumen adhesives for thermoplastic tiles and semi-flexible PVC tiles traditionally contained solvent naphtha. Gum spirit adhesive for linoleum and cork contained methanol/ethanol, and rubber solution adhesive for rubber flooring contained ketones. Because of the Control of Substances Hazardous to Health (COSHH) Regulations 1988[64], these adhesives are now being phased out and replaced with water based adhesives, but it is doubtful whether some of the replacements are technically as good as the products they replace.

Thermosetting resin floors contain solvents which release VOCs over a period of time. The most common solvents used with particular resins are:

- toluene with epoxies
- xylene with polyurethanes
- styrene with polyesters and acrylics

Seals and floor paints contain similar solvents and in addition some may contain white spirit.

Plasticised PVC flooring may contain a source of VOCs. Plasticisers are very high boiling point liquids with very low volatility. However, if flexible PVC flooring comes into contact with strong alkalis, either from the base screed or concrete, or from cleaning materials, the plasticiser can be saponified to form a volatile alcohol.

Traditional polishes and dressings contained volatile solvents. These were commonly white spirit and methylated spirit. Except for a few wax polishes available for timber flooring, solvent based materials have been replaced by water based emulsions. Natural products such as beeswax are emission-free.

Anti-static flooring and static electricity control

This is a very complex topic, and no more than an outline is given here.

In some circumstances, as where there is a risk of explosions, it is dangerous to produce sparks, whether by friction of hard materials on a hard finish, or by static electricity. To prevent friction sparks, a soft finish is required, and wood, special grades of asphalt, magnesite and latex cement have been extensively used in the past. More recently, rubber and other soft polymer materials have been used.

The prevention of sparks (arcing) and voltage discharges resulting from static electricity is crucial in some situations. People may acquire an electrical charge by moving across a floor, and this can be discharged as a spark when next in proximity to a lower potential. The electrical charge generated is the result of a number of factors, one of the most important being the resistivity of the flooring. To prevent the generation of static electricity, floorings are available with significantly lower resistivity than normal. These floorings are usually referred to as anti-static floorings and they have defined upper and lower limits of resistance. The upper limit is to control the generation of static electricity, and the lower limit to provide a safe working level in the environment against mains voltages. Resistance values in the range 10^7–10^4 ohms will often be quoted. Any value measured will depend on the total floor systems, the test method and the prevailing conditions.

Anti-static floorings may be specified for several reasons or circumstances, the following being the most common:

- where flammable solvents are used in industrial areas
- protection of microelectronics during manufacture, assembly or repair
- reduction of electrostatic charges in clean rooms to prevent attraction of particles to surfaces
- munitions and fireworks factories

Up to 1993 the use of anti-static floorings in operating theatres was obligatory, but since then these floorings have been optional as the risk from ignition of gases used in surgery has been considerably reduced (Figure 1.40).

Special anti-static floorings based on terrazzo, PVC, linoleum and various resin systems have been developed. These are dealt with under their respective headings in Chapters 5–7.

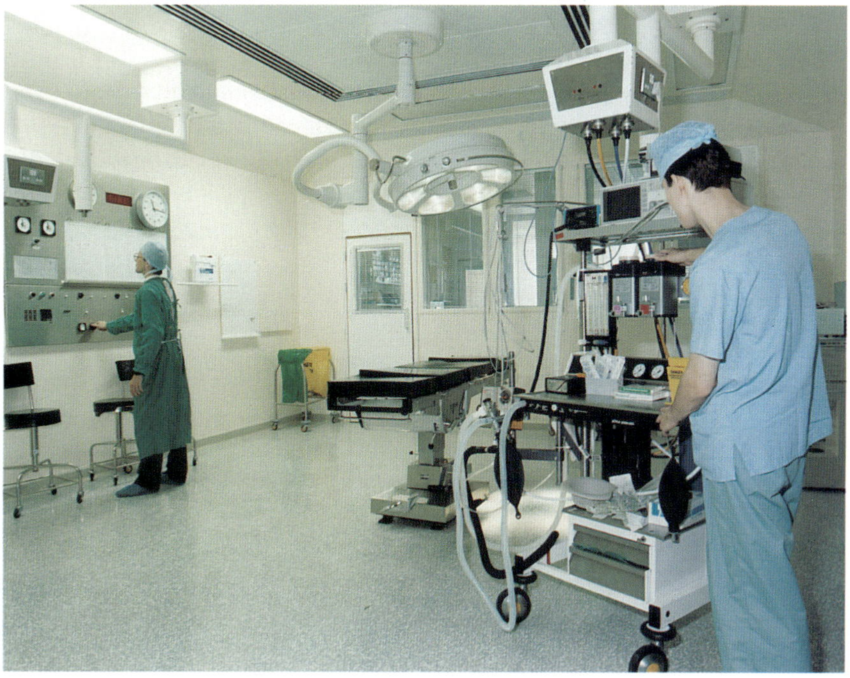

Figure 1.40
Anti-static flooring was formerly required in parts of hospitals where there was a risk of ignition of gases used in anaesthesia. The risk is considerably reduced now and anti-static flooring may not always be required

Levelness and accuracy

Any floor must be reasonably flat and level for safety reasons, and so that furniture standing on the floor is visibly plumb and level. The limits for local variations for newly built, nominally flat floors include three categories: up to 3 mm, up to 5 mm and up to 10 mm under a 3 m straight edge laid directly on the floor. Although these figures refer to concrete floors, levelling screeds and in situ flooring generally, there seems no reason why they cannot be applied to other floors as appropriate. The middle category is the one most often used in domestic and office buildings.

For some specific purposes, such as when specialised equipment is to be used on the floor, tighter specifications are required†. These can be expressed either as deviations under a 3 m straight edge or as maximum curvature over a length of 600 mm. Examples of the need for very flat floors are warehouses with very high racked storage where lifting trucks run in aisles or where access over the whole of the floor areas is required. Since the racks can be up to 13 m high, it is important that truck masts are kept vertical to avoid danger of collision with the racks.

TV studios are another example where greater accuracy is often required, albeit in small areas. For assessing very flat floors, electronically operated measuring devices are used.

Some difference in height across joints and cracks may also be tolerated. For joints in excess of 6 mm width, a variation in height of 2 mm is usually acceptable; but for cracks and joints under 6 mm width, the difference should not normally exceed 1 mm. However, for some specialist floors, no abrupt change in level can be tolerated.

In industrial situations where liquids are spilt or discharged onto the floor, adequate gradients should be provided to the finished floor surface to prevent ponding. If the surface is smooth and there is little spillage of aggressive chemicals, a slope of 1 in 80 should be regarded as the minimum with 1 in 60 specified to be on the safe side. Slopes greater than 1 in 40 can be dangerous in some circumstances, particularly where wheeled traffic traverses the floor. The direction of falls should be planned with traffic in mind so that the traffic will move across rather than up or down the slope.

Older buildings which have been in use for some considerable time often show significant inclines, particularly, say, in medieval timber framed structures (Figure 1.42). It will be a matter for client decision and professional judgement whether or not such slopes can be tolerated.

† Methods of measurement appropriate for floors built to higher standards are under discussion at the time of writing, and an International Standard is being drafted.

Figure 1.42
The horizontal picture frame betrays the slope in the joists and the floor above

Where existing floors are out of level by more than say 40 mm in 3 m, there will certainly be problems with furniture (Figure 1.43). It has been suggested that such floors should, if possible, be relaid to level (*Assessing traditional housing for rehabilitation*[7]).

Suspended floors will always have some degree of spring; if it is excessive it can cause problems with furniture rattling, ornaments falling off shelves in cabinets and with hi-fi systems. Floorboarding, particularly plain edge boarding, must be sufficiently secured and supported to resist differential deflections in adjacent boards, and also impacts, not only from foot traffic, but from point loads from furniture too.

In housing, small changes in floor level represent a hazard and will need to be eliminated or avoided where possible, particularly if young families or elderly people are likely future occupants. See the section on safety earlier in this chapter.

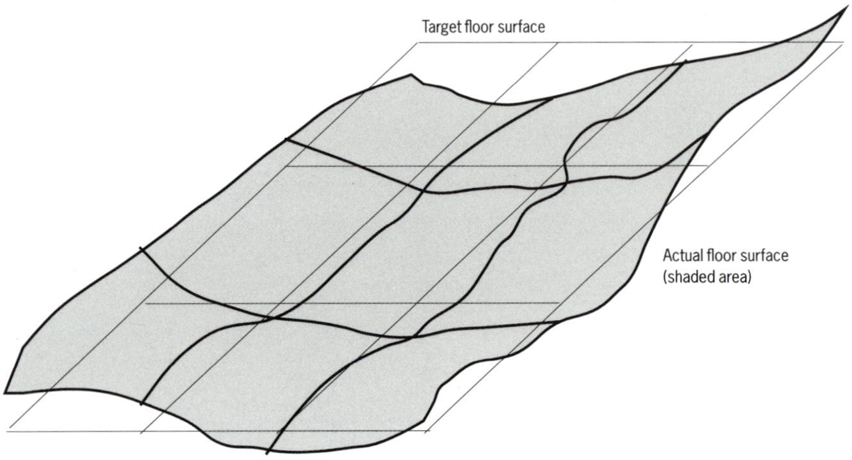

Figure 1.41
The undulations (exaggerated and shown shaded) in a floor surface compared with a flat target surface (shown as the uniform grid)

Hygiene

Certain floors are required not to harbour dirt and germs, and should not allow water to penetrate below the surface. Hot welding of, for example, flexible PVC and linoleum sheets with rods of compatible material, given adequate standards

Figure 1.43
Sloping floors are not confined, unfortunately, to medieval buildings. Even floors constructed more recently may show significant slopes. The reasons for distortion should always be investigated

of workmanship, can provide a continuous surface which offers considerable advantages in hospitals and certain industrial process areas. To provide a virtually tanked floor area, sheet flooring material can be formed with radiused corners at the floor perimeter and welded at the corners of the upstands. The integrity of the weld is of paramount importance.

Thermosetting resin floorings of epoxy or polyurethane can provide good resistance to a wide range of chemicals and be laid in situ to provide a smooth hygienic surfaces. These materials find wide use in food and pharmaceutical process areas and clean rooms.

Chapter 1.6 Fire and resistance to high temperatures

Of the 600,000 or so reported fires which occur annually in the UK and which are attended by fire brigades, about 1 in 5 happens in buildings, the rest in vehicles, industrial plants, forests etc. About half the fires in buildings occur in dwellings (around 65,000 in 1996) and half in other building types.

Although the number of fatalities in domestic fires has been falling since 1990, the number of casualties remains approximately the same. In this time the average number of fatalities in fires in dwellings has dropped to just over 500 per annum.

Separate statistics are not kept for fires which affect floors, although serious local damage to timber floors following a fire is not unusual (Figure 1.44).

In medieval times, the floors of farm houses and manor houses in Nottinghamshire and Derbyshire where gypsum was plentiful, often consisted of a paste of calcined gypsum and water spread on laths or reed laid on wood joists (Building Research Station National Building Studies Special Report No 27[65]). These floors could have a very good performance in fire (Figure 1.45).

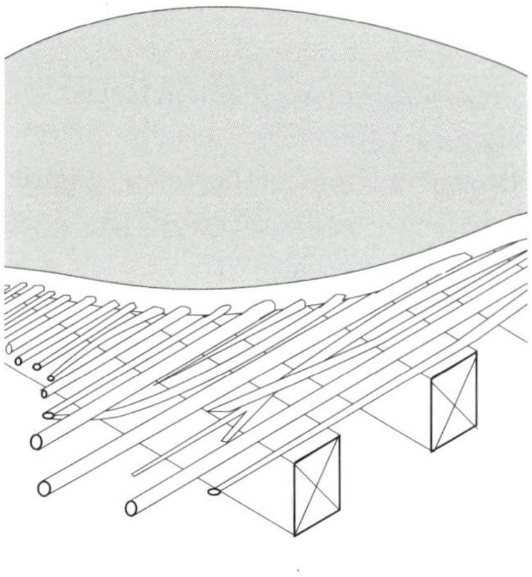

Figure 1.45
A gypsum floor carried on reed bundles, laths or hazel rods

Figure 1.44
The aftermath of a fire which has seriously affected a domestic timber floor. The fire has been confined to a relatively small area of the floor/ceiling cavity. Hardboard covered the plain edged boarding, the effect of which was to restore the performance given by tongues and grooves to the boarding. Both hardboard and boards had been cut away to assist with firefighting. The crocodiling seen on the sides of the joists and trimmer is caused by fire penetrating the weaker parts of the timber (eg caused by rot)

In cottages, the laying of gypsum floors continued into the nineteenth century. So long as the joists held, this type of floor has been said to be highly resistant to fire. The floors of typical multi-storey factory and warehouse buildings of the late eighteenth century, however, were of thick plank carried by solid timber beams resting on posts of timber. In a fire, massive timbers burn but slowly, and modern measurements show that, in standard furnace tests, from 15 to 25 mm of depth is charred on all exposed faces every 30 minutes, depending on the species of timber. If the fire is prolonged, the interior of the timber becomes steamy and weak, and ultimately the beams collapse and the collapsed floor structure adds fuel to the fire.

The first fireproofing liquid to be used on a significant scale, based on alum, was introduced in the UK by Wyld in 1735. Many developments took place in France where, for example, hollow pots were introduced about 1785. A development of these pots, made in London, was used in buildings of note in the London area between 1825 and 1838 (eg in Buckingham Palace, the National Gallery and the Treasury). Unfortunately these floors behave unpredictably in fire.

Brick arches carried on timber or wrought iron beams, with the spandrels filled with non-combustible materials and topped with stone flags, were introduced around the beginning of the nineteenth century. Although these floors were called fireproof, they were not in fact so since no protection was afforded to the lower flange of the beam (Figure 1.46). Any rivets also constitute a weakness in fire. Constructional aspects of these floors, and of vaults, are dealt with in Chapters 2.4 and 2.5.

Hollow tile floors came into more general use around 1905, and one of the earliest buildings to use Kleine hollow tile floors was the Imperial Hotel in Russell Square, London. So by the end of the nineteenth century, it was possible to construct large buildings with non-combustible materials. However, this did not prevent fires since the fire load of a building might include considerable quantities of combustible material as fittings and furnishings.

The objectives of fire precautions

The designer of a building or of alterations to a building, that is to say whether it is new or refurbished, will need to know a great deal about the conditions that will be imposed on the floors and their ceilings within that building when it is in service. Factors such as loading, environment and durability all have to be understood and assimilated into the design process, and considered in relation to behaviour in fire.

The objectives of floors as regards fire precautions include the following:

- providing adequate facilities for the escape of occupants
- minimising the spread of fire, both within the building and to nearby buildings
- reducing the number of outbreaks of fire

With respect to the first of these objectives, where roofs are used as means of escape they are considered to act as floors. They are dealt with in the companion volume *Roofs and roofing*[66].

Relevant practical aims arising from the second of these objectives are to limit the size of the fire by:

- controlling fire growth and spread
- dividing large buildings, where practicable, into smaller spaces

The various national building regulations and their supporting documents (eg Approved Document B[59] of the Building Regulations 1991) set minimum periods of fire resistance for floors according to the purpose group of the building and height above ground level, and it is not proposed to list these periods in detail here.

In the case of the lowest floor in individual premises, fire protection is not normally relevant. (Clause 7.4b of Approved Document B states that the lowest floor of a building is not considered to be an element of structure.) The London Building Constructional By-laws[67] did contain certain classes of building where no floors were required to be fire resisting, but outside the inner London area, certainly since 1965, all suspended floors above the lowest floor have needed in practice to be fire resisting to some degree. The Greater London Council 1972 By-laws also had similar provisions, but this changed in the 1979 Amending By-laws.

Non-combustible floors

As *Principles of modern building*[6] pointed out, non-combustibility does not of itself ensure fire resistance. Metals, for example, lose strength rapidly when heated above a critical temperature; any steel or aluminium used structurally must be insulated, and structural steel should also be protected by the use of fire protecting suspended ceilings. The fire resistance of solid reinforced concrete slab floors depends on the minimum thickness of the slab and on the thickness of cover to the reinforcement. With hollow floors, including precast concrete or clay units, it is the minimum total thickness of solid material which determines their fire resistance. The fire resistance of prestressed units, though influenced by many factors, depends mainly on the thickness of cover to the steel.

Compartment floors

Compartment floors, which are in essence needed in conjunction with walls to separate one kind of risk from another or where a portion of the building (a compartment) must

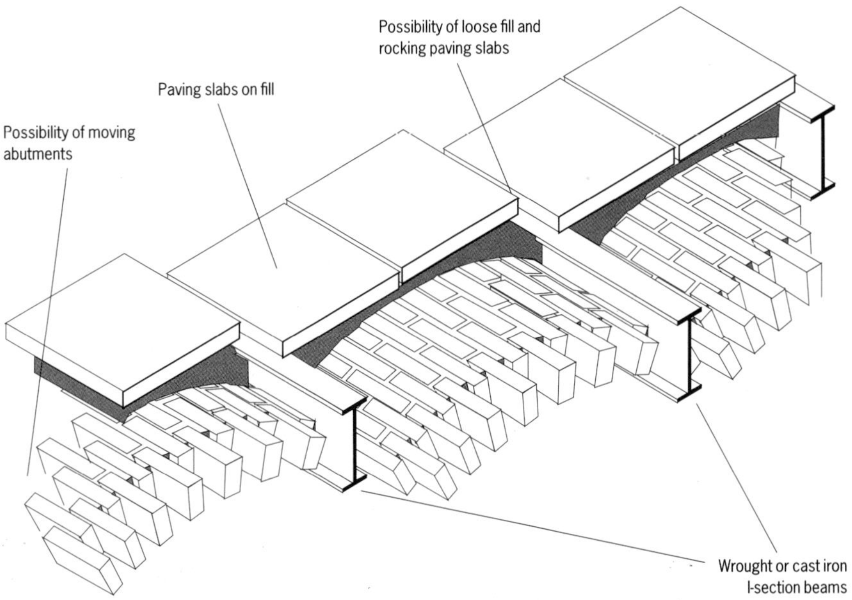

Figure 1.46
A brick arch, or jack-arch, floor

be restricted in size, are required to have enhanced requirements for performance in fire. To give an elementary example, every maisonette has to be a compartment; Approved Document B specifies that the floors above and below the maisonette should be compartment floors, but the internal floor, as an element of structure, simply has to be fire resisting (Figure 1.47).

The fire resistance of a floor is judged according to three attributes:

- loadbearing capacity
- integrity
- insulation

A fire resisting floor will also need to satisfy sound and perhaps thermal requirements, and upgrading an existing timber floor to achieve satisfactory values for these attributes may invalidate the fire resistance. An example is where mineral wool is inserted over, and is supported by, plasterboard alone without additional support. In a fire, the plasterboard will get hotter that much more quickly and is penetrated sooner. It is therefore important to provide the mineral wool with additional support, such as a wire mesh, so that it does not fall away when the plasterboard goes. Also it is important to note that the finished widths of timber joists to achieve adequate fire resistance must be a minimum of 38 mm; and there may be other kinds of problems such as calcium silicate board being insufficiently robust to carry extra weight. Furthermore, the underside of all floors exposed to rooms and corridors must satisfy provisions for surface spread of flame.

There is much else of relevance to floors and their fire protection in building regulations and their associated documents; for example in England and Wales Approved Document B, particularly in relation to domestic loft conversions (Clause 7.7, B1, Means of escape) and to compartmentation, concealed spaces and the need for cavity barriers (B3, Internal fire spread: structure). Notional periods of fire resistance for floors are given in *Guidelines for the construction of fire-resisting structural elements*[68].

Domestic loft conversions are not dealt with in this book, though it can be said that, so far as an existing loft ceiling is concerned, the usual plasterboard layer will not suffice from a fire point of view without additional work. Also, new joists will need to be fitted to carry the extra loadings (BRE Digest 208[69]).

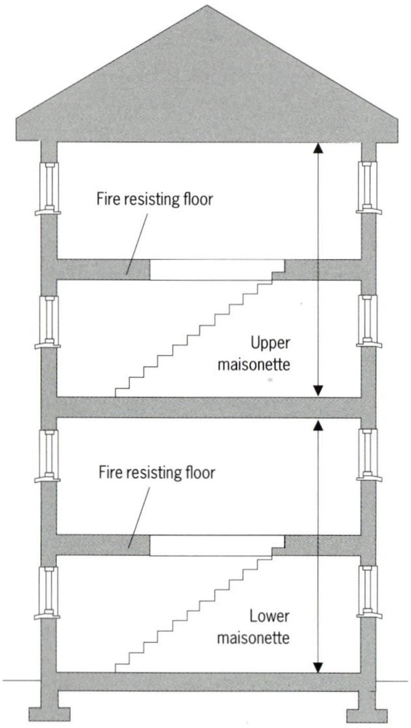

Figure 1.47
Floors above and below maisonettes need to be compartment floors, but those within just need to be fire resisting

Figure 1.48
The floors of galleries and balconies must be fire resisting

Openings

Where a floor needs to be a compartment floor, there are strict controls on openings in order to ensure that the effectiveness of the compartment has not been compromised. There are various ways of ensuring this; for instance by means of placing the opening within a protected shaft, or by means of a fire damper or shutter. However, where the floor is not a compartment floor, that is to say it is simply an element of structure, less stringent controls are applied depending on what penetrates the floor.

A protected shaft also requires the use of materials of limited combustibility and this will effectively rule out the use of certain categories of flooring. Approved Document B, though, allows combustible materials to be added to the upper surface of stairs except in the case of fire fighting stairs.

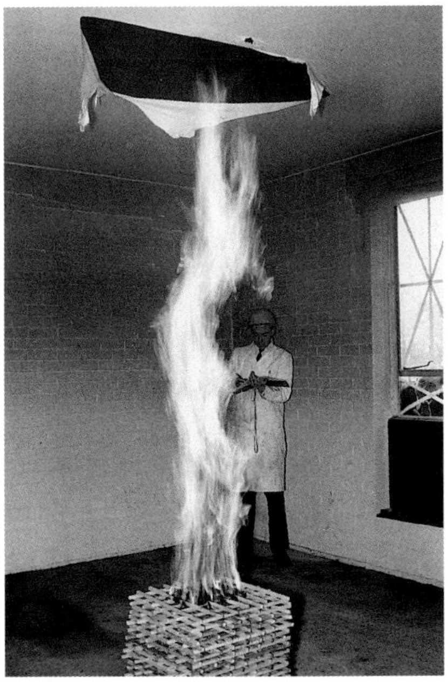

Figure 1.49
A fire test on a taut plastics sheet ceiling

Galleries and balconies

Galleries and balconies cannot be compartment floors since, by definition, they are incomplete floors. Fire and smoke simply bypass them. However, these floors need to be fire resisting and so resist collapse during fires (Figure 1.48).

Ceilings

So far as structural fire protection is concerned, it is the whole floor construction, including the ceiling if any, which determines the performance. The ceiling, whether suspended or not, may be required to contribute to the overall fire resistance of the floor or to prevent premature collapse of any beams which support the floor. An alternative to this is that, where the floor itself achieves adequacy of fire resistance, all that the ceiling is required to meet is the surface spread of flame requirement. A ceiling may not need to contribute to the fire resistance of a concrete floor but it may provide all or most of the protection needed for timber floors. Clause 9.2c of BS 5588:Part 5[70] recommends that the ceilings, stairs and landings within a fire fighting shaft should be constructed from materials of limited combustibility and not of the kind shown in Figure 1.49.

Most boards used for ceilings will be found to claim Class 1, though such claims are specific to each product or manufacturer. What is claimed for one particular product from one manufacturer is not necessarily applicable to a similar product from another manufacturer.

BS 476:Part 20[71] and Part 21[72] provide for fire test procedures for elements of construction†. So far as floors are concerned, these include the following components making up the floor element:

- floors (boards etc)
- beams
- suspended ceilings

Suspended ceilings can contribute to the fire resistance of a floor, or indeed may be needed to provide protection to the underside of a steel beam, but all ceilings are subject to control under building regulations for their surface spread of flame classification on the face exposed to the room below (Figure 1.50).

It is important to remember that, in relation to the floor as a whole, the fire tests are all carried out from the underside of the construction,

† Test methods on fire resistance are under development by CEN (Comité Européen de Normalisation) and these are expected to replace BS 476 test methods.]

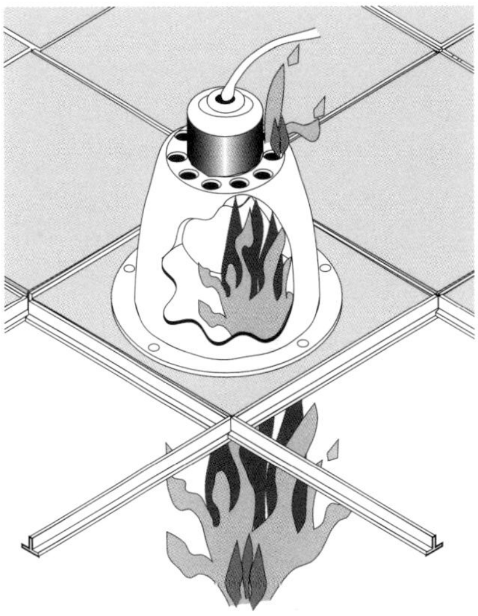

Figure 1.51
A downlighter which breaches a suspended ceiling makes the fire resisting qualities of the ceiling ineffective

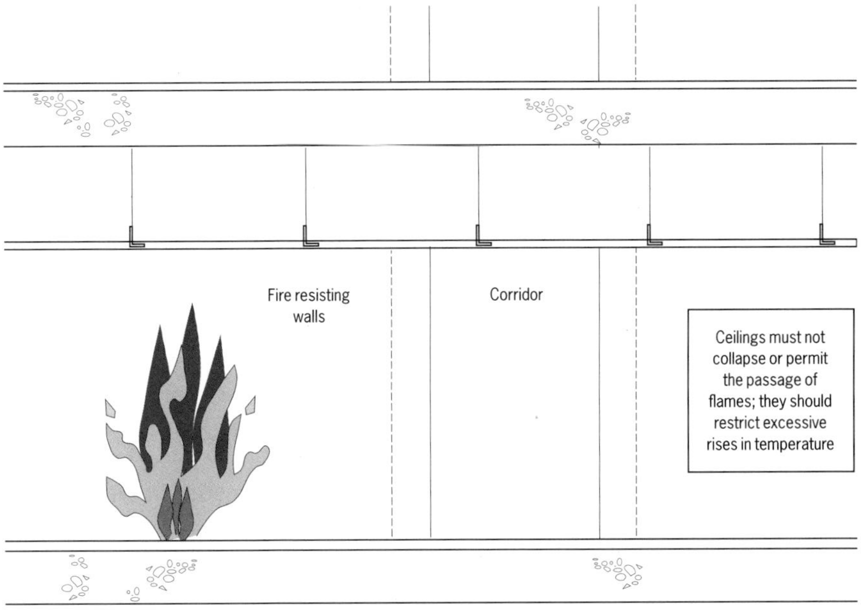

Figure 1.50
Suspended ceiling acting as a fire resisting membrane

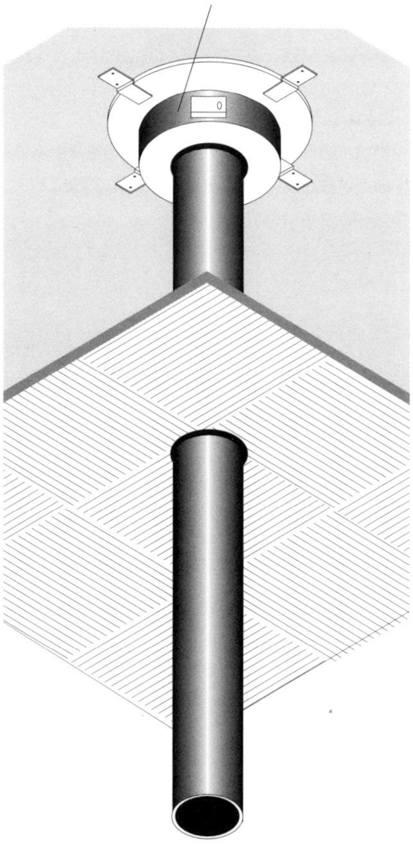

Figure 1.52
It is more effective for an intumescent seal to be contained within a box which is fixed to the soffit of a structural floor

and in consequence, suspended ceilings are not required to be tested for any resistance to fire already within the ceiling void. This is the reason why, in order to compartmentalise a fire, partitions flanking an escape route need to be carried up to the underside of the structural floor (BRE Information Paper IP 1/80[73]).

Recessed light fittings and ventilation grilles will, of course, puncture the ceiling, and it may not be possible to insert these therefore in refurbishment work (Figure 1.51 on page 47). Overlaying light fittings with fire protection material breaches the Institute of Electrical Engineers regulations (*Regulations for electrical installations*[74]).

Refurbishment

Where refurbishment is being undertaken on buildings controlled under the Fire Precautions Act 1971[75], the fire certificate issued by the fire authority will list conditions which will need to be maintained in the building provided the use has not changed. It therefore lays responsibility on all those concerned with refurbishment of buildings to make themselves aware of the contents of this certificate, and indeed with all other legislation affecting occupied premises.

There may also be insurance requirements under Loss Prevention Council rules (*Code of practice for the construction of buildings. Insurers' rules for the fire protection of industrial and commercial buildings*[76]).

Fire damage

Inspection of fire damaged structures should be carried out by those who are competent and experienced to do so. It should be noted that tests carried out to BS 476 are not intended to give guidance on the serviceability of such structures after exposure to a fire. Absence of spalling or absence of change of colour after a fire should not be taken as evidence that any reinforced concrete members are satisfactory since different aggregates vary in their behaviour in fire.

Cavity barriers and fire stops

Cavity barriers and fire stops can be the Achilles' heel of fire precautions. Where their importance is not fully appreciated by installers, work may be carried out only perfunctorily with potentially disastrous results. Typically, cavity barriers may be butted instead of being lapped, possibly leaving gaps between sections of the barrier. Care is needed both in briefing the installers and in supervision.

Cavity barriers within floor voids have an important influence on the growth and development of a fire within an enclosed space, though that is not their prime purpose under building regulations. In the case of a ground floor, cavity barriers will be needed where the height of a cavity beneath the floor is greater than 1 m or where crawl spaces are provided with access (Clause 9.11d of the Building Regulations 1991 Approved Document B). In the case of suspended floors above ground level, a cavity within the floor void needs to be limited in size, just as with all other elements of structure.

The majority of cavity barriers have been designed to BS 476: Part 20 criteria but a new test method is proposed in a new European standard.

In tests that were carried out at the Fire Research Station in the 1970s, it was confirmed that plastics pipes penetrating compartment floors constituted a route for fire to spread through the element, though bathroom services could be used safely in small diameters. Where services are required to penetrate compartment floors, the protection needs to be preserved. This is normally achieved by containing the services within a fire resisting builders-work duct (ie a protected shaft) or by using a proprietary sealing system. Other suitable methods for certain service pipes are described in Section 10 of Approved Document B. A common form of seal is intumescent in character, contained within a box, so that expansion due to the heat of a fire effectively seals the void (Figure 1.52).

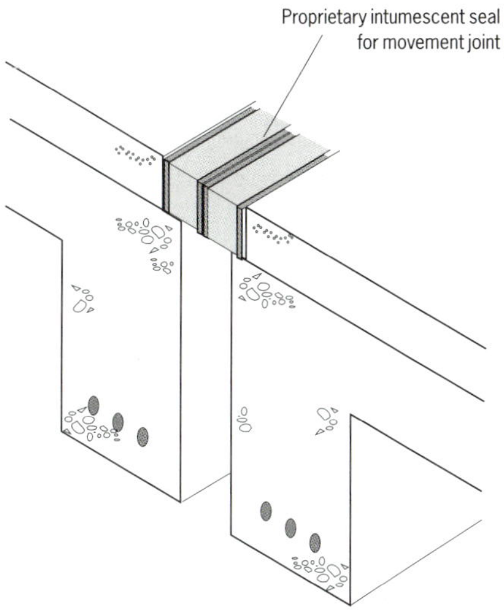

Figure 1.53
A flexible fire resisting seal in a movement joint

Movement joints
Flexible fire resisting seals are available which can provide continuity in the fire performance of movement joints of compartment floors (Figure 1.53).

Floor finishes
In the UK the rate of spread of fire on the upper surface of flooring is not provided for in building regulations as it is in other countries of the European Community. The relevant clause in the Building Regulations 1991 Approved Document B is Paragraph 0.42 of Guidance to Requirement B2 which notes that the upper surface of a floor is not significantly involved until a fire is well developed. Consequently it does not play an important part in fire spread in the early stages of a fire. It is these early stages that are most relevant to the safety of occupants. Remembering, however, that building regulations are concerned with personal safety, there may be circumstances (eg with insurance) where designers concerned with the integrity of a building and its contents may need to take a different view.

The *Guide to fire precautions in premises used as hotels and boarding houses which require a fire certificate*(77), issued under the Fire Precautions Act 1971, calls for floor coverings to be included in the overall assessment of the suitability of surfaces to protected routes. (The same requirement also includes premises covered by the *Guide to fire precautions in existing places of entertainment and like premises*(78), and to factories, offices, shops and railway premises in the appropriate guides issued under the Fire Precautions Act). Where new textile coverings are being fitted, they should comply with BS 5287(79) as conforming to the low radius of fire spread (up to 35 mm) when tested in accordance with BS 4790(80).

Fire properties of industrial floorings are not specified for testing to BS 476, though it is understood that some countries in Europe have spread of flame and smoke emission tests.

While the resistance of floorings to softening by the high temperatures of some processes in industrial premises will be a factor in the choice of materials for floorings, it is relatively unusual for these floor finishes to contribute to fires on a significant scale. Points relevant to each kind of finish are made in Chapters 5–8.

Stair treads and landings within firefighting shafts are required to be constructed from materials of limited combustibility (Clause 9.2, Section 2 of BS 5588:Part 5).

Floor heating
Heated floors are not a recent innovation. The Romans developed ingenious systems for passing the products of combustion from a furnace through ducts under mosaic covered floors built of brick or stone called hypocausts, though the thermal efficiencies they achieved were inevitably low. When the Romans left, hypocausts largely fell out of use because the necessary building skills were no longer available.

Many centuries later, systems for heating floors were reintroduced, consisting for the most part of methods of heating the surface of the floor rather than its structure. The temperatures reached in these systems require careful control.

Where a floor finish is to be renewed over a heated floor, consideration will need to be given to the temperatures reached.

Two basic kinds of underfloor heating laid in screed or slab have been used since the 1960s:

- low pressure hot water in steel, copper or, more commonly, polypropylene pipes (Figure 1.54)
- electric cables laid either directly or in conduit

In the case of hot water based systems, the temperatures reached in the pipes may be as high as 54 °C.

Figure 1.54
Low pressure plastics hot water heating pipes exposed by removing the screed

In situ floor finishes, such as terrazzo and granolithic, on newly laid floors which incorporate heating elements are more likely to curl and crack because of the increased drying shrinkage and the thermal gradient imposed across the section of the floor. Clay, concrete and terrazzo tiles usually present no problem provided that the base has been dried out by applying the underfloor heating before the tiles are laid. However, the heating must be switched off and the floor cooled while the tiles are being laid.

Sheet and tile finishes of linoleum, cork, rubber and PVC, and the adhesives normally used with them, behave satisfactorily provided that the base is dry. Floorings such as asphalt, pitchmastic, thermoplastic tile and vinyl tile are more likely to suffer permanent indentation on a heated floor than on an unheated one, the more especially so where the flooring is covered with a textile covering.

It is generally recommended – for a variety of reasons including indentation, shrinkage, and degradation of flooring and adhesives – that the surface temperature of a floor should not exceed 27 °C.

Chapter 1.7 Appearance and reflectivity

User satisfaction with floors depends largely on the condition of the top wearing surface. After installation, the surface should be clean and hygienic, and capable of being maintained in this condition. It should be suitable to receive a floor covering if one is not provided under the building contract. In areas where water may spill onto floors, both the effect of moisture on the floor and the ease with which the floor may be cleaned and dried, together with its subsequent rate of deterioration, will need to be considered. Fashion and taste will sometimes dictate rehabilitation measures that lead to the removal or covering up of otherwise quite adequate surfaces such as older tiled floors.

Matters which should receive considerable attention are the needs of visually handicapped people for safe access and movement within buildings (Royal National Institute for the Blind's *Building sight*[81]); these matters include appropriate contrasts at changes of level.

Appearance

The appearance of a floor finish is often considered to be of paramount importance, both in its initial selection and its acceptability for continuing use. This is probably true irrespective of situation and building type with the possible exception of those buildings used for certain industrial and storage purposes. The visual acceptability of the various kinds of floor finishes, in the final analysis, is a matter for individual taste and judgement. However, some of the factors which affect appearance, the quality and distribution of lighting in interiors, and the effects they might have on the choice of various materials can be identified (Figure 1.55).

Comments on particular finishes are contained in Chapters 5–8. Safety aspects of appearance have already been dealt with in Chapter 1.5.

Reflectivity

The reflectance of the flooring will affect the amount and distribution of light in a space. After all, in a reasonably sized space most of the light from windows, roof lights and light fittings will fall on the floor. A light coloured floor will increase the illuminances on walls and, especially, the ceiling, making the room look less gloomy and reducing glare from overhead light fittings. In a daylit space the light levels away from the windows, and hence the overall balance of light in the room, will be improved if floor reflectance is increased. Therefore reflectance values will always need to be considered when choosing floor coverings.

The CIBSE *Code for interior lighting*[82] recommends an average floor reflectance of between 0.20 and 0.40. It recognises that in some industrial buildings a reflectance of

Figure 1.55
The highly reflective composition tile floor finish improves the level of lighting in this sports hall

0.20 may be difficult to achieve; in these cases steps should be taken to keep the floors clean so that the average reflectance is maintained at 0.10 or above. Reflectance values for some floorings are given in Table 1.8.

Other values are given in later chapters appropriate to the flooring under consideration. BRE Digest 310[83] explains how to calculate the effects of floor reflectance on the daylight levels in a space. The *Code for interior lighting* does the same for artificial lighting.

The part played by the reflectivity of a floor can be demonstrated by calculating daylight factor. Average daylight factor *DF* is used as the main criterion of good daylighting – the ratio of indoor to outdoor daylight illuminance under the standard overcast sky. A daylight factor of 5% or more means that the space is well daylit (Figure 1.56). A daylight factor of less than 2% means that artificial lighting will be used all the time. Good housekeeping would ensure, anyway, that windows, and especially laylights in ceilings, are kept clean. The transmittance of the glazing will normally be around 0.85 but can be as low as 0.6 if the glazing is dirty.

Table 1.8
Typical reflectance values for floorings

Portland cement screeds, cream PVC tiles and light coloured carpets	0.45
Maple, birch and beech floorings	0.35
Light oak, marbled PVC tiles, medium shades of carpet	0.25
Other hardwoods and cork	0.20
Red and brown quarries, dark tiles and carpets	0.10

Figure 1.56
This interior of an airport building demonstrates the value of a light coloured floor in achieving a good level of lighting

Chapter 1.8 Sound insulation

Where buildings are in multiple occupation, it is crucial that a floor provides adequate insulation against noise generated either above or below it. The ability to absorb airborne sound may also be required of a flooring, though soft furnishings will normally play the more significant role in reducing reverberation times of rooms.

Sound insulation of floors against internally generated noise comprises two aspects which have to be measured separately (Figure 1.57):

- airborne noise
- impact sounds

Sources producing sound waves in air, such as from conversation or television, will in turn produce vibrations in floors. On the other hand, floors may also be vibrated by direct impacts such as footsteps or objects dropped onto floor surfaces.

The basic principles of sound insulation for floors and the terminology used are covered in BRE Digest 334[84].

Noise

Noises differ in their level (loudness) and frequency content (pitch); these two terms are called noise descriptors and they may vary with time. Consequently, different units have been developed to describe different types of noise.

Noise level is described on a logarithmic scale in terms of decibels (dB). If the power of a noise source is doubled (eg two compressors instead of one) the level will increase by 3 dB. Subjectively, this increase is noticeable but not large. An increase of 10 dB(A) doubles the perceived loudness of a sound.

The ear can respond to sounds over a wide frequency range (roughly 20 Hz to 20 kHz), but most environmental noises lie between 20 Hz and 5 kHz. The ear is more sensitive to sounds at some frequencies than others, and is particularly sensitive between about 500 Hz and 5 kHz. The 'A weighting' now in wide use is an electronic circuit built into a sound level meter to make its sensitivity approximate to that of the ear. Measurements made using this weighting are expressed as dB(A). Noises with tonal components (such as fan noise) are particularly annoying, and sometimes this is recognised by adding 5 dB(A) onto a measured level.

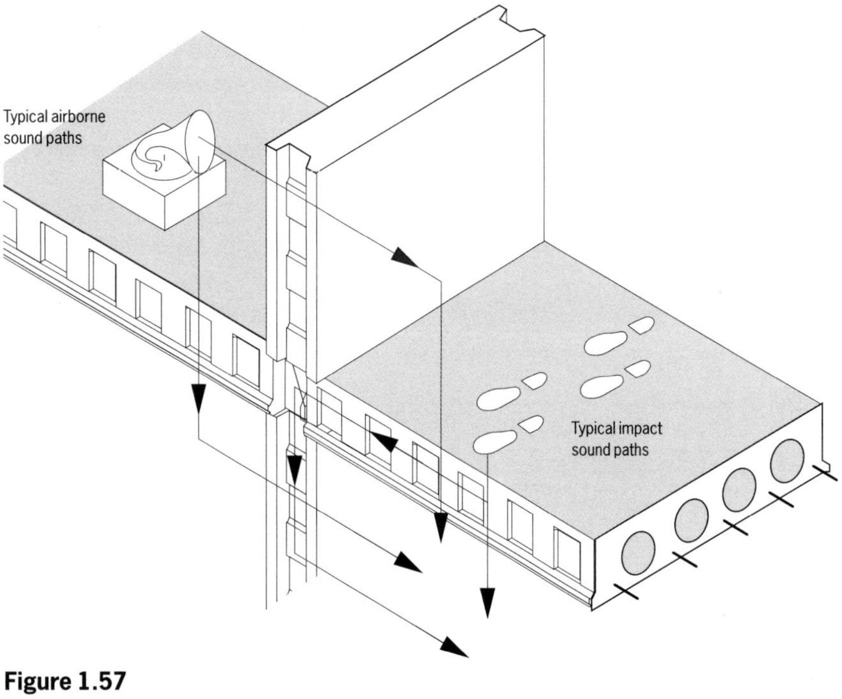

Figure 1.57
Typical airborne and impact sound paths

Airborne sound insulation

Airborne sound insulation of floors depends largely on the type of material used for the supporting or structural deck. Approximate attenuation values are as follows:

- timber structure supporting timber boards (no ceiling): 20–25 dB
- timber structure supporting timber boards (with ceiling): 30–35 dB
- concrete deck not less than 365 kg per square metre: around 55 dB

Airborne sound insulation of concrete deck floors ranges from 45 to 55 dB, mainly depending on the superficial mass. Some improvement may be achieved by adding, for example, a screed, but only if the existing structure can be shown by calculation to be strong enough to support it. On the other hand, proximity to a duct which passes through a floor adjacent to an inadequate wall can lead to flanking transmission (Figure 1.58).

Sound insulation levels between rooms separated by a suspended floor having a proprietary lightweight ceiling have sometimes been shown by test to be lower than expected.

The properties of the material providing the resilient layer in a floating floor are important in determining the overall performance of the floor. Techniques for improving the sound insulation of existing floors are given in appropriate later chapters. BRE Digest 293[(85)] offers general guidance.

Impact sound insulation

Impacts on the surface of a floor can be transmitted through the structure and ceiling into the room below. They can be minimised by either of two basic methods:

- using a floating screed or board isolated from the substrate by means of a resilient layer
- using a resilient layer for the flooring itself

Graphs showing impact sound reduction at different frequencies are normally available for the various kinds of flooring on various kinds of substrates.

A reasonable level of performance is required under the building regulations for England and Wales[(13)] (E3) – spelt out in Approved Document E[(86)] – H3(2) in Scotland[(14)], and Technical Booklets G[(87)] and G1[(88)] of the Northern Ireland regulations for separating floors in flats and maisonettes. For England and Wales, the performance standard for the approved procedure that enables an existing construction to be replicated is currently 61 dB $L^{1}_{nT,w}$† when it is tested to BS 2750:Part 7[(89)] and BS 5821: Part 2[(90)].

† This is an index of impact sound transmission defined by BS 5821:Part 2

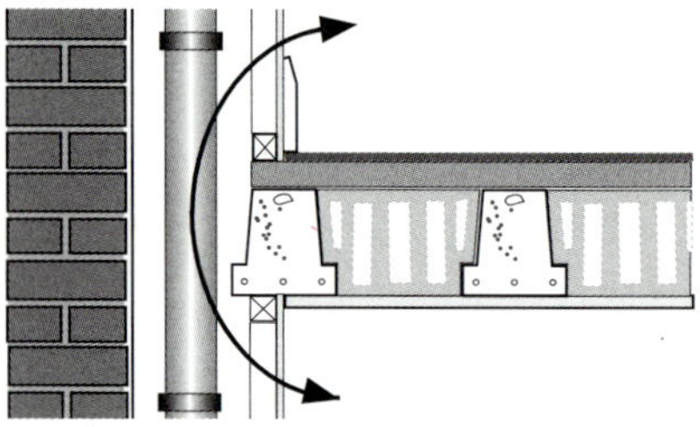

Figure 1.58
A typical sound path via a service duct

Figure 1.59
A standard impact sound testing machine for floors

Surveys of sound insulation of existing floors

In a BRE survey of flats built during the 1970s involving some 500 individual tests on floors, nearly half the floors tested failed to achieve the performance standard given in the building regulations for insulation against airborne sounds and two thirds failed to achieve the standard for insulation against impact sounds (Figure 1.59). Furthermore, one third of floors were found to have a poorer performance than those in similar investigations carried out in the 1950s (National Building Studies Research Paper No 33[(91)]).

Although it is suspected that matters have improved in dwellings built since the 1970s, sound insulation still appears to be a significant problem in buildings that have been converted ([(92)], BRE Information Paper IP 5/81[(93)], BRE Current Paper CP 3/79[(94)] and BRE Information Paper IP 9/79[(95)]).

In a study of noise from neighbours, and the sound insulation of separating floors and walls in purpose built flats, 70% of the residents interviewed heard noise, either from neighbours living immediately at the side or above, or from other parts of the building. Despite the widespread incidence of noise reported as present in all parts of the buildings and their immediate surroundings, relatively few respondents were seriously bothered by neighbours' noise. The most intrusive noise was that coming from neighbours living above, chiefly of footsteps and articles falling on the floor. Significant relationships were found between airborne sound insulation and residents' ratings of insulation quality, both of floors and sound insulation generally[96].

Another BRE survey looked at the factors which influence response to noise in large houses converted into flats. Overall, 50% of respondents thought the sound insulation between their flats was poor and some found the need to be quiet very restricting. Subjective response can be reliably measured, with a workable degree of separation of noise from above, noise from below, and airborne and impact noise. Noise exposure is more difficult to assess, but the number of different types of noise heard may provide a useful proxy for physical measurements. In other words the most important difference between response to noise from above and from below is that impact noise, perhaps not unexpectedly, seems to be the dominant component of noise from the flat above.

BRE has also conducted a survey among organisations that have experience of house-to-flats conversion work. The main practical problems occurred with floor modifications and were caused by:

- the increase in height of the finished floor level when a floating floor was installed
- the reduction of room height when an independent ceiling was fitted

One factor which seriously affects the sound insulation of floors in flats is the relationship of the external and separating walls with the separating floors. There may be flanking transmission along lightweight walls, bypassing the sound insulation provided by the floors. Flats in the centres of blocks generally have heavy separating walls on both sides. However flats at the ends of blocks may have lightweight gable end walls which provide flanking paths neutralising the effect of the sound insulation. Where centre block flats have separating walls which are as light as the gable end walls, the performance is more like that of gable end flats. Impact insulation is affected to a smaller extent. Where separating walls are similar in construction to inner leaves, differences in floor performance are not significant.

In studies conducted since the 1980s[97], it was found that, by and large, people do not complain without good reason, though there are some people who are dissatisfied even where their homes meet the intended standard. The Building Regulations 1991 do not make provision for post-construction testing and, in difficult situations, further specialist advice may be needed.

Quietness

If there is too much reverberant sound in a room, the components of speech may be difficult to distinguish. The sound absorbent qualities of flooring will certainly help to reduce reverberant sound in rooms, though arguably the influence of the ceiling and the walls is significantly greater than that of the floor (BRE Digest 192[98]).

Chapter 1.9 Durability

This is a general chapter which includes information relevant to all kinds of flooring – information relevant to a particular kind of flooring, such as typical wear rates on a particular surface, will be given in the appropriate later chapter.

Floor finishes may be considered to provide for the following main functions:

- protection of the structural floor
- better appearance
- increased comfort and safety

The relative importance of these functions varies according to circumstances and budgets. However, one of the most important attributes is the maintenance of the functions over time; in other words, the finishes must be durable. Because of its importance, durability has often been regarded as a basic property of a floor finish, but it represents only the length of time that the chosen properties persist and depends as much on the conditions of use as on the properties of the finish[99].

Many factors influence the life of a floor finish:

- wear
- water and other liquids
- indenting loads and impacts (Figure 1.61)
- sunlight
- insects
- moulds and fungi
- high temperature

as well as the fundamental properties of the materials and adhesives used, and the compatibility of these with other parts of the structure and its behaviour in use.

Figure 1.60
The durability of any flooring is governed ultimately by the kind of use it gets, particularly by heavy wheeled vehicles

Figure 1.61
This industrial floor has suffered very heavy wear in spite of the robust finish. It has been subjected to piecemeal replacement. The wheel tracks of mechanical handling equipment can be seen

The European Union of Agrément (UEAtc) have given priority, in their Method of Assessment and Test covering innovative floorings, to assessing the conditions of service for floorings using what has been called the UPEC system. UPEC is the acronym for wear (usure) index 1–4, indentation (poinçonnement) index 1–3, water (eau) index 0–3, and chemicals (chimiques) index 0–3 (UEAtc Method of Assessment and Test No 2[25]). Given the existence of an Agrément certificate for a particular product, therefore, a surveyor should be able to make a judgement on its performance. UEAtc and the British Board of Agrément (BBA) have also issued further guidance on the assessment

Figure 1.62
This paint finish is breaking up. There is little substance to resist continuous wear

of plastics floorings (UEAtc Methods of Assessment and Test No 23(100) and No 36(101), and BBA Information Sheet No 2(102)). These are referred to in greater detail in the appropriate later chapters. Experience has shown that there is a relationship between wear and indentation of PVC floorings; these floorings, invariably, have a C index of 2. For PVC floorings, it has therefore been found possible to omit the C rating, and to combine the U and P ratings into a single G classification (1–5).

Of all the performance factors described above, wear is perhaps the most influential because:

- it is unavoidable in normal use
- it applies to all kinds of flooring
- in many cases, it can be very obvious to users when it occurs

Expected life of floorings

There is a British Standard on durability, BS 7543(103), which applies to floors and flooring. It gives general guidance on required and predicted service life, and how to present these requirements when preparing a design brief.

The service life of a floor covering depends as much on the conditions in the building and the degree of use as on the inherent properties of the material and the techniques adopted in laying it. However, even bearing in mind the comparative ease of replacement of many thin floorings, building users should have a reasonable expectation of a minimum service life for a properly selected flooring, taking all circumstances into account. What that minimum might be is a matter for debate, but UEAtc have decided on 10 years, and BRE would not dispute that figure(25).

Durability is affected by the care and skill with which the floor is laid, and the behaviour of the subfloor, as well as by the choice of material itself. Failure to provide adequate dampproofing and ventilation, and a sound screed, can lead to serious deterioration in finish; for example as buckling of wood block floors or loss of adhesion in tiled or painted finishes (Figure 1.62).

Resistance to wear

All floor surfaces wear to some degree when subjected to foot or wheeled traffic. There may also be other factors including the movement of furniture, which may scrape or cut the surface, and the movement of heavy loads such as in warehouses. However, using the definition of wear as the progressive loss from the surface of a body brought about by mechanical action, it is not just the quantitative loss of material which is important but also the qualitative assessment of the condition of the worn surface. Where changes in appearance are involved, the assessment becomes more and more subjective.

Wear and damage can be considered to arise from a number of causes:

- mechanical action (leading to progressive loss of substance)
- abrasion caused by fine solid particles
- cutting by the action of vehicles (eg trolleys) and furniture
- corrosion from spillages
- degradation from soluble salts rising from the subsoil or entrapped moisture in concrete
- fatigue in the surface material
- movement or rocking from uneven or friable bedding causing chipping of arrises of slabs or tiles

There has been considerable concern since the 1960s with the degree of wear of floors in heritage buildings. This is dealt with in detail in Chapters 5.2 and 5.5. In some cases, depending of course on the material involved, it is evident that wear of around one millimetre in ten years is occurring(104).

Accelerated testing for wear

Over the years there have been many attempts to design apparatus which reliably simulates the wear that takes place in normal use and which ranks particular materials according to criteria which reflect

Figure 1.63
Measuring the forces applied by foot traffic to flooring. Although these tests were conducted in the late 1950s, the results still apply

Figure 1.64
Research on flooring at a London Underground ticket office: testing on a section of material placed at the point where turning and scuffing produces maximum wear

normal use. International cooperation in the 1950s indicated that correlation obtained up to then was unsatisfactory[105].

In the UK, these efforts were given added stimulus by the need to develop new flooring materials for use on concrete floors when restrictions on the use of timber ground floors were applied after the 1939–45 war.

Research work by BRE[106] has led to a much better understanding of the actions which lead to wear in floorings, particularly wear cause by foot traffic (Figure 1.63 on page 57). Four separate forces are involved:

- vertical load on sole
- vertical load on heel
- horizontal forces on heel and sole
- torque on sole

So far as pedestrian traffic is concerned, it is generally recognised that wear is most severe where changes of direction occur and where people are moving in a confined area; for example moving slowly through a ticket barrier. Research work carried out by BRE in 1960–61 at a London Underground Station (Figure 1.64) demonstrated that wear was directly related to the numbers of people passing over the surface, and established which floorings available at that time were unsuitable for areas of heavy use. Indeed, one material (cork) did not survive sufficiently long enough to be walked on by 500,000 people. When leather was the predominant soling material for shoes, tests using it on the sole plate of the artificial wear machine showed a good correlation with performance in use with a range of different flooring materials, including thermoplastic, linoleum, flexible PVC, vinyl asbestos, rubber and cork.

Indentation

Indentation arises from the effects of furniture on relatively soft floorings, impacts from articles dropped onto the surface, and damage caused by footwear, whether by small heels or protruding nails. Tests are available for indentation by heavy furniture

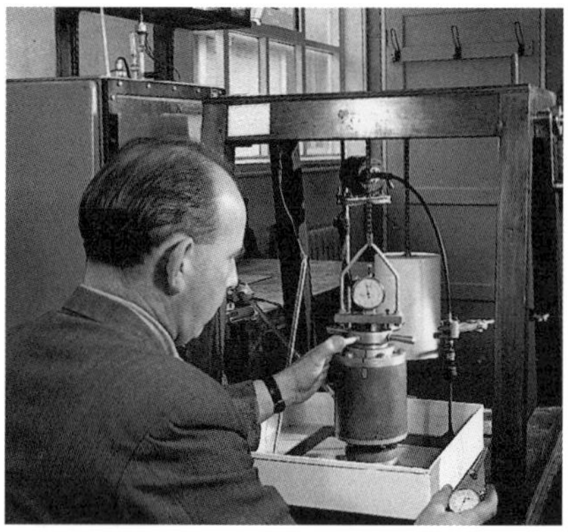

Figure 1.65
An indentation test on thermoplastic tiles. Although the photograph dates from 1954, the principles of this test are still valid

and chair legs, both stationary (Figure 1.65) and moving. Apparatus has been developed for similar purposes by UEAtc[25].

Avoidance of wear

Wear can be prevented only by denying access. Nevertheless there are various housekeeping measures than can be employed to reduce wear, such as the avoidance of unprotected metal on chair legs and trolley wheels. Also the effects of wear on some floorings can be mitigated by good maintenance; for example by applying seals and polishes to timber, cork and linoleum flooring. Good cleaning procedures and barrier matting will keep grit (which is often implicated in abrasion) to a minimum.

Floorings for the heaviest wear situations

BRE is often asked to examine the floors of shopping malls, railway and airport concourses, and stores. In these buildings the volume of pedestrian traffic can be very intense and large cleaning machines are often used to cover the vast areas involved in the shortest times. A wide variety of floorings has been seen, none of which has proved to be ideal and some have performed particularly badly. Only robust floorings of the highest quality laid to a good standard give good service. Materials which have performed best are relatively thick ceramic tiles, high quality decorative terrazzo

Table 1.9
UPEC E indices[25]

Index number	Maintenance	Tolerance to wetting
E_0	Dry methods only	Short duration
E_1	Occasional wet methods	Accidental standing water
E_2	Wet methods and washing	Not prolonged standing water
E_3	Swilling	Prolonged standing water

concrete tiles with hard aggregates, and hard wearing natural stones like granite and quartzite. High quality bedding and jointing with all these floorings is essential for good service. High quality flexible PVC and thermosetting resin screeds have been used less often, but usually with good results.

Floors for heavy industry are, of course, the areas which can potentially receive the highest wear rates, depending on the types of industry and the wear processes involved. Because of the possibility of contamination by mineral oils, the choice of finishes for general manufacturing areas is restricted in many cases to various forms of concrete, wood blocks, clay paviors and thermosetting resins.

Protection to floorings

Protection of floorings, particularly in cases of rare and historically important floors (eg of medieval tiles), should be undertaken only after advice. Textile coverings can serve simply to trap grit, which is then ground into the surface of the underlying flooring by small movements in the covering. Secure fixing of protective matting is crucially important.

Probably a first covering with a sheet of plastics material might serve as a cushion; impermeable sheets should be specified for this task only if there is no risk of condensation underneath. Breather membranes provide a better solution, and will be serviceable for preventing grit from reaching the surface below, though they will need to be renewed or at least shaken from time to time to remove grit lodged in the fibres. Natural rubber and foam backed textiles should be avoided.

A boarded floor on ventilated battens probably provides the best longterm protection, if necessary with the incorporation of suitably glazed panels for viewing purposes. Barrier matting at the entrance to a building with sensitive (ie easily damaged) flooring will help to remove grit before it can do any damage.

Resistance to water

Principles of modern building[6] pointed out that the effects of moisture probably cause more damage to floor finishes than any other agency, even abrasion. This is still thought to hold true, as studies of the workload of BRE Advisory Service have shown. However, in contrast to a commonly held view, moisture rising from the ground through a groundbearing slab above a defective DPM is not one of the most frequent sources of dampness in floors. In the majority of cases it is the complete absence of a DPM and excess construction water in the slab or screed which has not been given sufficient time to dry before floorings are laid which lead to dampness. Excess construction water has become more of a problem with 'fast track' construction.

Other sources are spillages, overzealous or inappropriate use of water for cleaning purposes, and condensation on ground floors. Some finishes may only be able to tolerate accidental wetting for short durations, while others are immune even to prolonged wettings.

The UEAtc E indices are given in Table 1.9.

With respect to plastics floorings, the G classification applies, as explained earlier in the chapter. Floorings classified as G1–5 have limited tolerance of wet conditions, the 'w' suffix (in Gw) indicates that a product can tolerate wet conditions, and the 'ws' suffix (in Gws) that it can tolerate standing water[101].

Test methods are available covering exposure to water and assessment of sensitivity, spread of water under a puncture, resistance to standing water, and dimensional variations when wetted. The behaviour of individual floorings is described in Chapters 5–8.

Osmosis

This is a phenomenon which affects some kinds of flooring.

It is only since the mid-1970s that osmosis has become evident in failures of some sheet and tile floorings. BRE has heard of about 50 cases of osmotic blistering and has examined nearly half of them. Even though much more is now known about blistering caused by osmosis, it remains a rare and unpredictable event. It is described in more detail in the feature panel on page 60.

Resistance to chemicals and effluents

There is a risk in all kinds of premises that chemical compounds will be accidentally spilled on floor surfaces. In housing these typically may include food products, pharmaceuticals and cleaning products (Type A). The products may stain the flooring, or cause its softening or other forms of deterioration. In other buildings such as factories and laboratories, special chemical products may require to be resisted (Type B). One of the most difficult of building uses to accommodate satisfactorily is the dairy. Milk residues and the cleaning agents used in dairies are notoriously corrosive to particular floorings. Advice given in *Dairy Floors*[107] is still relevant.

An explanation of the UEAtc classification is shown in Table 1.10 on page 61.

Test methods are available covering the resistance of flooring to fatty stains of vegetable and animal origin; non-fatty stains such as tea, coffee and fruit juices; chemical

Osmosis and how it has been observed in flooring

Osmosis is a well known thermodynamic property of solutions. It occurs as the spontaneous flow of water into an aqueous solution or of water from a dilute to a concentrated solution where the two solutions are separated by a semi-permeable membrane (Figure 1.66); this membrane allows the free passage of water but not the dissolved solute. The cement paste in the concrete base can act as this semi-permeable membrane. The movement of the solvent (ie the water) through the membrane into the concentrated solution can generate very high pressure known as osmotic pressure.

Three conditions are required before osmotic blistering of flooring can occur:

- the presence of a concentrated soluble salt or soluble organic material at or near the surface of the concrete base
- a semi-permeable membrane at or near the surface of the flooring
- a source of water in the substrate

A **salt** can concentrate at or near the surface of concrete in a number of ways (eg from acid etching or the application of surface hardening agents). It is considered that salts derived from Portland cement are usually implicated as they can migrate to the surface and concentrate there as the concrete is drying. However, contamination from sources outside the concrete or screed are not usually involved.

In some cases investigated by BRE where blistering has occurred, the concrete bases were old ones laid on the ground without any DPM. Over the years, and before the resin flooring was applied, small amounts of water passing through the slabs could have carried soluble salts, derived either from the ground or the cement in the concrete, to the surface. Here they could concentrate in a manner similar to efflorescence on drying brickwork. Old concrete floors can also be contaminated with soluble salts during previous use – for instance deicing salts could have been tracked into a building by vehicles.

Analysis of fluid taken from blisters at a number of sites has shown that the fluid always contained a mixture of inorganic salts – mainly sodium and potassium sulfates, carbonates and hydroxides – and often organic constituents contained in the epoxy resin hardening systems. Total soluble material concentrations between 4 and 28% by mass have been measured. These concentrations would have been higher before dilution by osmosis occurred.

In some cases of blistering it has been suspected that the resin constituents of the flooring have not been thoroughly mixed together, although there has been no direct evidence for this. Incomplete mixing could leave concentrated pockets of soluble hardening agents adding to the soluble salt.

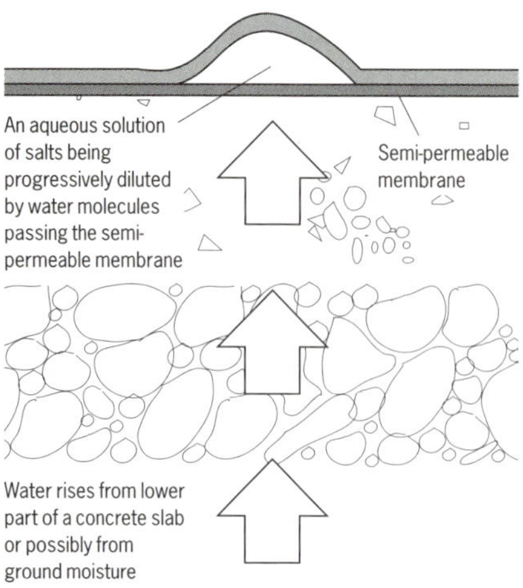

Figure 1.66
The mechanism of osmosis

The **semi-permeable membrane or layer** is one which allows water-size molecules (but nothing larger) to pass through. Good quality concrete can act in this way. Even if the pore size in the concrete is not small enough to form a semi-permeable membrane, it is possible that the application of a primer coat could reduce the pore diameter at the surface sufficiently for this to happen. Epoxy films have been shown to behave in the same way as semi-permeable membranes.

A **source of water**. It is often specified that the moisture content of concrete should be below 5% before thermosetting resin floors are laid. However, the osmotic force is a very powerful one. There is probably sufficient water, even in so-called dry concrete of 3–5% moisture content, for osmosis to occur. The minimum moisture content for osmosis to occur is not known.

Therefore the conditions for osmosis, and the consequential formation of blisters, can frequently exist in a concrete floor covered by an impervious flooring material.

The size of the blisters depends on a number of factors such as the initial concentration of salts, the quality of the semi-permeable membrane formed, and how well the flooring is stuck to the concrete.

Usually the blisters are first noticed some one to six months after the flooring is laid. They can range from a few millimetres up to 300 mm in diameter, but commonly from 10 to 50 mm with a height of 2 to 5 mm. They can continue to appear and increase in size for up to two years. The frequency of formation and increase in size slow down and stop when the ionic activity on each side of the semi-permeable membrane becomes equal due to the dilution of the salt concentration and its transfer back into the bulk of the concrete. If blisters are opened after some years, they are often found to be dry: all aqueous fluid and soluble material has probably returned to the concrete through an imperfect membrane.

Blisters can be punctured due to internal pressure or the effects of traffic and fluid within the blister then leaks to the surface. Initially the fluid is a pale straw colour produced by the organic components. The water quickly evaporates, depositing the inorganic salts as crystalline material, but the organic components appear to oxidise to a much darker, sometimes nearly black, sticky material. Further solutions may exude from a broken blister until the ionic activity on either side of the semi-permeable membrane becomes equal. This will happen for the reasons already given and also because soluble material is being lost to the surface.

Blistering of resin floorings caused by osmosis has not been well researched and, although BRE has examined about 20 cases, it cannot predict when osmosis will occur or the precise conditions necessary for the formation of blisters. Nor can it predict that a replacement floor will not blister again; this has happened in a number of cases.

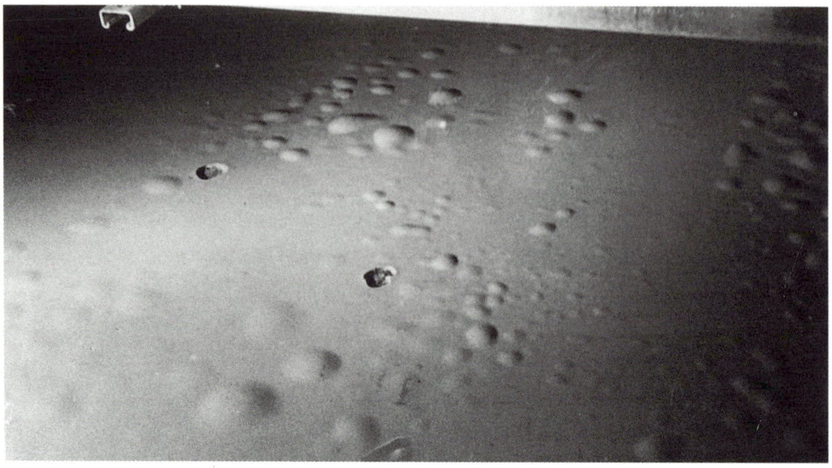

Figure 1.67
The very first case of osmosis investigated by BRE in 1971

stains (eg from cleaning products such as ammonia, carbon tetrachloride and white spirit); and burns (eg from cigarettes). The behaviour of individual floorings is described in Chapters 5–8.

BRE Advisory Service has received a large number of enquiries concerning the effect of spillages of central heating oils on concrete and other absorbent hard finishes. The common response has been to say that these oils have no effect on the strength and integrity of concrete or of sand and cement screeds, but, because they are porous, oil will impregnate them. Bitumen and pitch based DPMs and DPCs will be solubilised; polyethylene is likely to be unaffected. Where moderate spillage has occurred, it is prudent to remove any affected screed, but, unless the spillage has been large, some impregnation into the concrete base can be tolerated. An affected base concrete should be covered with a 300 μm (1200 gauge) polyethylene sheet before renewing the screed.

Slopes for drainage of liquids

Any industrial or agricultural process involving a liquid may require a chemically resistant floor. The problems which arise seem to centre on those floors which are also regarded as catchment areas for the drains, and on which large quantities of liquids are deliberately discharged. Such use of a floor is to court trouble. Waste liquids should be discharged to a drain and the floor finish should be required to deal only with accidental discharges. If floor finishes are expected to receive discharges, they should slope to the drain. Recommended slopes vary between 1 in 80 for plain water to 1 in 40 for the most corrosive liquids. Slopes steeper than this may present a safety hazard to users unless further precautions (eg on creating non-slip surfaces) are taken.

Four methods of arranging for drainage have been in common use and, provided slopes are adequate, all have proved satisfactory:

- continuous slope from the walls of the longer sides to discharge towards a central channel running the length of the space
- crowned slope to discharge from the centre of the space to channels along the walls of the longer sides
- short slopes in bays to discharge to channels running transversely across the space, then to channels or a concealed drain running the length of the space
- shallow rectangular saucers each discharging to a central gully, then to a common concealed drain and, ultimately, to any suitable discharge point

Care should be taken that drainage channels are at a suitable distance from walls, and that maximum discharges, for example from accidental spillages, do not overtop the channels. Where bases for plant and machinery interrupt the drainage slope, the bases should be oriented so that a spillage does not accumulate (Figure 1.68).

Resistance of various types of floor finish to various liquids is dealt with in later chapters where appropriate.

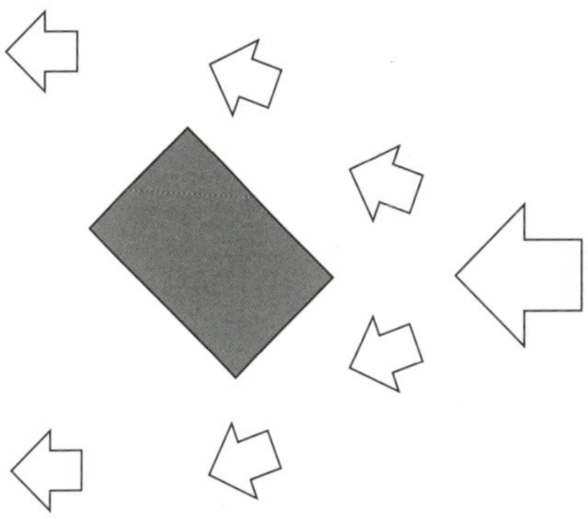

Figure 1.68
Machinery base (shown in plan) angled to the direction of fall so that drainage is not impeded

Table 1.10
UPEC C indices[25]

Index number	Type of chemical	Degree of resistance to attack
C_0	Type A (eg in fats, non-fat products, common pharmaceuticals and chocolate)	Low (stain removal causes damage)
C_1	Type A	Some (leading to slight staining)
C_2	Type A	High (impervious to staining)
C_3	Type B (household cleaning products and solvents)	High

Indenting loads and impacts
These have been dealt with in Chapter 1.1

Resistance to ultraviolet and sunlight
Many flooring materials fade on prolonged exposure to sunlight, including some woods and cork, and some plastics become yellow; not all materials, though, are affected. Whether colour changes are important depends on circumstances. Material specifications do change, and inspection of old samples is not necessarily a good guide to future performance. It may be possible to carry out laboratory assessments. Manufacturers' advice can also be sought.

Insects
Two types of problem occur in floors as a result of insect action: degradation of materials, and infestation and harbourage.

The degradation of timber and some timber based board materials can result from the action of wood boring beetles. *Anobium punctatum* is the most common species encountered in the UK. Though damage is potentially of structural significance, the risk of significant damage in floor structures is very low because the normal designs and conditions generally militate against initiation and development of attack.

In a survey carried out in 1993, BRE found that there had been a marked decrease in infestations of *A. punctatum* in UK properties up to 30 years old, though just under one third of buildings older than this can still be affected, with slightly greater numbers recorded for properties built before 1900. A total of 2,212 survey reports were received which showed that 33% of the infestations occurred in intermediate floors and 23% in ground floors (the remainder was in roofs). Rural locations showed rather higher infestation rates overall compared with urban and coastal locations. The data do not support a case for preservative pretreatment of building timber against *A. punctatum* alone.

House longhorn beetle (*Hylotrupes bajulus*) infestations can cause severe structural damage to softwood floor timbers, but are mainly concentrated in a few areas to the south west of London with the highest levels occurring in properties built between 1920 and 1930. Areas of England where treatment is required are defined in the Building Regulations 1991.

Guidance is available in *Recognising wood rot and insect damage in buildings*(108) and *Remedial treatment of wood rot and insect attack in buildings*(109).

Infestation by pest insects can occur in suitably sized voids in floor constructions. Where these voids have openings, for example between walls and floors, this can make pest control operations both complex and expensive. Cockroaches and pharaoh's ants are usually only found in larger voids whereas book lice (psocids), bed bugs and silver fish utilise small crevices on internal surfaces.

Detailed requirements for the exclusion of some pests, and general methods for reducing infestation by exclusion, reduction of habitat and food supply, and design and construction requirements for excluding pests and for facilitating the treatment of infestations, are to be found in BRE Digest 238(110).

Mould and fungus
There are two kinds of building fungi – those that cause wood rot, and those that do not (Figure 1.69).

Wet rot in floors occurs mainly at the bearings of timber joists in external walls. Wet rot decay of timber and timber based board materials can take place only where these are maintained in persistently damp conditions. Initiation of attack generally results from microscopic airborne spores, but can also occur where pre-infected timber has been

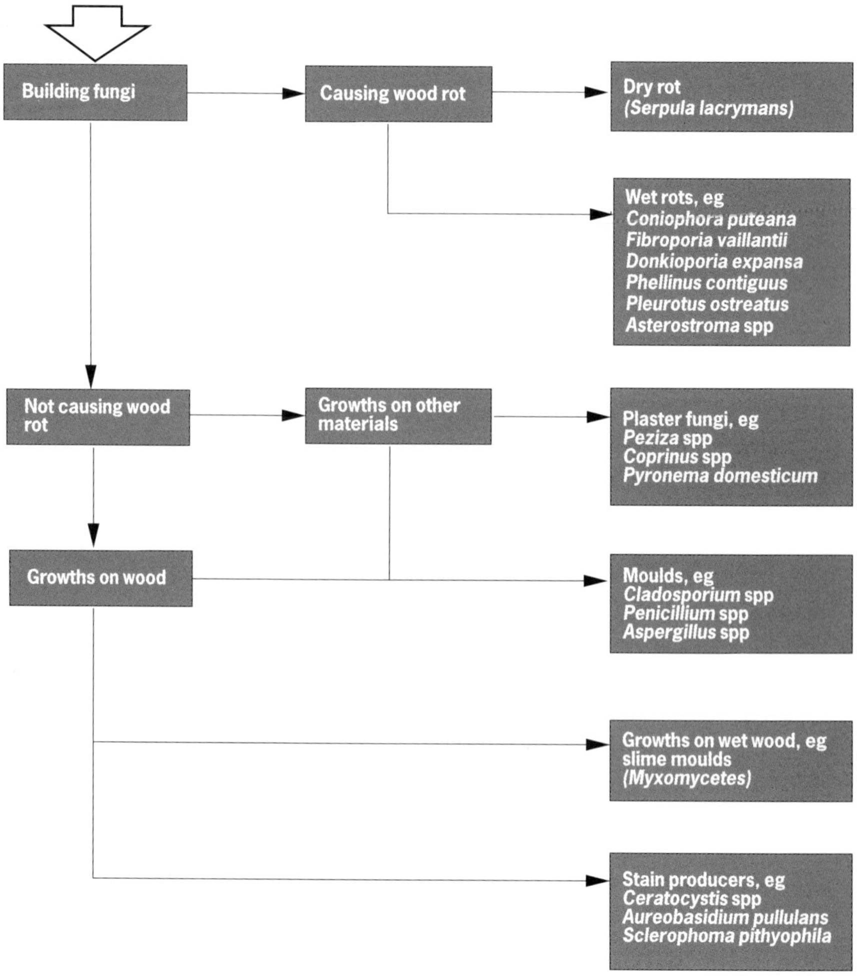

Figure 1.69
Categories of fungi and moulds found in buildings

used; in this situation, very rapid decay can occur in new construction. Under certain conditions damage may be rapid and severe, and therefore of structural significance. The dry rot fungus, *Serpula lacrymans*, which can also occur in floor timbers (Figure 1.70), is more devastating though less common than wet rot. There are many species of wet rot.

Surface moulds can be found on external and internal building surfaces, including ceilings, usually preceded by persistent condensation or some other form of wetting (Figure 1.71). Externally, moulds are unsightly and may also cause premature failure of paint films. On internal surfaces, as well as being unsightly, mould is thought to cause respiratory problems in susceptible individuals.

Guidance is available in *Recognising wood rot and insect damage in buildings* and *Remedial treatment of wood rot and insect attack in buildings*.

Adhesives

In a study of 100 BRE Advisory Service site investigations of failures involving site applied adhesives, nearly half involved flooring. Errors in design or specification, or faults outside the control of site operatives, were identified in three quarters of the failures, while a contributory factor of poor site workmanship in making the adhesive bond was identified in about half[111].

Adhesives for floorings are dealt with in BS 5442:Part 1[112].

Suitability for underfloor heating

Notes are given for each flooring under separate headings in Chapters 5–8.

Quality assurance

Inspections by means of subsurface radar techniques are being used increasingly for the investigation and assessment of defects in buildings and other forms of construction. The method is being used not only to provide information about construction details and the presence of anomalous features such as under floor voiding, but also in some instances about the condition of these elements[113].

One of the more important determinants of durability is the quality with which the original work was carried out†. Specific points to bear in mind on workmanship are

† With respect to quality assurance of the complete floor and its covering, the situation is complicated by the existence of separate agencies, contractors, subcontractors and suppliers. While it may be relatively straightforward to warrant the quality of materials, it is far less straightforward to warrant the quality of installation processes. Suitable initiatives by the industry and its major clients are awaited.

Figure 1.71
Moulds occur on surfaces such as walls and ceilings. They are normally caused by persistent condensation which, in turn, is due to moist air in unventilated conditions or in contact with cold bridges

Figure 1.70
This floor has been devastated by dry rot. Collapses of this severity occur comparatively rarely

listed later in the book under headings appropriate to each kind of floor but the point is best made here that no floor, even if of the highest quality materials, will give good service if crucial aspects of workmanship are not properly carried out and verified (see Chapter 1.10).

Chapter 1.10 Inspection and maintenance

It is inevitable that floors which are subjected to pedestrian and wheeled traffic will wear, and will also show other types of deterioration (Figure 1.72). It is therefore essential, particularly for heavily used floors, that a regular inspection and repair schedule is drawn up which will ensure that they are kept in a satisfactory and safe condition.

Survey methods

The identification of a particular kind of flooring under inspection is dealt with in Appendix A. The flooring, by definition, is open to view, but finding out exactly what kind of floor construction supports that flooring is often not at all easy. A suspended ground floor may often be accessible from the underside via a trap and a crawl space, or the underside of a suspended upper floor may be accessed by removing a ceiling tile. But where the floor is fully enclosed, such as a groundbearing concrete floor, and a drawing or specification of the original construction is not available, some destruction in the interests of diagnosis of a problem may be inevitable. It will be a matter for professional judgement in balancing the severity of problem with the consequences and costs of making good any damage.

One way of minimising damage is to use an optical probe. Optical probes may be useful in some circumstances, but BRE experience is that they need to be used with caution since the field of view is restricted and may not show all features or problems. See Figures 8.15 and 8.16 in Chapter 8.5.

Installation of floorings

Points which require particular consideration in respect of different kinds of floorings when they are being replaced, or, for that matter when being laid new, are covered in Chapters 5–8 as appropriate. New installations are covered generally by standards such as BS 8203[35].

Inspection and maintenance schedules

Intervals between inspections will vary according to the type of building and type of traffic. It is reasonable to suggest that busy public access areas such as shopping malls, railway and bus stations, and airport concourses should be examined at least every year, and preferably every six months. Defects like areas beginning to wear, cracked tiles or missing grout can be rectified before further deterioration takes place. A straight edge over traffic routes should give some indication of wear.

To summarise the above, the maintenance of flooring includes four tasks:

- dealing with faults at early stages of their development
- regular servicing
- repair of unexpected damage
- replacement of worn out areas

The effort involved in any of these tasks is directly related to the correct selection and timely installation of the replacement materials.

Maintenance of sheet and tile floorings is covered in BS 6263:Part 2[114] †. Other installation codes also

† At the time of publication of this book, Part 2 is the only one to have been issued.

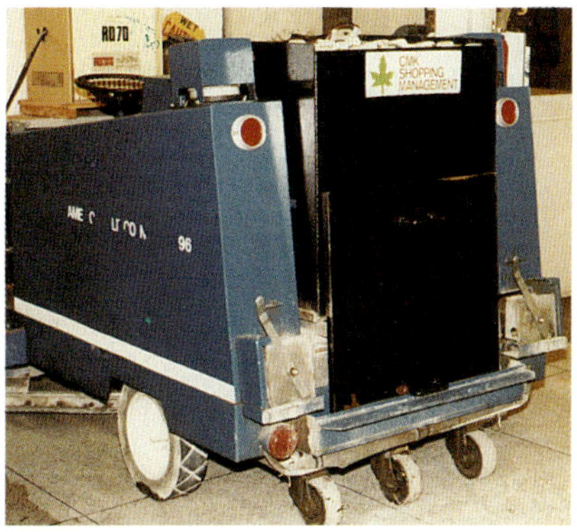

Figure 1.72
An industrial floor cleaning machine which transmits heavy loads to the flooring

have sections on cleaning and maintenance.

Treatment of a surface may be possible to consolidate friable material. Care should be taken to remove or ameliorate the original cause of damage before treatment, such as the lifting of the floor and laying of suitable DPCs where deterioration has been caused by soluble salts carried by rising damp, is undertaken.

Further points on maintenance are covered in Chapters 5–8.

Ease of cleaning

Ease of cleaning is important for all floor surfaces, but especially so for domestic areas in all kinds of buildings – kitchens, bathrooms and WCs – where a washable floor surface is desirable. Loose grit and other forms of ingrained and surface dirt add to wear rates on all kinds of floorings where it is scuffed by foot and wheeled traffic. Regular cleaning will prolong the life of the flooring, and using suitable surface treatments such as seals and polishes will not only ease the cleaning process but will also assist in maintaining its appearance. Polishes, however, can make a flooring slippery; advice is given in later chapters on the degree of slip resistance attributed to each kind of finish.

Detergents used for cleaning floorings should be neutral in character if at all possible. Solvent based cleaners should be avoided, in general, since they can cause damage to vinyls and rubbers. Alkaline detergents may occasionally be necessary to remove old layers of polish, but caustic detergents should not be used on floorings.

Any abrasives used should normally be of the fine or medium grades since coarse abrasives will scratch most flooring leaving grooves where dirt can accumulate. Steel wool should be avoided as the residues may induce rust stains which become difficult to remove. Nylon abrasive pads are available in a variety of grades.

There is one point that can be made about surface dressings. These polishes, used for the protection of floorings, have in the past tended to be of two basic kinds:

- water based emulsions
- solvent based pastes

The water based emulsions contain varying proportions of waxes and polymers – the higher the proportion of wax the higher the gloss when buffed, but also the greater its slipperiness.

The second category is now being phased out in line with environmental concerns. However, inappropriate use of these solvent based pastes could have led to progressive deterioration of certain types of flooring, including mastic asphalt, pitchmastic and rubber based latexes, as well as certain resins used for jointing.

The most vulnerable area of any flooring is near the entrances to buildings where rainwater and grit fails to be removed from shoes by the entrance matting (see Chapter 4.6). Regular inspections to identify breakdown of any seals, and rapid stripping and renewal of the more vulnerable surfaces, is essential if appearance is to be preserved. If at all possible, some form of grid should be incorporated outside all entrance doors to encourage dirt to fall outside rather than inside the building.

Ease of cleaning is considered for each flooring material in Chapters 5–8 under the general heading of maintenance.

Conservation of flooring

Since the range of floorings used in the past is so wide, it is impracticable to give advice on all aspects of their conservation. Not only will materials sometimes be difficult to identify but their condition will vary too, whether simply from the ravages of time or from changes in the environment leading to chemical or biological attack. However, inappropriate conservation practices may easily accelerate deterioration. Amongst the more serious mistakes made have been:

- the practice of re-bedding old tiles in cement-rich mortars. Subsequent cracking of the bedding also cracks the tiles (Figure 1.73)

Figure 1.73
Tiles at Titchfield Abbey which show some cracking resulting from the use of a cement-rich mortar when they were relaid

- painting or sealing old floorings laid on substrates which have been inadequately dampproofed. Salts, rising to the surface with moisture, crystallise within the material leading to flaking of the flooring material or its detachment from the base

Topics on which specialist advice may be necessary include wear and other aspects of durability, timber specification for repair and replacement, behaviour of the fabric in fire, and monitoring of structures for movement. The advice may be available from the heritage organisations.

Structural surveys of floors

It will be a matter of judgement when the building surveyor or architect needs to call in the assistance of a structural engineer to carry out a full structural survey. All very large span floors, or those which carry abnormal loads, should be regularly inspected; building owners should be advised to adopt arrangements for regular and systematic inspections.

Any preliminary examination and assessment of a floor in order to determine whether or not to call in a structural engineer will need to consider at least the following:

- the age of the building
- the use of the building (in particular, access by the public)
- any features which make the building particularly vulnerable (eg span, local exposure, previous history of damage etc)
- obvious signs of distress
- the degree of accessibility to the main structural members
- jointing techniques used for the main structure of the floor (eg welding, glueing etc) and any evidence of cracking, sloppy fit of bolts or snapped off rivets
- the condition of the supporting walls or frame members
- evidence of rain penetration into the bearings and where the rainwater collected
- the presence of corrosion (in the case of metal structures) deeper and more extensive than just surface corrosion
- surface damage, cracking or spalling (in the case of reinforced concrete structures)
- the presence of fungal or insect attack (in the case of timber structures)

Where it is decided to call in an engineer to carry out an appraisal of a structure, it will normally be undertaken in accordance with the framework described in *Appraisal of existing structures*[115] published by the Institution of Structural Engineers.

The report from the engineer to the client or his other professional advisers should consider at least the following points:

- items (numbers of faults in structural members) requiring immediate repair or replacement for reasons of safety, stability or serviceability of the structure, either as a whole or substantial parts of it. (The assessments will need to be supported by calculations in appropriate cases)
- items likely to require attention within the short term (say two years)
- items requiring attention within the medium term (say five years)
- a suitable maintenance regime, if one is not already in place
- a date for the next structural survey and appraisal

Chapter 2 Suspended floors and ceilings

This main chapter deals with all kinds of suspended floors in a wide range of materials and structural forms. Balconies are included since they are often formed as extensions to floors, the complete units cantilevering over intermediate structural supports. Where these balconies are external to the building, they are also included, but only so far as they need special treatment for thermal insulation. Weathertightness aspects are not dealt with.

Ceilings, of course, occur with suspended floors, mostly above ground level. It would be impossible to consider comprehensively the performance of a suspended floor without its ceiling. (The great majority of these floors will have ceilings fixed to their undersides.)

However, the variety of ceiling designs is legion, especially suspended ceilings, and all that can be done in this chapter is to consider the types most often associated with the particular floor construction being described.

Figure 2.1
Even apparently solid floors often prove to be suspended rather than groundbearing, especially those dating from Victorian times

Chapter 2.1 Timber on timber or nailable steel joists

This chapter deals with both timber ground floors and timber upper or intermediate floors. Although the conservation of medieval timber floors is not described, some of the observations made in relation to aspects of performance of newer construction will apply also in varying degrees to older construction.

Hearths of concrete or brick construction, or flagstones, bearing directly on timber joists with no boards underneath in otherwise timber floors, are dealt with in Chapters 2.2 and 2.5 respectively.

Basic structure

Ground floors

In housing, and small cellular type buildings, three types of timber ground floor are commonly encountered (Figure 2.2) and can be categorised thus:

- single span (ie fully spanning) where perimeter walls provide support
- multi-span (ie partially spanning) where sleeper walls or beams reduce effective spans
- ground-bearing (ie non-spanning) where the floor deck is fixed to battens resting or partially set into a slab or screed of concrete or other solid material, and which is dealt with in Chapter 3

Sleeper walls were built in honeycomb bond to permit through-ventilation to airbricks placed in the external wall (Figure 2.3). Traditionally, either square edged or tongued-and-grooved boards were fixed to the joists, but chipboard, plywood and OSB (oriented strand board) increasingly have replaced these materials. Similarly, joist hangers instead of

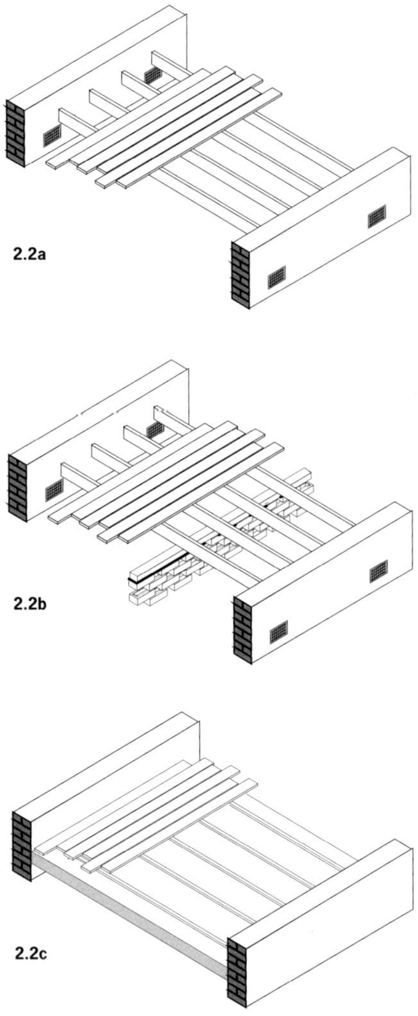

Figure 2.2
Types of timber ground floor: fully spanning (2.2a), partially spanning (2.2b) and non-spanning (2.2c)

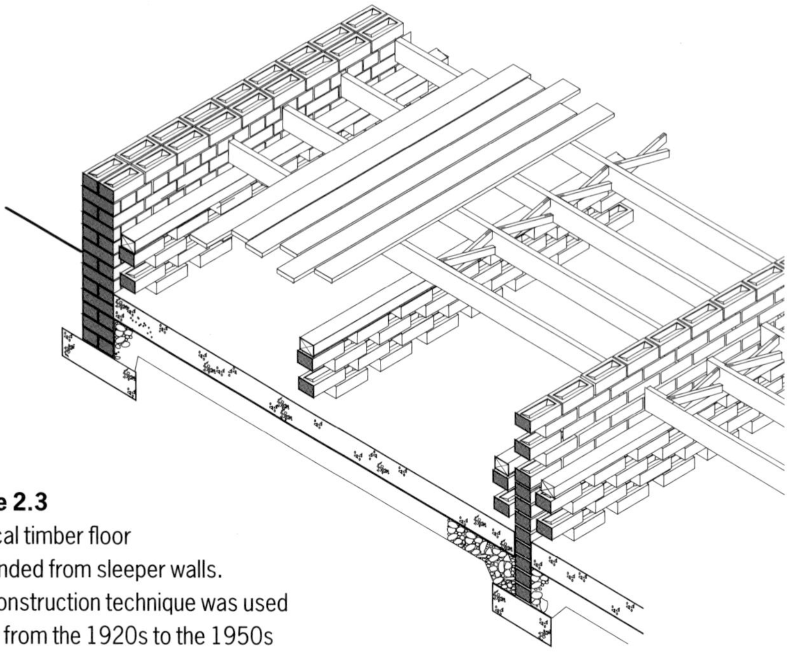

Figure 2.3
A typical timber floor suspended from sleeper walls. This construction technique was used widely from the 1920s to the 1950s

sleeper walls or pockets are used to support joists from perimeter walls.

Before the mid-1930s, it was common practice in a few parts of the UK to place a concrete oversite under suspended timber floors. In the next 30 years it became more universal, following the requirements and exceptions introduced in the Model Byelaws under the Public Health Act 1936.

Between 1967 and 1985, building regulations required a concrete oversite to be installed under suspended timber ground floors. Since 1985, it has been possible to substitute a layer of inert gravel laid over polyethylene sheeting.

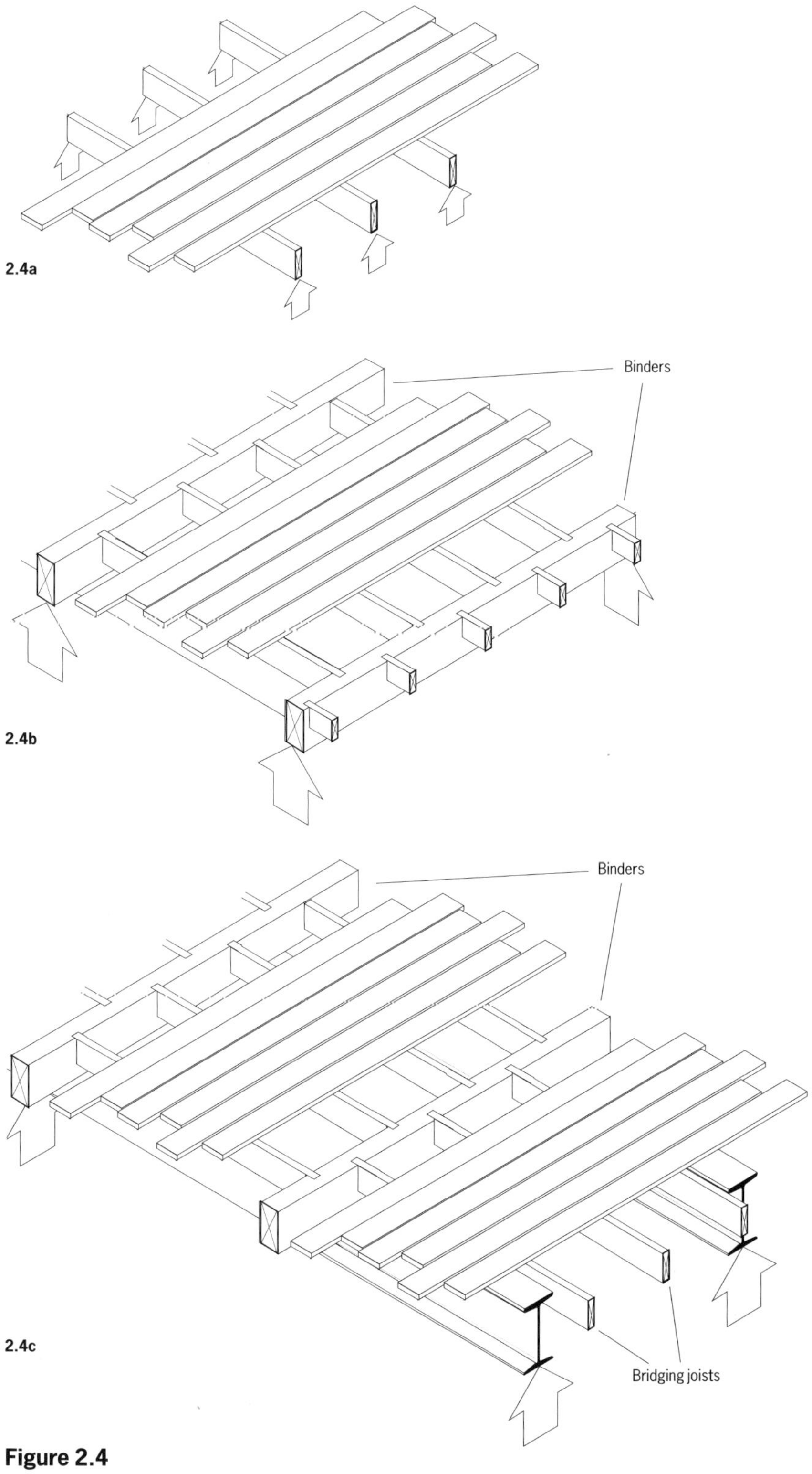

Figure 2.4
Types of timber upper floor:single (2.4a), double (2.4b) and triple (2.4c). Ceilings are not shown

Upper floors

Timber upper floors for simple cellular buildings usually comprise joists spanning between walls, with stairwells trimmed to run either with or across the direction of span. Often these joists have intermediate support provided by a beam or internal wall. Occasionally in post-1960s construction, plywood decked floors will be encountered in which the design of a diaphragm floor takes significant account of the contribution to strength provided by the floor deck, usually as plywood glued and nailed to the joists. Extra care is required when adapting these floors, and expert advice may be needed.

For other non-cellular buildings of greater spans the floors will often be composite in character. In former days these were conventionally defined as single, double and triple (or framed) floors, though these terms are now less frequently used than they used to be.

Single timber floors are those floors where the span usually is less than about 5 m – which is about the maximum for easily available lengths of timber for joists – and where the joists are carried directly on loadbearing walls or indirectly by hangers or corbels. This type of floor has been used for many centuries for simple buildings of comparatively short spans, right up to the present day. In some cases the old local authority byelaws prohibited timber joists to be built into separating walls. However, many examples of so-called rationalised traditional construction housing in the 1950s and 1960s employed cross-walls at about 5 m centres, with joists spanning directly onto the walls (Figure 2.5 on page 72). In order to provide adequate separation for sound and fire, the joists were usually staggered and the bearings limited to about 60 mm.

The main criteria for performance of the traditional suspended single timber floor have been stiffness (lack of which gives rise to deflection) rather than strength (lack of which leads to collapse). The tabulated sizes given, for example, in the Building Regulations 1991 Approved Document A[116] will provide more than sufficient strength and stiffness for normal use in domestic construction of the kind shown in Figure 2.6.

However, even if the joists are of adequate section for the loads to be borne, there remains the problem of rotation. Joists normally need to be blocked at bearings in shorter spans, and intermediate points also need to be strutted at least:

- once at mid-span (Figure 2.7b) in spans of 2.5 to 4.5 m
- twice in spans over 4.5 m (Figure 2.7c)

Joist bearings should always be either built into their supporting walls with appropriate beam filling, or blocked or strutted appropriately. Plasterboard by itself will not provide sufficient restraint against rotation.

Double and triple timber floors have been used where the span exceeds the economic span for single joists, or where extra loading needs to be taken into account. In those cases seen by BRE, the secondary span is normally around 3 m or less, carried between downstand beams (sometimes also called binders) which are usually found to be in steel but perhaps in massive timber sections or flitched timber or in wrought or cast iron if the building is old. The joists (in double timber floors sometimes called bridging joists) of course span in the direction of the larger dimension of the space, so the boards run in the direction of the lesser dimension of the space. Triple floors (sometimes called the framed floors) – where binders are carried on girders, often of steel, and the ordinary timber bridging joists span around 2 to 3 m between the binders – can be found in buildings with very large clear spans.

In medieval and later heritage buildings, the variety of different types of floor is vast. Bridging joists were sometimes tenoned into mortises in the binders at one bearing and pivoted into notches at the other, or merely housed in notches let into the top surface at both bearings. Some bearings can be found pegged. In very old buildings, closely spaced bridging joists may also be found which are much wider than they are deep with wide floorboards set parallel to the joists and let into rebates running lengthways along the joists (Figure 2.8). These floors were designed either to have no ceiling at all or to have a ceiling tight to the bridging joists.

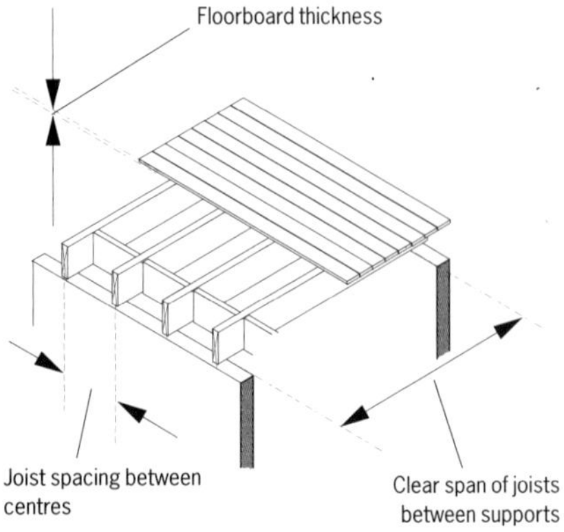

Figure 2.6
The measurement criteria specified by Approved Document A of the Building Regulations 1991 for the elements of floors

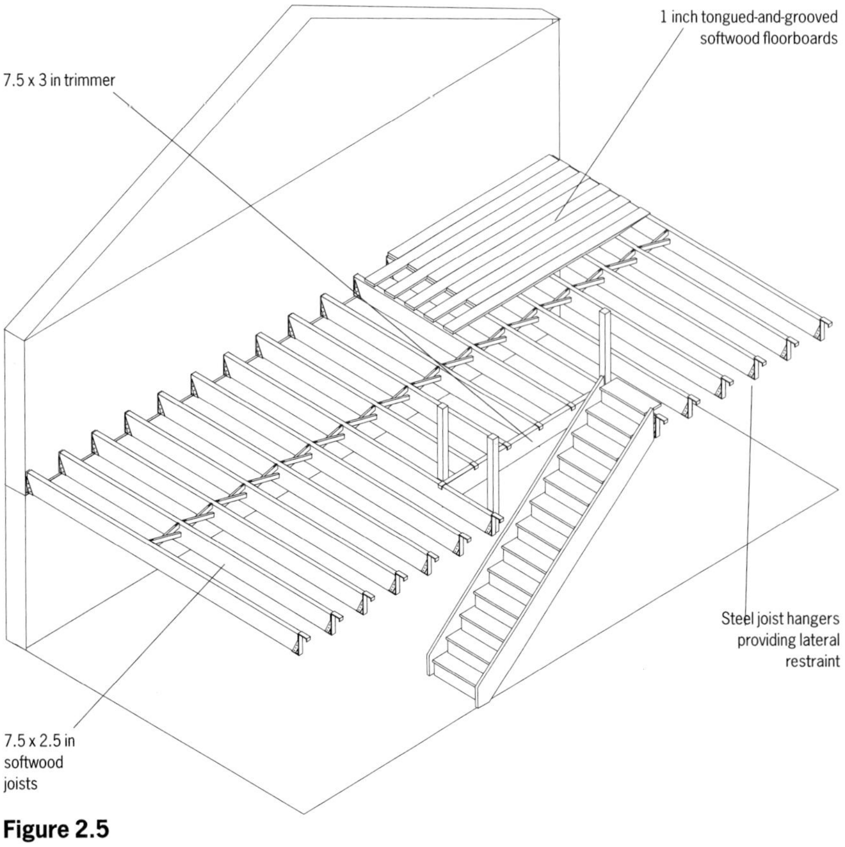

Figure 2.5
The floor construction of a rationalised traditional cross-wall house built in about 1960

Suspended timber floors used at ground level

As explained in the introductory chapter (Chapter 0), in the section on house condition surveys, a substantial number of dwellings have suspended timber floors at ground level. These floors are almost universally of softwood boards on solid softwood joists. The boards may be tongued-and-grooved or they may be square edged. The use of square edged boards was common in older properties leading to high upwards ventilation rates into habitable rooms above. It is only since the 1970s that sheet materials have come into prominence as the preferred decking material.

Floor panels – a combination of prefabricated joists and decking, and designed to be assembled with wall panels – may be found in some timber framed or even masonry buildings (Figure 2.9).

There may be some cases, probably not in housing, where the

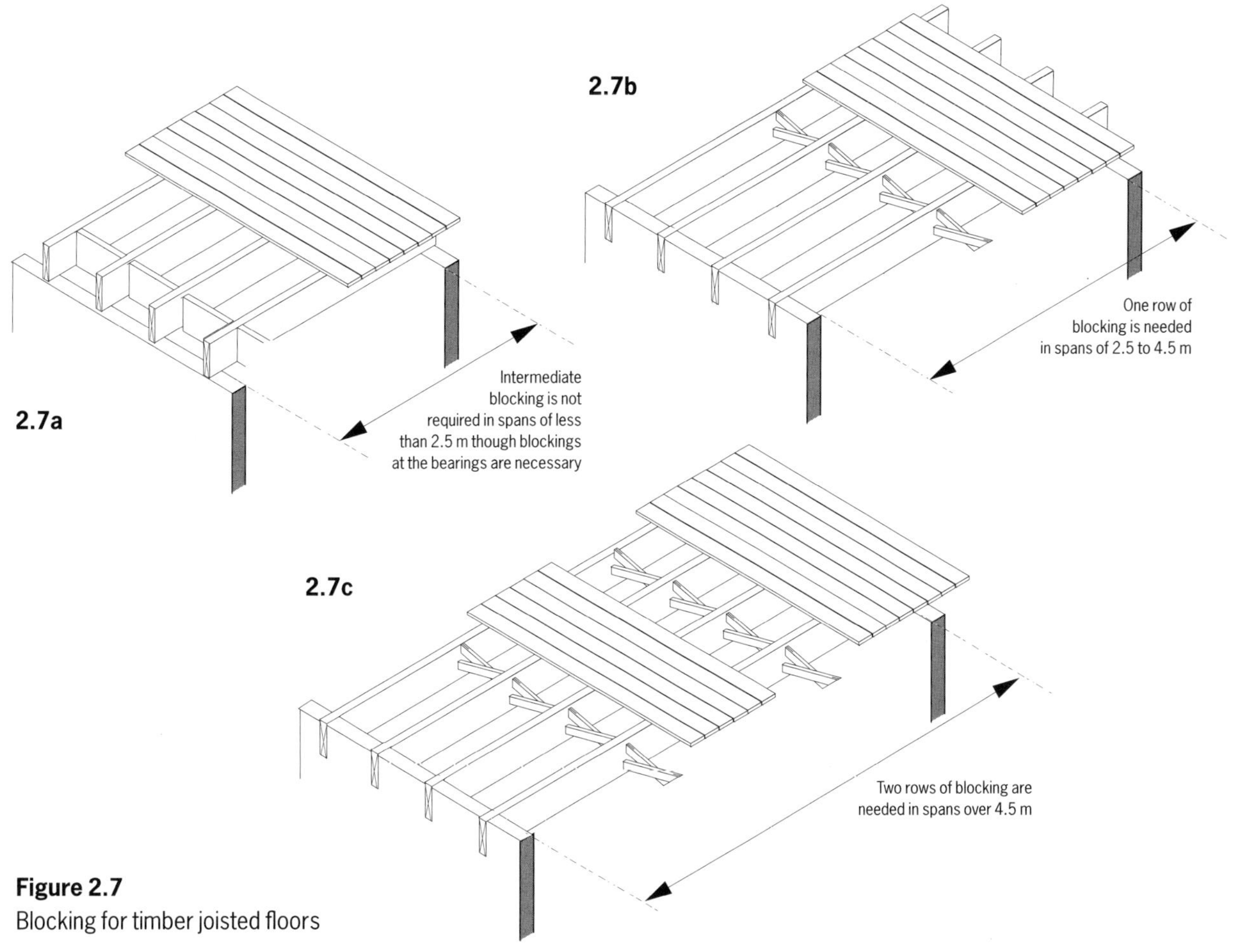

Figure 2.7
Blocking for timber joisted floors

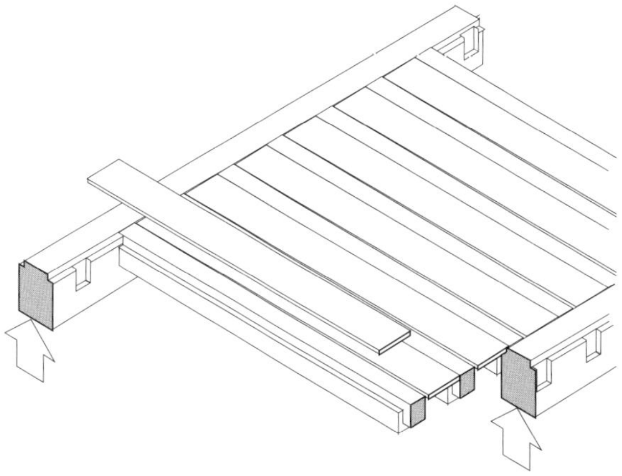

Figure 2.8
Some very old timber floors have boards laid lengthways in the same direction as the joists

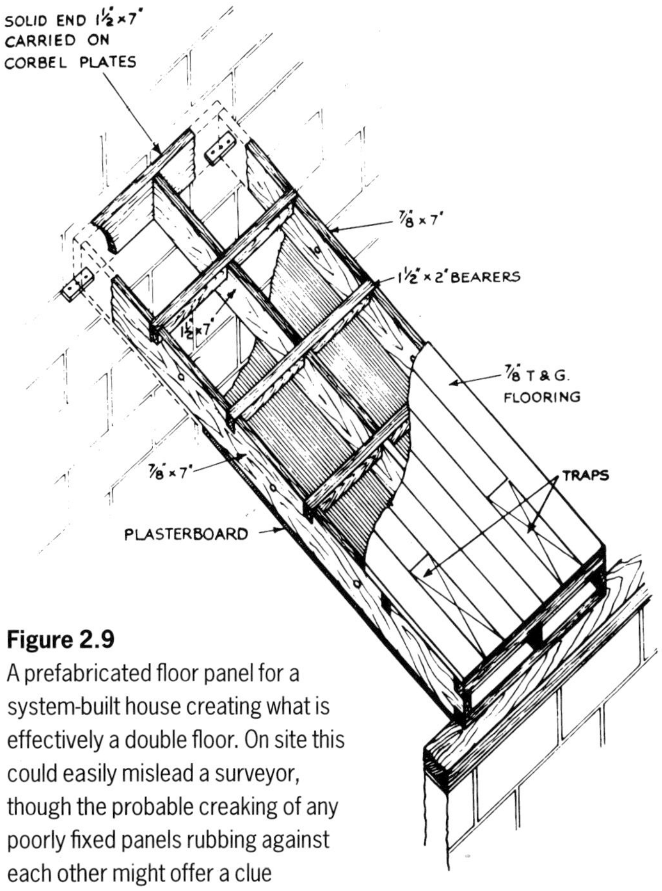

Figure 2.9
A prefabricated floor panel for a system-built house creating what is effectively a double floor. On site this could easily mislead a surveyor, though the probable creaking of any poorly fixed panels rubbing against each other might offer a clue

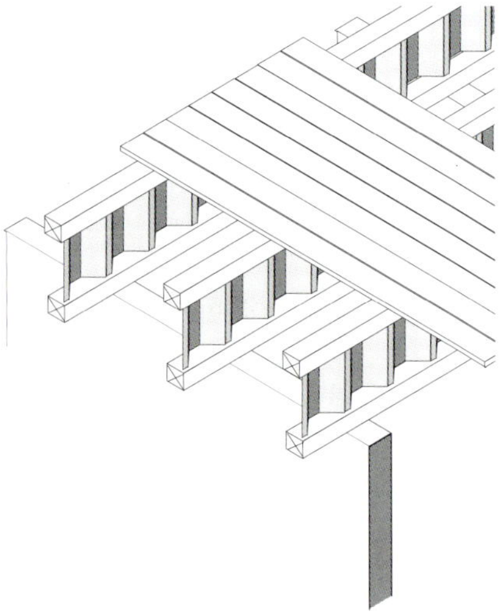

Figure 2.10
Ply-web beams which are prefabricated joists made from timber upper and lower flanges, and a suitable web material (eg fabricated plywood)

Figure 2.13
Joist hangers round a stairwell

Figure 2.11
Open web, nailable steel lattice joists for housing

Figure 2.14
Both the joist hangers and the floor shown here are of relatively unusual design

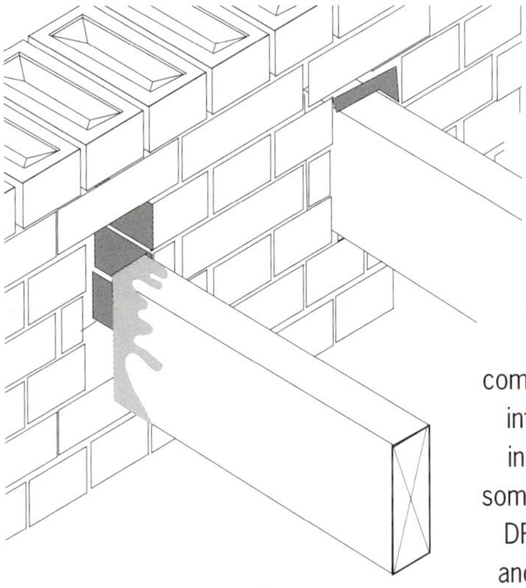

Figure 2.12
In older properties it was quite common to build joists into pockets directly in the external walls, sometimes without the DPCs between joists and bearing surfaces

Figure 2.15
Although ventilation may have been provided for this floor, builder's debris and detritus have made it completely ineffective

joists are fabricated from solid timber top and bottom flanges, and webs of plywood (Figure 2.10) or welded steel (Figure 2.11).

Bearing details

Ground floors

The Building Regulations 1991 Approved Document A gives the minimum bearing length at supports as 35 mm. The bearing should be 90 mm where the floor provides lateral restraint to the wall.

In older properties, say built before the 1920s, it was quite common to build joists directly into pockets in the external walls, sometimes without a DPC. Slate DPCs are occasionally seen (Figure 2.12). Joists without a DPC frequently rot and many have had to be replaced. A technique for replacing rotted joist ends is shown in Figure 2.20.

Upper floors

Minimum bearing dimensions for joists in upper floors are the same as those already given for ground floors. Since there are openings to be found in floors above ground, for example for staircases, trimmers will be found which bear on other joists as well as walls. Steel joist hangers also provide a convenient means of supporting joists from masonry and for trimming round openings (Figure 2.13). However, there are many varieties (eg Figure 2.14) and the choice of the correct one for the circumstances is vital to good performance (BRE Good Building Guide GBG 21[117]).

Treatment of the surface of the ground

Many, if not most, houses built before the 1939–45 war had no oversite concrete placed following removal of topsoil. In England and Wales this could involve around 5 million houses which have suspended timber floors over the whole or part of the ground floor area (*English House Condition Survey 1991*[2]). English and Welsh local authority building byelaws issued after the Public Health Act 1936 called for adequate drainage and the covering of the site with a 4-inch layer of concrete on hardcore, or asphalt on a suitable base, *'unless the exceptional condition of the site or the exceptional nature of the ground renders this requirement unnecessary.'* In Scotland, it was traditional to treat the solum with coal tar pitch or a similar product (BS 2832[118]).

These local authority byelaws also contained provision for a minimum space *'affording through ventilation'* of 3 inches if the ground was covered with one of the specified coverings, and 9 inches if it were not covered.

The Building Regulations for England and Wales, which came into force on 1 June 1972, made provision in Clause C4 for similar deemed-to-satisfy construction; for example a requirement for the space between the upper surface of the concrete and any wall plate to be not less than 75 mm, and, to the underside of any suspended timbers, 125 mm (Figure 2.15). Whether these subfloor voids received adequate ventilation is dealt with under the section covering ventilation later in the chapter.

Protection against radon and methane

An outline of the problem of radon was given in Chapter 1.5.

Where there is a problem of very high radon levels in a building with suspended timber floors, there is a question as to which method of ventilation will be the most effective means of achieving a reduction – natural or mechanical.

Increasing the natural ventilation rates in the subfloor space is the preferred solution for suspended timber ground floors. The easiest way of doing this is to make sure that the existing ventilation provision is working properly and that airbricks, for instance, are clear of debris. Where properties have airbricks installed on only two adjacent faces of the perimeter walls, it may be possible to install extra vents in the other faces. This is a more cost effective solution than resorting to mechanical ventilation.

Where natural ventilation is insufficient to provide a solution, mechanical ventilation must be provided. This then raises the question of whether air input or extraction is the best solution.

The answer depends on the relative airtightness of the floor and the permeability of the soil or oversite surface. If the floor is relatively tight, supply ventilation may be more effective, whereas if the floor is relatively leaky or there is oversite concrete, extraction may be better. In the latter case it is better to keep the underfloor pressure low.

When mechanical ventilation is provided to the underfloor space, increasing the number of airbricks will be useful.

Precautions may be required to prevent sparking in areas where there is also a risk of methane.

On no account should vents (eg for combustion appliances) be inserted in the suspended ground floors of buildings in areas where there is a radon risk.

Main performance requirements and defects

Choice of materials for structure and deck

Joists

Timber which is used to sustain loads in buildings is said to be structural. The term includes joists and associated trimmers and trimmed members which support ceilings and floors. If timber is correctly specified at the design stage, it is more likely to be supplied and fitted without trouble, and will give better service throughout the life of the building. Those aspects which should be considered in a specification for structural timber joists include:

- the adequacy of section size for the loads envisaged
- the risk of being wetted in service and, hence, the need for preservative treatment
- the strength class of timber
- the adequacy of bearings
- acceptable limitations in deflection

Aspects which apply to existing floors include:

- change of use or changes in loadings
- deterioration through insect or fungal attack

Structural timbers will be mainly softwoods (Coniferae). Floors consisting of structural members cut from hardwoods (Dicotyledonae) will be rare and mainly confined to floors built before and during the eighteenth century, though decking in hardwoods, for example oak, will be found in floors built later. Most joists will be of redwood (eg pine) or whitewood (eg spruce).

Floorboarding

Most floorboarding seen will be tongued-and-grooved softwood, though floors in historic buildings may be of a native hardwood such as oak, normally square edged. It is only since the 1960s that floor decks of plywood or chipboard have become common. Some of these early floors, largely through the inexperience of specifiers, did not perform well (eg Figure 2.16), but floors built since the 1970s will usually give little trouble.

For wood based panel products used as decking or overlays, see Chapter 8.1. See also the inspection list in Chapter 8.5.

Strength and stability

Maximum clear spans for timber floor joists have been controlled by rules of thumb for centuries. However, these rules have gradually been codified, and construction text books over the years have given recommendations which differentiated between various building types and, hence, floor loadings.

New and replacement construction is covered by BS 6399: Part 1[19] and BS 5268:Part 2[119].

Some old floors may well be inadequate for current floor loadings (Figure 2.17) – from storage heaters for example. Although it is sometimes recommended that jumping on a floor will reveal its structural inadequacy, for obvious reasons this must be done with circumspection. Judgement may be required on the degree of shortfall that can be accepted, bearing in mind that modern home electrical equipment, such as washing machines and hi-fi, may induce or be susceptible to oscillations in floors. Moreover, conversion work can involve additional loads; for example new cast iron baths or deep freezers, or even expose the need to upgrade roof joists to floor joists (Figure 2.18).

A timber floor is referred to as

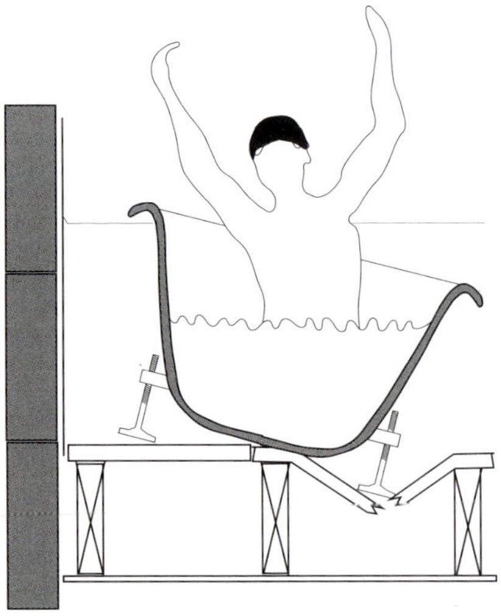

Figure 2.16
Chipboard which is incorrectly specified or protected will be vulnerable to water leakage and subsequent failure

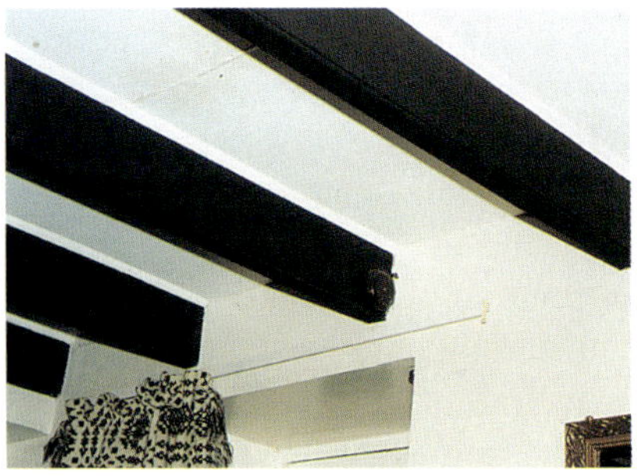

Figure 2.17
This domestic floor is quite inadequate for the span and the loading. The joists are approximately 190 x 50 mm at about 600 mm centres, spanning around 4.5 m

Figure 2.18
This floor was being upgraded from a flat roof to provide a further storey to the building. Additional joists were installed to support the new floor while the original joists continued to support the ceiling. However, the ends of the old joists from the adjacent dwelling still perforated the separating wall, magnifying the fire risk and increasing sound transmission

overspan if it is stressed too highly, or deflects more than the 0.003 of span permitted by BS 5268:Part 2. But overstressing may not be the only cause of structural inadequacy. Other faults revealed in a largescale BRE survey of rehabilitation included insect and fungal attack, excessive notching or drilling, and inadequate blocking and strutting.

The sizing of timber joists for domestic situations (and other situations where domestic loads are not exceeded) can be checked by reference to tabulated information contained in the Building Regulations 1991 Approved Document A1/2. Tables A1 and A2 are appropriate for checking normal construction where imposed loads do not exceed 1.5 kN/m^2. These Tables do not include any allowances for partition loads. Where timber cannot be inspected, it should certainly not be assumed to be of higher quality than strength Class 3, and perhaps may be lower.

Where floors suffer from excessive deflections (eg of more than 14 mm), it may be feasible to insert extra joists of the same depth as the original (to avoid difficulties with ceiling heights) or, in less severe cases, to introduce strutting or flitching of the existing joists.

It will be a choice of least overall damage as to whether to remove floorboards or ceilings. Where the ceiling is removed, most of it will have to be sacrificed, though it may be possible to preserve cornices. An alternative sometimes used is to insert a beam at midspan beneath the ceiling; this does have serious implications for the appearance of the room beneath, though sometimes it can be made to appear as an intended feature of the ceiling. Even where joists are of adequate strength, a floor may still deflect unacceptably because:

- joist bearings are not blocked to prevent rotation
- blocking or strutting between the joists has worked loose or is absent
- floorboards are inadequately fixed to the joists
- loads exceed those allowable

Supporting sleeper walls are sometimes constructed on inadequate foundations, or have been altered during earlier work (Figure 2.19), and may settle causing floors to slope. Holes though sleeper walls for services, or to provide access to crawl spaces have been found without lintels to support joist loads above.

Excessive spring may be due to undersized supporting timbers.

Where replacement joists are to be supported by metal hangers, inaccuracies in cutting and fitting the joists can lead to excessive deflections or settlement.

The use of inadequate or degradable packing material under joists can also result in settlement of joists and floors.

If strutting is thought to be inadequate, flitching by means of a cripple joist coach-screwed to the side of the existing joist may be considered. Not every joist will need to be treated this way.

Rotten joist bearings can be replaced. Figure 2.20 gives one method which has proved to be satisfactory. Where the old joists have been supported from the external wall directly, however, it may be preferable to block the old joist holes and to insert steel shoes to carry the replacements (BRE Defect

Figure 2.19
Deficiencies in sleeper walls sometimes lead to inadequate performance of the floor

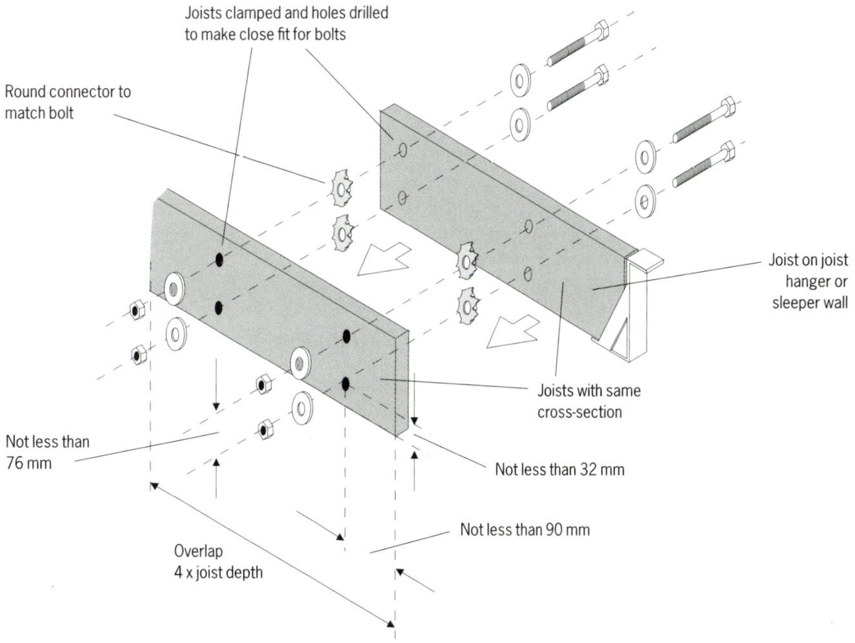

Figure 2.20
Repairing a deteriorated timber joist with a new section

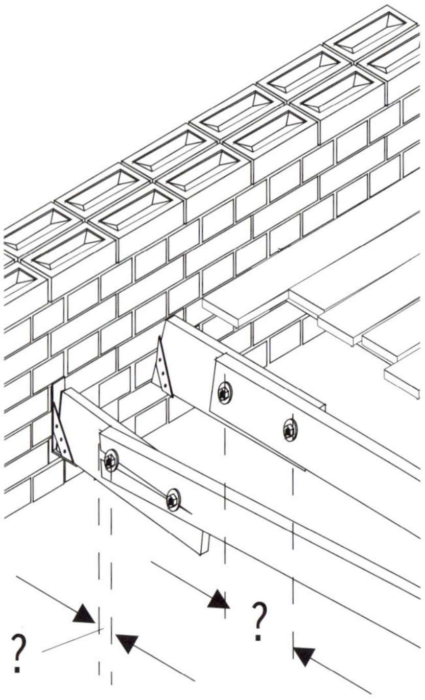

Figure 2.21
Inadequate repairs to joist bearings which have failed or have the potential to fail

Figure 2.22
A joist hanger which has been poorly fitted to a wall: the reason why some modern timber floors are springy

Action Sheet 74[(120)]). It is most important that rotted wood is cut away and removed together with any debris; adjacent timber must be then correctly treated (*Remedial treatment of wood rot and insect attack in buildings*[(109)]).

Where a timber floor has already been repaired, and shows signs of distress, it could be that the replacement ends have been inadequately bolted to the sound parts which have been retained; or that the repaired joists have rotated (Figure 2.21); or that new timber has shrunk during drying out and the bolts need retightening.

In some circumstances specifiers have found it necessary to replace suspended timber floors with solid concrete; this is dealt with in Chapter 3.2.

Specialist techniques are available for repairing structural timbers in the floors of heritage buildings. (Advice can be sought from the national heritage organisations or the BRE Heritage Support Service.)

Upper floors

The installation of services can seriously weaken a timber floor if excessively deep or excessively wide notches in joists are made in the middle third of the span, particularly if blocking or strutting has been removed as well.

Floors may be overloaded by the weight of concrete hearths, partitions and, sometimes, roof loads and water tanks carried on partitions.

Apart from ensuring sufficient bearing at supporting walls, particularly where they are required to give lateral support to the walls which carry them, it is the trimming of openings round stairwells which seems to give the most cause for concern, and which have figured significantly in the site observations carried out by BRE. Trimming round stairwells may have been poorly executed, with end-nailed or notched joints used instead of the stronger tusk-tenoned or cleated joints. In older buildings the trimming round upstairs hearths and their associated flues also often needed attention.

Particular points included trimming joists which were of the same width as the trimmed joists and therefore inadequate, and the use of flimsy and inadequately nailed straps and shoes to hold both trimmers and trimmed joists. Joists supported by steel hangers may lack strength

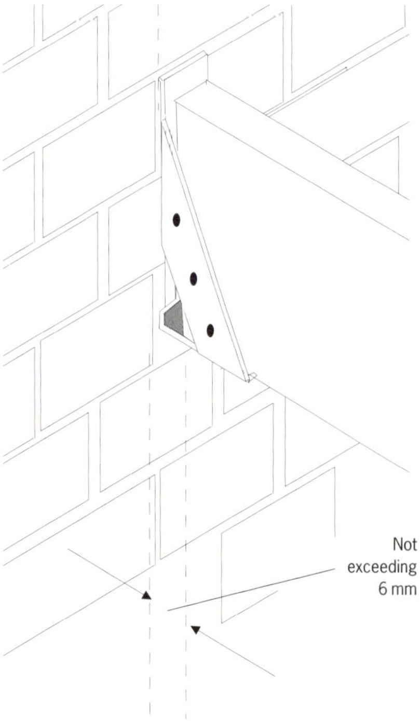

Figure 2.23
Any gap between a joist and the back plate of its hanger should not exceed 6 mm

Figure 2.24
Installing services in an already overcrowded space. The joists are well over the notching limits

because of the rotation of hangers, poor fitting of joists into hangers (Figure 2.22), hangers which are not fully embedded in the wall or are set on packs that deform under load, or because the hangers are not of a design appropriate to the crushing strength of the wall material.

Where new or replacement heating and other service pipes are to be run within existing floor depths, and the joists need to be cut to accommodate them, the recommended sizes and positions of holes and notches should not be exceeded (Figure 2.24). The appropriate conditions are illustrated in Figures 2.25 and 2.26 and BRE Defect Action Sheet 99[121].

Chipboard decked floors need to be nogged in certain cases (see Chapter 8.5).

In houses built since the 1939–45 war, it has been frequently found necessary to provide lateral restraint support to external walls from floors, especially in those dwellings of cross-wall design with light cladding infill front and rear, and in conventional masonry boxes where little support from partitions and flues was available. BRE site inspections have shown that many examples of lateral restraint were inadequately carried out, though the results of such inadequacies will rarely show as distress in the fabric unless extreme winds have been experienced. Figure 2.27 gives a reminder of the most usual provisions for providing lateral restraint to an external wall from a timber floor.

Information from Housing Association Property Mutual indicates that housing built in the early 1990s has suffered, too, from deficiencies in lateral restraint strapping. Of the schemes inspected, 1 in 7 did not make adequate provision for lateral restraint (Figure 2.28).

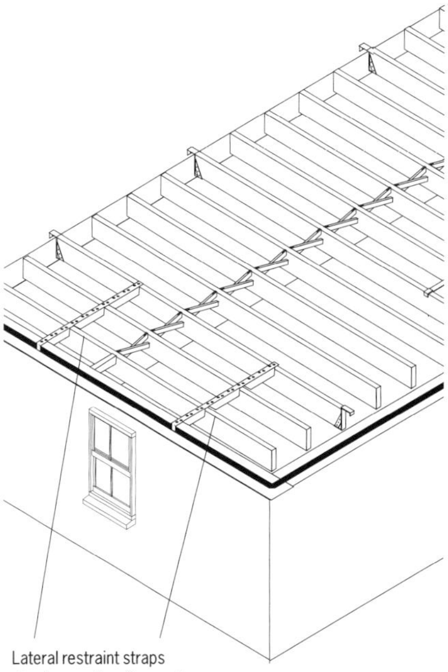

Figure 2.27
Lateral restraint strapping on the first floor of a house

Figure 2.28
A badly fitted lateral restraint strap which would offer no support to the external wall. Any movement in the wall will not be restrained by the strapping

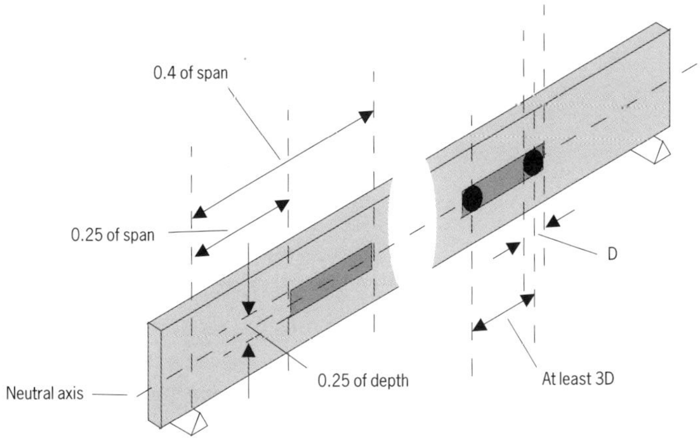

Figure 2.25
Allowable positions for drilling holes in timber joists

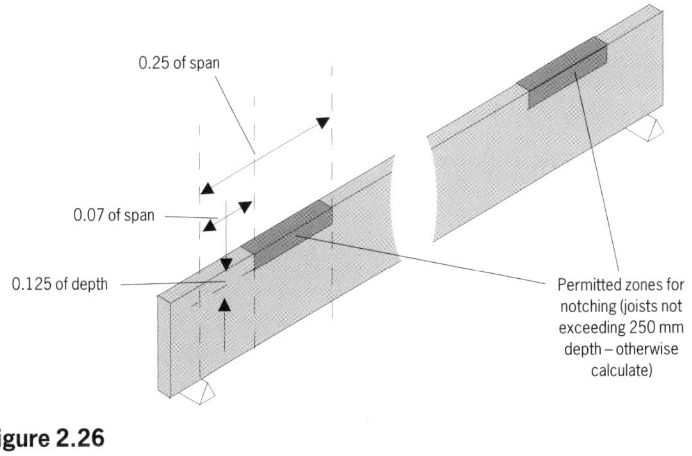

Figure 2.26
Allowable positions for notching timber joists

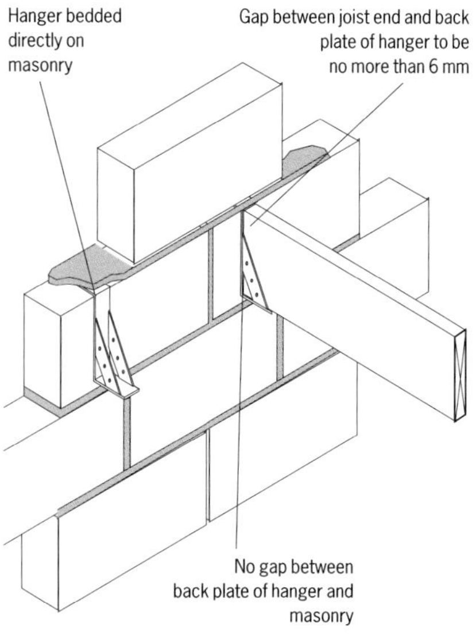

Figure 2.29
Maximum gaps permitted for joist hangers and joists bearing on them

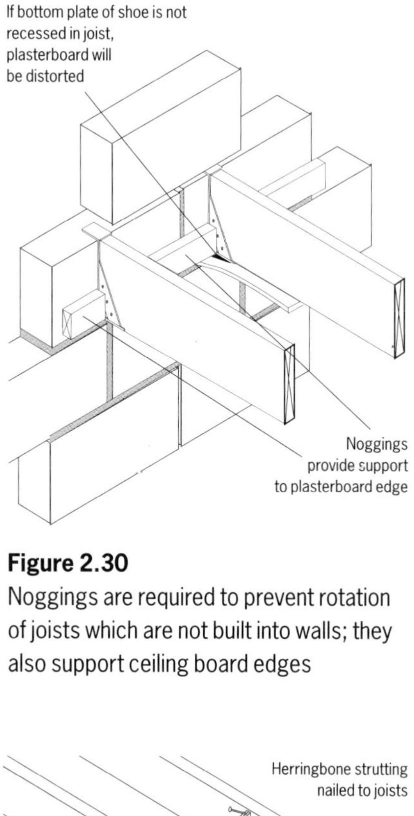

Figure 2.30
Noggings are required to prevent rotation of joists which are not built into walls; they also support ceiling board edges

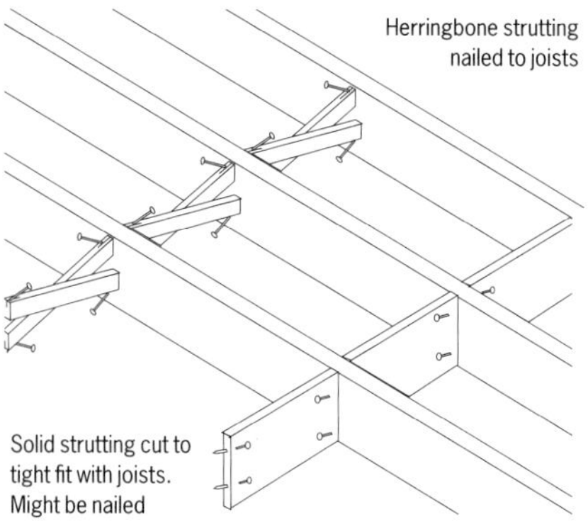

Figure 2.31
Strutting (or blocking) a floor. Herringbone is often used to prevent rotation

Cases have also been encountered where joists have been lapped at mid-span without adequate fixings, leading to deflections.

Joists carried on hangers have not always been installed correctly and may give rise to unsatisfactory bearings for the joists. A reminder of some of the main criteria for joist hangers is given in Figures 2.29 and 2.30.

It is the simple tenoned bearings in bridging joists in older double floors which can suffer some of the worst movements since the depth of timber in the tenon can be very small indeed. Such connections may need remedial strapping.

Dimensional stability, deflections etc

Coefficient of linear thermal expansion per °C:
hardwoods and softwoods:

- with grain 4 to 6 × 10^{-6}
- across grain 30 to 70 × 10^{-6}

Reversible moisture movement:

- hardwoods 0.5–4.0%
- softwoods 0.45–2.6%
- plywoods 0.15–0.3%

Further information on moisture movements is contained in Table 8.1 in Chapter 8.1.

There have been problems of excessive vibration with some long span timber floors in housing. Vibration can be caused by foot traffic as well as domestic equipment such as washing machines. Joists supported by joist hangers seem to be more prone to this problem than joists which bear directly on walls. The result of rotation is to reduce the loading capacity of the joist. Adequacy of blocking to restrain lateral rotation, is important (BS 6472[122]).

Deep timber joists need to be adequately restrained from rotating under load at mid-span where spans are greater than 2.5 m (Figures 2.7b and 2.7c). However, missing noggings at bearings can also lead to lateral rotation (Figure 2.30).

Herringbone strutting between timber joists is the traditional method of reducing or preventing rotation of the joists (Figure 2.31). Such struts may be of protected steel or of timber. Solid strutting is sometimes used, though it is more difficult to achieve a good fit to the joists than with herringbone; as the joists shrink, the strutting loosens. Herringbone strutting does not loosen to the same extent when joists shrink, though it may perform less well than properly fitted solid strutting.

Shrinkage

Softwood floorboarding can vary in its condition in existing buildings (Figure 2.32), particularly when, over the years, it has been lifted for access to concealed heating pipes. There is further discussion of typical conditions in Chapter 8.2.

Where double timber floors are carried on rolled steel members, and insufficient allowance has been made for timber shrinkage under the effects of too rapid buildup of central heating, some apparent humping of the flooring surface may become apparent (Figure 2.33). In these cases the only practical remedy is to re-lay the floor and pack up to level after the timbers have dried down to their longterm moisture content.

Where floor joists have been

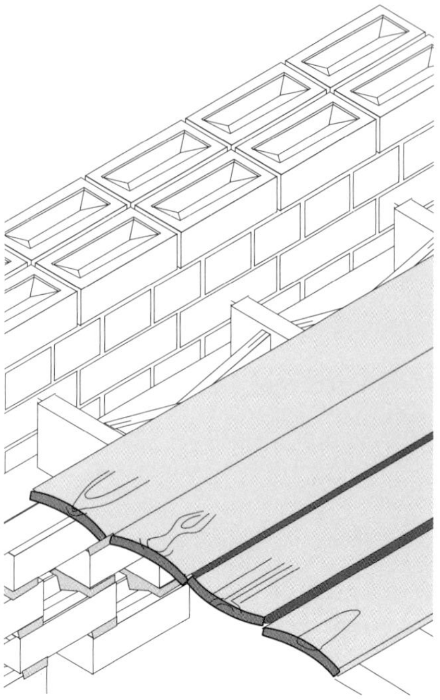

Figure 2.32
Cupping of floorboards

supported on metal joist hangers, a defect seen by BRE investigators on site has been localised crushing of lightweight blockwork where the hangers used have been of unsuitable design. Such crushing tends to happen fairly early in the life of a floor and is unlikely to occur where the floor has been in use for some time, unless local heavy floor loading has been introduced. Old joist hangers may not be marked with appropriate loadings, though hangers to BS 6178:Part 1[(123)] † are marked as suitable for minimum block strengths of:

- 2.8 N/mm^2 (very lightweight)
- 3.5 N/mm^2 (light to medium density)
- 7.0 N/mm^2 (medium density or heavier)

Several cases of inadequate bearings to joists have also been seen – in one case only 10 mm. Such small bearings can lead to local crushing of wood fibres with consequent settling of the joist and springing in floorboards.

The amount of out-of-levelness of floors which can be accepted is often a subjective matter. Floors which are out of level by more than, say, 40 mm, will often present difficulties in placing furniture. See Figure 2.34 and also the section on levelness and accuracy in Chapter 1.5.

† At the time of publication of this book, Part 1 is the only one to have been issued.]

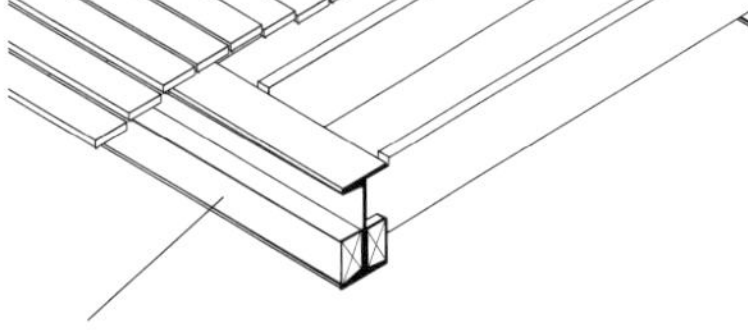

Figure 2.33
Floor humping over steel joists

Thermal properties and ventilation

Rot in ground floors

Voids under suspended timber ground floors need to be ventilated to ensure freedom from condensation. Existing floors often have less ventilation than is currently recommended, and the ventilation openings provided may be obstructed by outside ground, pavings, debris or vegetation. Areas of solid floor may impede cross-ventilation (ie to kitchens sculleries and halls). Building rubbish left under floors can also obstruct ventilation.

Recommendations on the amount of free ventilation area which should be provided have varied over the years. Nevertheless, it should be recognised that there is little scientific evidence available as to how much ventilation is required to ensure freedom from deterioration of floor timbers. It would appear that the provision of airbricks in subfloor voids in late Victorian times was governed entirely by rule of thumb. The earliest Building Research Station (BRS) recommendation, just prior to the 1939–45 war, was 1.5 in^2 open area per foot run of wall. This was confirmed in BRS Digest 1, First series, in December 1948[(124)]. *Principles of modern building*[(6)] in 1961 also repeated this provision and, at the same time, recommended that vents should be provided in at least two external walls on opposite sides of the building, if possible in all walls. Pipes or ducts were also recommended to provide for air movement around obstructions caused by solid floors and hearths (Figure 2.35).

Although it was customary to insert airbricks under the suspended timber floors in a building built under the old Model Byelaws introduced under the Public Health Act 1936, the only specific requirement was provision for 'free air space' (ie the space between the underside of the joists and the top of the slab or ground); this was afforded through a space of 3 inches if the subfloor was concreted or covered with asphalt, or 9 inches if not concreted or covered. The

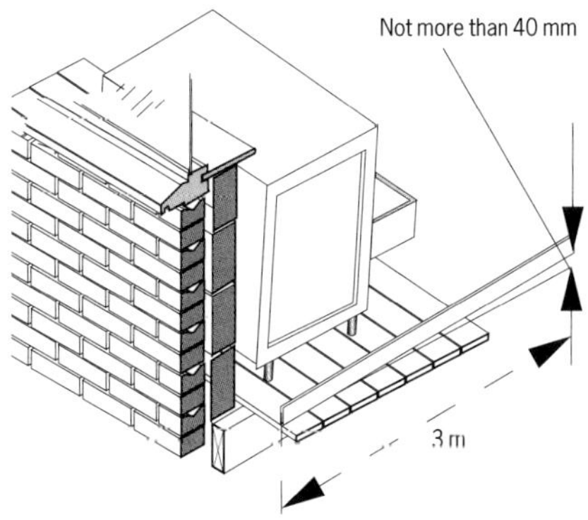

Figure 2.34
Floors which are out of level present difficulties in placing furniture

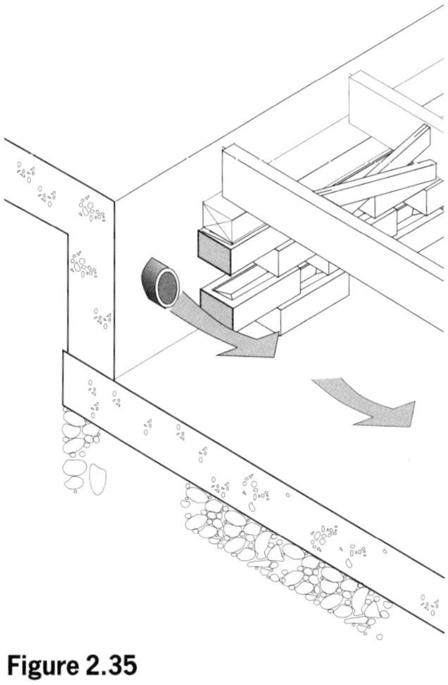

Figure 2.35
Vents inserted to provide cross-ventilation in a part solid, part suspended ground floor

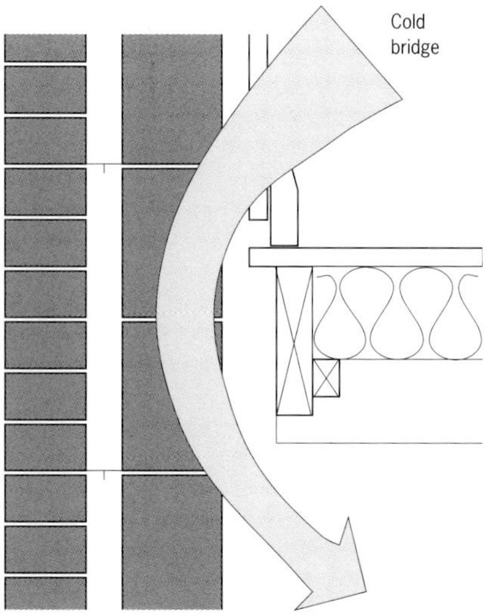

Figure 2.36
Where the perimeter of the floor is not insulated, there will be a cold bridge

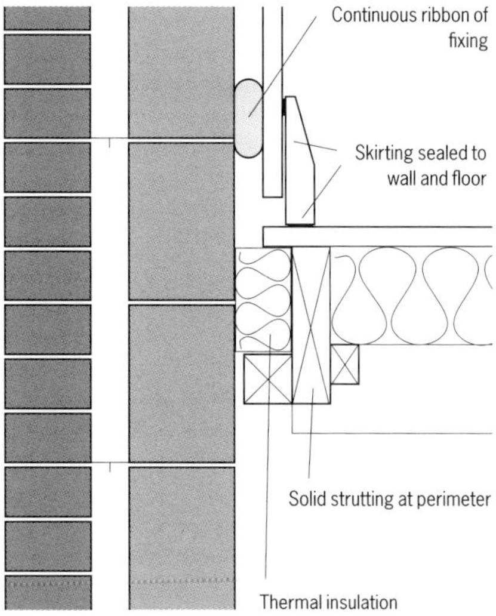

Figure 2.37
Overcoming the cold bridge shown in Figure 2.36

Building Regulations for England and Wales 1972 (Statutory Instruments No 317:1972)[125] increased the void depth requirement for covered subfloors to 125 mm, but still the free area requirement for vents in the external wall was defined only as adequate.

BS CP 102[40] gave a requirement of 3200 mm^2 per metre run of external wall, whereas Clause C4 of the Building Regulations 1991 Approved Document C[42] gives in Paragraph 1.10(b) a figure of 1500 mm^2 per metre run. BS 5250[126] gives a figure of 1500 mm^2 per metre run or 500 mm^2 per square metre of floor area, whichever is the greater.

The ventilation rates actually achieved under suspended ground floors naturally vary according to the positions of vents and the prevailing conditions. However, it must be assumed that provisions in accordance with the building regulations should achieve adequate ventilation rates. In tests carried out by BRE, the ventilation rates actually achieved under a timber suspended floor were measured over a period of two to three days. The tests included a range of subfloor airbrick locations although the total ventilation area remained at the level recommended in the then current building regulations. The subfloor ventilation rates measured for this example fluctuated widely, ranging from 2 air changes per hour (ACH) to over 18 ACH. It was also found that the subfloor temperature remained very constant through the day and night, and that wind speed had only a limited effect on the subfloor ventilation rate. Temperature differences between subfloor and external, and internal and external, were far more significant.

Gaps between boards in older floors provided additional means of ventilating the underfloor space. Decay often resulted if these floors were covered with an impervious material which prevented this ventilation. At one time this was widely known as linoleum rot.

Radon

Dwellings with timber suspended floors and high radon levels have been the most difficult to resolve, particularly those which do not have a concrete or other relatively impermeable oversite. Increased natural ventilation, and mechanical supply and extraction, have all been used successfully for moderate concentrations of radon. Sump systems installed into oversite concrete have been very successful in reducing high radon levels in houses where such oversites exist. In those buildings where no oversite exists and radon levels are very high, solutions are likely to prove more difficult and costly.

Energy issues at ground floor level

Most suspended timber ground floors in buildings constructed before the 1980s, and not upgraded since, will have no added thermal insulation. Moreover, since ventilation of the subfloor void was needed to ensure freedom from decay, draughts through the joints in floorboards and under skirtings would negate even what little insulation was available from the wood itself.

Where recently built dwellings have had thermal insulation incorporated, BRE inspections on

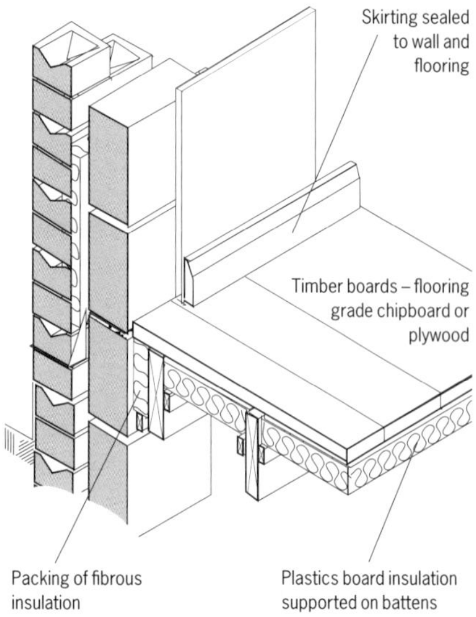

Figure 2.38
Boards on battens retrofitted to the underside of a suspended timber ground floor

site have shown:

- insulation missing from perimeters
- lack of sealing at floor/wall interfaces (eg at skirtings) (Figures 2.36 and 2.37)
- insulation draped over joists before laying decks leading to movement between boards and joists
- ill fitting boards leaving large gaps

It is frequently practical to insert thermal insulation under suspended timber ground floors as a retrofit operation. If there is a cellar, installation of insulation should be a relatively simple matter; or if there is a crawl space which is sufficiently deep, it may not be necessary to remove more than a small area of boards to provide an access trap.

The retrofitting of board insulation supported on nails or light battens nailed to the sides of the joists (Figure 2.38) may be found to be more practical than quilt on netting (Figure 2.39). Quilt on netting is a technique which may be more useful in new-build or when complete replacement of the floor becomes necessary. Insulation boards can be pushed up to be in contact with the underside of the floorboarding and should be cut to a tight fit.

Of course, if the boards need to be removed for any reason, it is a comparatively simple matter to insert insulation.

Upper floors

Where a floor is entirely contained within a heated building it is not usually necessary to provide thermal insulation. However, if a floor separates dwellings with very different heating standards, provision of some insulation may be desirable and it will be necessary in the case of flying floors (floors of rooms directly over passageways or garages open to the external air).

Cases may be seen where quilting has been incorporated into suspended upper floor voids, though normally for reasons of sound insulation rather than thermal (Figure 2.40).

Control of dampness and condensation

Ground floors

Softwood timber is generally used for floor construction and must be maintained in a dry condition to avoid rot. Where these floors are at ground level, timber is normally supported on masonry or concrete and protected from rising damp by impervious membranes (DPCs or DPMs). If the membranes are bridged, or have fractured or deteriorated, rising damp may reach the timber. Additionally, timber floors may become damp if joists have been built in contact with solid external walls, particularly if these are located in an area of high exposure to driving rain (Figure 2.41) or where builders' rubble has accumulated (Figure 2.42).

Although there is a small risk of condensation developing in a suspended timber ground floor which has been upgraded with thermal insulation, it is not necessary to introduce a vapour control layer since any small amounts of condensation which form should be vented safely away by air currents

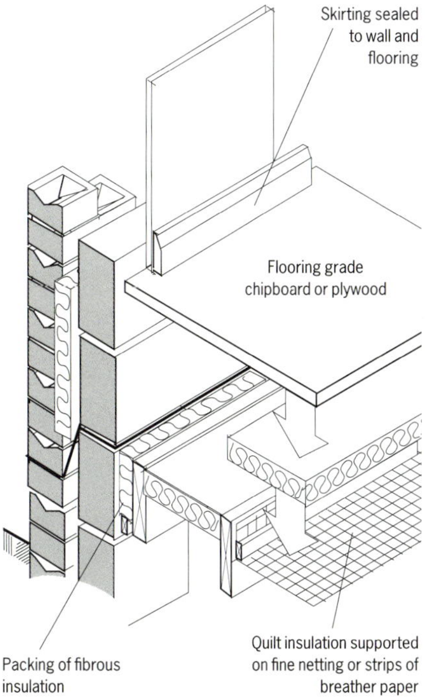

Figure 2.39
Quilt on netting: a less appropriate form of retrofitting thermal insulation. It is feasible only if the old boards are being replaced

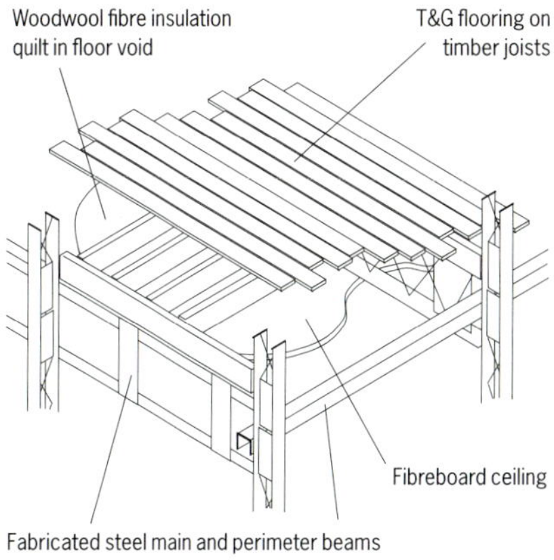

Figure 2.40
A suspended first floor in a prefabricated house from the 1950s. This example is from the Birmingham Corporation system

Figure 2.41
This floor joist has rotted after being in contact with wet external brickwork

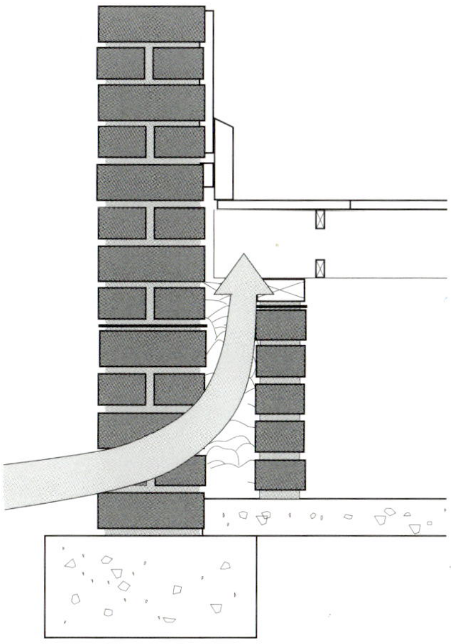

Figure 2.42
Debris or builder's rubble may bridge the DPC leading to dampness and rot in the bearing

Case study

Massive condensation from drains leaking into underfloor void

A 1930s semi-detached house had been occupied for decades without experiencing any problem of condensation or rising damp. Part of the front garden was compulsorily purchased by the local authority for a road widening scheme, and the boundary wall adjacent to the road was replaced. Some weeks later, as winter approached, the house began to suffer from extreme condensation problems. The occupants patiently wiped it up but it got worse.

Soon rising damp began to show above the skirtings on plastered internal brick partitions. Then, one day, the leg of a chair broke through a floorboard. Inspection revealed that the contractors, in rebuilding the boundary wall, had inadvertently cut a storm water drain from a gully taking roof rainwater. The water had found its way into the underfloor void to a depth of several inches and the heavy clay soil of the site had prevented its escape. The remedy was to provide a new drain for the rainwater gully, and a land drain to the underfloor void as an insurance policy against recurrence. The floor joists and wall plates over the sleeper walls had not been provided with a DPC, although the external walls had a DPC of two courses of engineering blue bricks. Consequently, joists, wall plates and boards had suffered from wet rot, and had to be renewed over most of the ground floor area of the house. Plaster had to be cut away from the partitions and replaced, and extensive redecoration needed to be carried out.

under the deck. A vapour control layer also might provide a catchment tray for accidental spillages of water.

Bridging of a DPC, or even flooding of underfloor voids, may occur if the ground or paving around the building is raised so that rain or floodwater reaches masonry above the DPC. The ends of any joists or boards in contact with solid external walls will need to be protected or replaced with less vulnerable construction. Replacement DPCs may need to be provided at these points below wall plates or joists on sleeper walls, and drainage may need to be provided for the subfloor void.

Fully spanning or partially spanning timber ground floors are not usually regarded as prone to condensation problems. Warmth from a dwelling will normally maintain the floor timbers above dewpoint and conventional timber floors do not normally include any vapour control layers where moisture may collect at low temperatures. However, the use of impervious floor finishes, such as PVC flooring, above the timber deck may create a risk of moisture accumulation under certain conditions. Shortfalls in under floor ventilation can result in high humidity in the subfloor void and increase the risk of condensation. The passage of water vapour from the ground into the void should be limited by the provision of a concrete oversite, asphalt or DPM. The presence of water in a subfloor void due to flooding or longterm groundwater problems will lead to particularly high humidity and associated condensation on cold surfaces, and will require specific action, either to prevent water passage through the structure or to improve drainage or lower groundwater levels.

Experience from buildings in Scotland suggests that where buildings are located in a frost hollow, condensation can occur on the underside of timber floors near to airbricks on north walls.

The durability of timber ground floors may be ensured if the timbers are maintained at a moisture content below that at which rot will occur (ie at about 20–22% or below). In addition to the risk of becoming damp through the omission or bridging of DPCs and DPMs, or due to the accumulation of moisture from condensation, timbers may achieve a high moisture content in service merely by reaching an equilibrium with moist air which can be present in a subfloor void. Humid air in the subfloor void occurs as a result of migration of moisture from the ground, the rate of migration being dependent upon the permeability of the ground surface treatment, which ideally should be concrete or asphalt, and the amount of naturally occurring ground moisture. The humidity in the subfloor space will then depend on how well it is ventilated. Correct provision of ventilation openings and unrestricted cross-ventilation are both important considerations.

Upper floors

Rooms directly over passageways open to the external air or above garages – the already mentioned flying floors (Figure 2.43) – should be insulated at floor level. Where they suffer from condensation, it is usually as a result of inadequate floor insulation. The risks associated with this situation are covered in *Thermal insulation: avoiding risks*[31]

Voids in upper floors may provide an unwanted route for air movement from ventilated wall cavities or tile hanging to living areas, particularly if floorboards are plain edged and have gaps.

Figure 2.43

A 'flying floor': most are not thermally insulated

Fire

As *Principles of modern building* stated, the fire resistance of a timber floor depends on the size of joists, the thickness and type of jointing of the floorboards, and the type of ceiling, if any, provided below. Except where the timber members are exposed, joists which are suitable for their normal loadbearing function are adequate for fire resistance, provided their widths meet the need for ceiling fixings. Flames will penetrate through the boards before the joists collapse.

Fire protection to a timber floor is a function of the complete floor including any ceiling (Figure 2.44). The performance appropriate to the various categories of use of the building are defined under national building regulations; for example, for England and Wales, in Approved Document B[59]. Adequate fire resistance can seldom be achieved solely from above the structure and it is usually the ceiling which contributes most.

Existing timber floors may lack a prescribed fire resistance. Assessment of the fire resistance of an existing construction would normally be made by comparison with the tables in the supporting documents to the national building regulations. However, lath-and-plaster ceilings may present a problem, as also may pugged floors and plain edged boards. (Pugging, called deafening in Scotland, is described in the section on sound insulation later in this chapter.)

In general, the requirements call for a modified half hour resistance for two storey houses, a full half hour for three storey, and a full one hour for dwellings converted into more than one unit or for floors over inhabited basements.

Older lath-and-plaster ceilings where the keys have broken are most unlikely to perform well. Upgrading is therefore likely to be required where conversion work is proposed. Improving the fire resistance of timber floors usually involves adding extra weight and

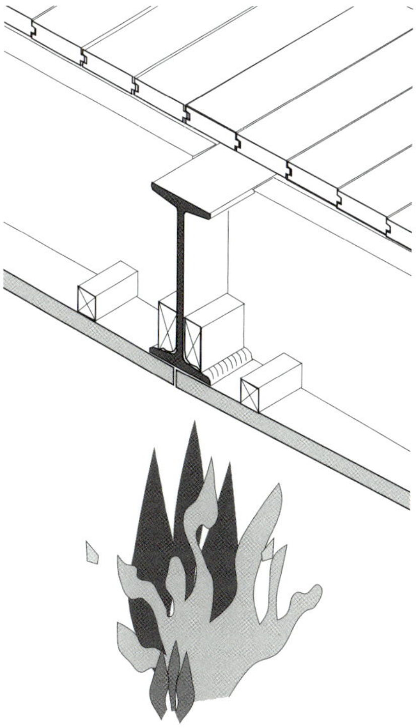

Figure 2.45
Steel beams within floors may need additional protection

Figure 2.44
The aftermath of a fire in domestic premises. The floorboards and ceiling boards have been broken away to assist with firefighting

therefore the structural implications must be fully considered.

It used to be considered practicable to underdraw an existing lath-and-plaster ceiling with wire mesh and plasterboard. However, a lot of time needs to be spent searching for the joist centres which could mean more total effort than removing the lath and plaster and starting again. Cornices and roses cannot be underdrawn in any circumstances.

Fibre building board ceilings do contribute to fire resistance in the sense that any building material can, but whether that contribution is worthwhile is another matter. Calcium silicate board or plasterboard may lack noggings behind the joints – noggings are necessary for full fire protection as well as for structural reasons. Beams of steel or timber may be included within or underneath a floor but may lack adequate fire protection. If these beams support loads in addition to the floor, for example partitions, then a higher level of fire resistance could be required. It is also just possible that asbestos board casings, or even whole ceilings, will be found which must be removed by approved contractors.

Where timber joists are cut to bear on the lower flange of a narrow steel beam, the bearing of the timber on the flange is normally minimal (Figure 2.45 on page 85). In a fire the timber at the steel interface will burn away quickly, distorts, loses its bearing, and the floor falls without achieving its notional period of resistance.

Where the floorboarding is plain edged, it is the ceiling which has to give the necessary fire resistance. By definition therefore, since all floors above the ground have to be fire resisting, it is not feasible to have a suspended timber floor with plain edged boarding without a ceiling of some kind.

It is very unlikely that any timber joisted floor with joists spaced at centres exceeding around 600 mm apart can be economically upgraded for periods in excess of one hour – the protection would have to be so massive that the loadbearing capacity of the joists might be compromised. However, timber floors are known where 300 mm × 225 mm deep baulks of timber have been installed shoulder to shoulder (Figure 2.46). Such floors, on test, have been shown to give longer periods of fire resistance with charring rates in standard furnaces of around 25 mm per hour (Building Research Station National Building Studies Special Report No 27[(65)]).

Tongued-and-grooved boarding would give around 8 to 10 minutes resistance, and in the case of such a floor without a ceiling this would be all that the floor would achieve. In these cases, it might be practicable to insert a ceiling between the joists, leaving the soffits of the joists exposed, though this does demand scribing the ceiling boards to close tolerances (Section 4.2 of BS 5268: Part 4[(127)]).

For most timbers with densities exceeding 400 kg per cubic metre, which includes most commonly used softwoods and hardwoods, the surface spread of flame to BS 476: Part 7[(128)] is Class 3. This may be uprated to Class 1 or Class 0 by appropriate methods of treatment. Section 4.1 of BS 5268:Part 4 gives the rate of charring of various species of timber. Chemicals used for

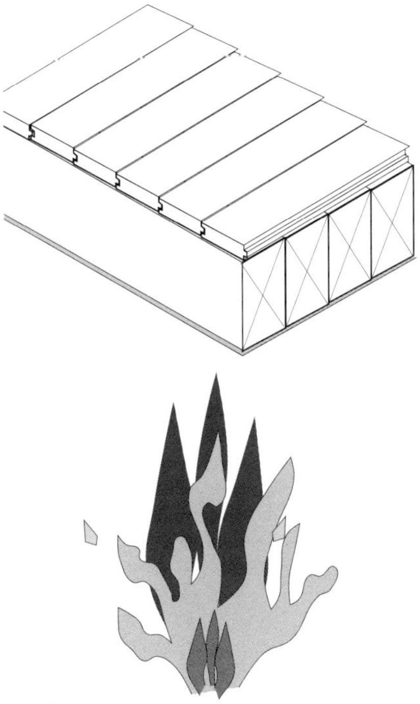

Figure 2.46
Timber baulks placed shoulder to shoulder to create a floor which is relatively fire resisting

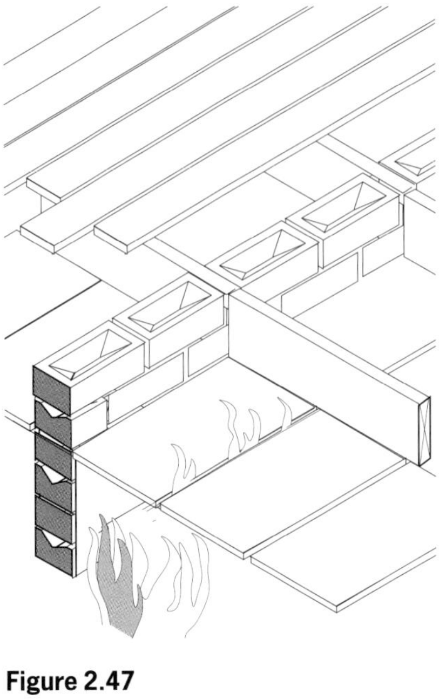

Figure 2.47
Plasterboard needs to be nogged on all edges to provide fire resistance

preservative treatment of timber may not be compatible with fire treatments.

There is a potential problem with requirements for the protection of steel supports to timber joists in floors. BS 5268:Part 4 suggests that a 12 mm plasterboard ceiling beneath 1 mm steel should achieve 30 minutes, whereas 31 mm of plasterboard would be needed to achieve one hour. BS 6178:Part 1 requires a minimum thickness of steel of 2.5 mm. This suggests that joist hangers would need further protection where floors of more than 30 minutes are needed.

Where joist hangers are used, the performance of the floor in a fire can be reduced by up to 25%. The point of failure is reached when the joist pulls out of the hanger rather than through failure of the steel.

Upgrading existing floors

Rehabilitation work will often need to include improvements to the fire resistance of suspended timber floors, especially where a change of use is being made.

Where there is an alteration or extension to a building, or a material change of use, the period of fire resistance of an existing timber floor may need to be increased (BRE Digest 208[69]). Periods of up to one hour may be achieved by the addition of protection to the underside of the ceiling, over the floorboarding and between the joists. There may, however, be problems in increasing fire resistance when the joists are exposed to view from below. Precise improvement depends on the details of construction and condition of the existing floor. In older double timber floors, where separate ceiling joists span between the main beams, these ceiling joists may be insufficient to carry the weight of extra material, and could be in danger of sagging.

Upgrading the fire performance of existing floors can be achieved using a variety of additional materials for the ceiling; for example 13 mm of gypsum plaster on expanded metal lathing would give half an hour, and 19 mm would give a full hour. Thicknesses of this order, of course, need to be built up in multiple coats. A single layer of 12.5 mm thick plasterboard with joints taped and filled, and backed or nogged by timber (Figure 2.47), would give a modified half hour; similar construction using a thicker board would give half an hour. Two layers of plasterboard to a total thickness of 25 mm, laid to break joint on 47 mm thick joists, would be required to give the full one hour.

Existing ceilings may be adequate, but lath and plaster is very variable in likely performance and may not be in sound condition (Figure 2.48).

Some ceilings may not achieve (indeed, some have not achieved) even the modified half hour requirement. Provided, though, the joists are sufficiently strong, these ceilings can be underdrawn with plastered metal lathing or with plasterboard. However, as already considered in Chapter 1.6, whether this is worth doing is open to question.

The floorboards themselves contribute to the overall performance of the floor in fire, but boards which have been raised to gain access to services, and hence have had their tongues removed, may not be adequate for fire resistance purposes (Figure 2.49). There will also be a need to check that the spaces between joists are

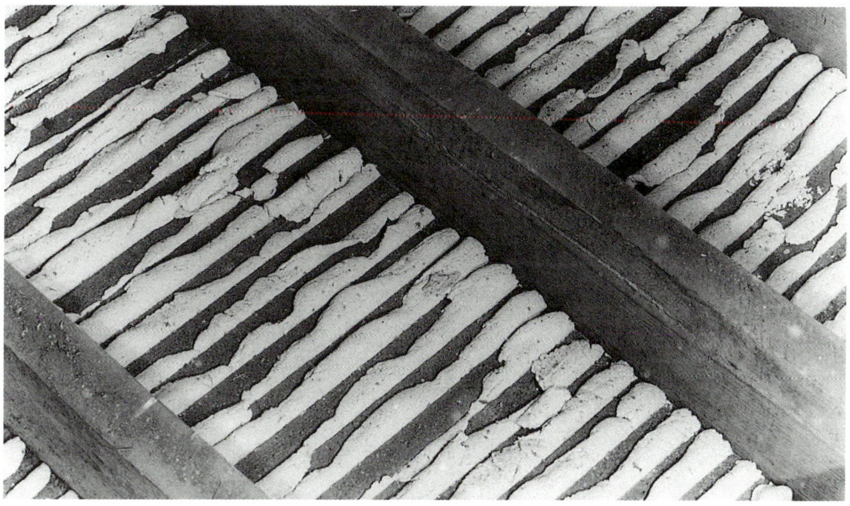

Figure 2.48
A lath-and-plaster ceiling from above, with the floorboards removed

Figure 2.49
The effects of penetration by fire from below, through the joints in the floorboards, can be clearly seen on the singed carpet

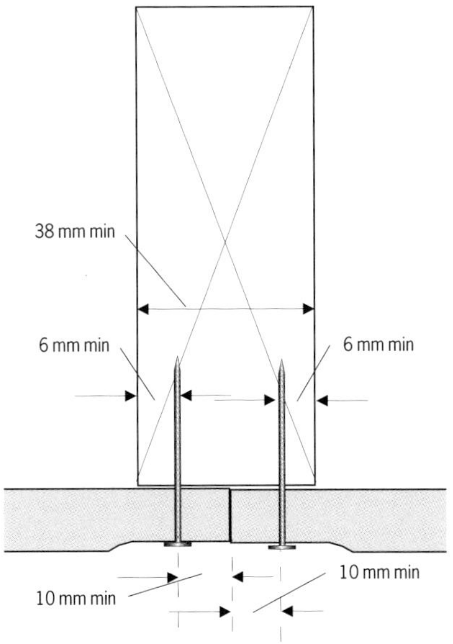

Figure 2.50
Normal fixing for plasterboard ceilings

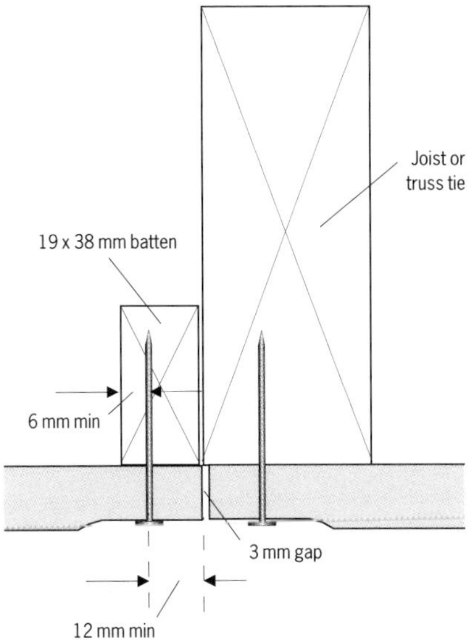

Figure 2.51
A batten is needed for support where fixing centres have strayed

fully blocked where the wall carrying them performs a fire resisting function.

In some non-traditional blocks of flats, there may be a potential fire risk where joints between precast balcony units and the floors of the flats are filled with a form of polystyrene which could allow rapid flame spread in the event of fire.

Pugged or deafened floors may be a problem to assess for likely behaviour in fire. In order to make this assessment, the floor needs to be examined closely, involving the removal of at least some of the boarding. Many such floors will consist of a sand pugging on boarding on battens. Once the support has been breached by fire, say in 10 to 15 minutes, the pugging will fall out and no extra benefit is gained. It is probably better to remove the pugging and seek alternative fire resisting measures.

Fixings

For ceilings expected to give additional protection to the floor to enable it to meet appropriate performance requirements, the provision for fixings is most important and, in the experience of BRE investigators, the least likely to be complied with in all respects.

With plasterboard, fixings should be at not less than 150 mm centres using the following lengths for galvanised nails:

- 9.5 mm board: 30 mm
- 12.5–15 mm board: 40 mm
- boards of 19–25 mm total thickness: 60 mm
- boards of 30–35 mm total thickness: 65 mm

Noggings are a crucial requirement for preventing distortion of a plasterboard ceiling under normal conditions as well as in a fire. Particularly vulnerable is the perimeter of a floor where noggings should be fixed to the wall as shown in Figure 2.30.

The recommendations for fixing plasterboard include providing support to all edges, particularly cut edges. The normal joint centred on a joist is shown in Figure 2.50, and the situation where a batten needs to be used where the joint has strayed from the centre of the joist is shown in Figure 2.51.

For expanded metal lath, fixings should be at not less than 100 mm centres using 38 mm galvanised nails or 32 mm staples. End laps should be not less than 50 mm and side laps not less than 25 mm, and in both cases the overlaps to be wired together at not less than 150 mm centres.

Downlighters

Downlighters are set within the floor void, but, in puncturing the ceiling, they destroy the integrity of the ceiling from the point of view of fire resistance (Figure 1.51).

Sound insulation

In many cases, existing timber floors, consisting of timber boards, joists, and plasterboard or lath-and-plaster ceilings, do not provide good resistance to the passage of sound, and certainly not to the standard required in new construction for floors separating dwellings which may be required under building regulations.

In a study carried out by the BRE Housing Defects Prevention Unit in the mid-1980s, it was found that a number of conversions from single dwellings to multiple occupancy produced floors which were not acceptable for sound attenuation purposes.

There are no mandatory requirements for sound insulation in upper floors within dwellings. However, there have been many cases where a floor is particularly inadequate in circumstances in which quietness can be required or reasonably expected. There may also be situations in industrial premises where sound insulation needs to be improved.

The floor shown in Figure 2.52 originally had a lath-and-plaster ceiling tight to the underside of the bridging joists. When the ceiling was removed in order to expose these joists for aesthetic reasons, the sound insulation through the floor worsened dramatically. The ceiling boards have been scribed to try to fit the sides of the joists, and many gaps are in evidence. The floorboards also are not tongued-and-grooved, giving further gaps through which airborne sound can travel.

Timber joisted floors are much more vulnerable to flanking sound transmission via walls than are concrete floors, and extra care needs to be taken that general measures to improve sound insulation are not negated by this source.

Where a timber joisted floor was required to have improved airborne sound insulation, for example where single dwellings were being converted into flats or in purpose built flats, it was common practice to introduce pugging of the interjoist spaces with suitable material (Figure 2.53). This pugging would either be carried on a layer of rough boarding between the joists or on the ceiling suspended from the joists. In the latter case the ceiling would need to be of heavy construction, and the floor also would need to be specially constructed to carry the extra loads of the pugging; for example using heavier joist sections and expanded metal lathing to carry the plaster finishes. Typical pugging usually consisted of 50 mm dry sand, but practice did vary considerably. In the New Town of Edinburgh pugging typically consisted of a mixture of lime and ashes. Pugging, by filling the spaces between joists with mineral wool, does not give a worthwhile improvement – broadly speaking, the heavier the material, the better the improvement. Pugging has fallen out of favour as a solution to unwanted noise with preference being given, where ceiling heights permit, to double construction. Pugging is no longer recommended as providing a worthwhile improvement to resistance to airborne sound transmission.

An unpugged wood joisted floor with a light plasterboard ceiling will give noise reduction of around 43 dB, whereas one with a two or three coat lath-and-plaster ceiling can be expected to give around 48 dB.

Where reduction in impact sound transmission is needed for timber joisted floors, a floating floor surface isolated from the main joists has been a common method of providing the necessary isolation (Figure 2.54). Alternatively, where there is room, a new ceiling spanning from the surrounding walls underneath the old ceiling can give a very worthwhile improvement in sound insulation

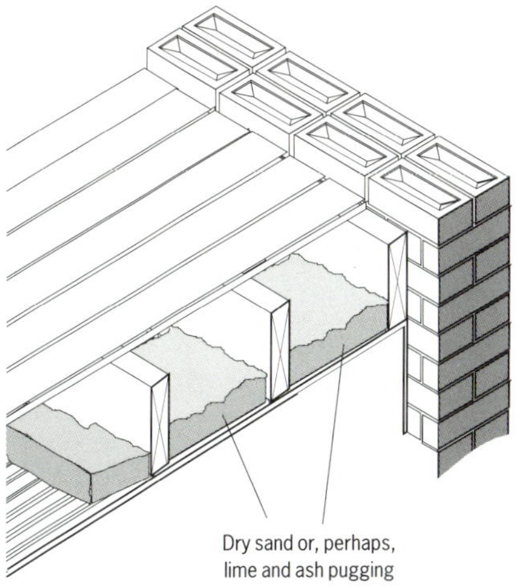

Figure 2.53
Pugging in a traditional joisted floor to improve sound insulation. Although it may provide a worthwhile improvement to airborne sound transmission, there are several negative factors such as the additional loading on the floor joists and foundations, the unknown effect on the fire resistance of the floor, and the possible calamity if the floor becomes wet and collapses

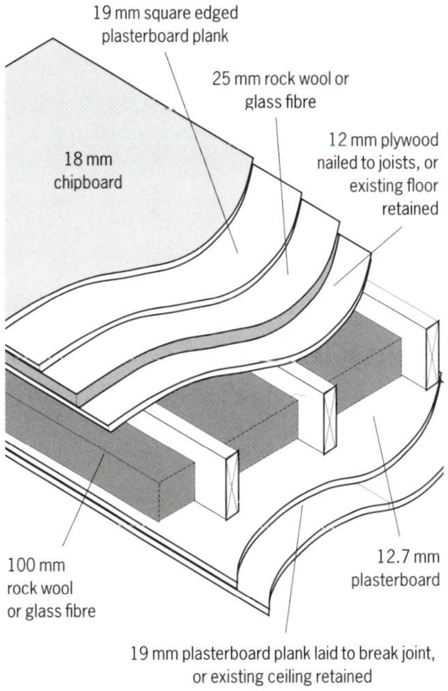

Figure 2.54
A floating floor installed over an existing timber joisted floor

Figure 2.52
A double floor dating from around the sixteenth century. The soffits of the bridging joists show the marks of a former lath-and-plaster ceiling, removed for aesthetic reasons

(Figure 2.55). This is the solution which BRE recommends is used wherever possible.

Tests carried out in the early 1990s have shown that the majority of floors with independent ceilings have given an airborne sound insulation of more than 52 dB $D_{nT,w}$.

Certain double timber floors with ceilings suspended on hangers were purpose built to provide adequate fire and sound separation in flats. Some, for example, in urban areas of Scotland, have typically included binders of around 170 × 350 mm carrying 70 × 170 mm bridging joists with a false ceiling carried on 32 × 80 mm branders (slender ceiling joists) slung from the bridging joists by means of hangers. This solution, depending as it does on connections between the two leaves of the floor, is not so effective a sound insulating floor as one where the ceiling is carried only by the perimeter walls (ie it is not slung from the joists).

Noise has been a problem in some conversions and studies have been carried out by BRE to assess its scale. Subjective response can reliably be measured, obtaining a workable degree of separation of noise from above, noise from below, airborne and impact noise. Noise exposure is more difficult to assess, but the number of different types of noise heard may provide a useful substitute for physical measurements. This measure, together with sensitivity to noise, accounted for nearly 60% of the variance in response to noise in one particular study carried out by BRE[129].

In BRE tests carried out on timber separating floors, not bridging the construction with nails was found to be important (BRE Current Paper CP 46/78[130]).

Noise from washing machines

Noise from washing machines has been identified as disturbing for occupants of adjacent dwellings. The effects of washing machine location, added mass and vibration isolation were investigated by BRE in a series of experiments. It concluded that it was not possible to define an absolute level of noise from washing machines which would not be a problem as this is dependant on a number of factors such as the background noise level. Direction of orientation and minor changes in position would appear to have no significant effect. Location of the washing machine by a very lightweight wall instead of a masonry wall showed a small reduction in noise; however, in many homes the scope to do this is not available. Location of the machines on concrete slabs alone is generally detrimental. The use of readily available materials as anti-vibration mountings can decrease the transmission at mid-frequencies but may increase it at low frequencies. Overall, any improvement is bound to be small.

Durability

Ground floors

Timber is an organic material which, under damp conditions, may be at risk of decay by wood rotting fungi (Figure 2.56). Different timber species have different degrees of natural resistance to decay fungi but the softwood species most commonly used in modern building construction all have a low natural resistance to decay. It is important, therefore, that the moisture content of timber in service remains below the critical threshold for decay. To be immune from attack, the measured moisture content of timber must be maintained below about 22%, and preferably below 20%. Occasional brief periods of superficial wetting, due say to condensation, are unlikely to allow the decay process to start. This threshold value of 22% applies to decay by the dry rot fungus *Serpula lacrymans* as well as the wet rot type which results from attack by a number of other fungi.

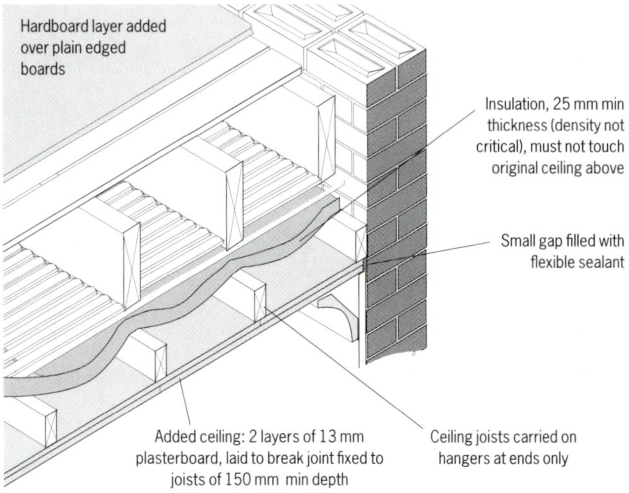

Figure 2.55
A separate ceiling carried on the perimeter walls: a solution preferred to the timber joisted floating floor

Figure 2.56
The lath-and-plaster ceiling to a cellar has been removed to reveal significant deterioration caused by rot of this timber floor

Checks on the moisture content of timber members will give a guide to the extent of any problem and the need for remedial measures to prevent further wetting and promote drying.

It is the bearings of joists built into external walls which often give problems due to dampness and subsequent rot. The replacement of such damaged joists has been dealt with in the section on structure earlier in this chapter.

Rotten or infested timber left under suspended timber ground floors from previous construction or repairs can lead to further outbreaks of rot or wood borer attack. Ground beneath the floor deck may lack oversite concrete or other impervious material, which is particularly true of pre 1939–45 war construction, and if it is below the external ground level there will be a risk of flooding. Wood boring beetle infestation can also reduce floor durability; while attack may occur anywhere, damp timbers are more vulnerable than dry timbers. It is common to find damp floor timbers damaged by both fungal decay and wood boring beetles. Dampness originating from within dwellings may also cause wood rot; it is often found in kitchens or hidden behind panelling, particularly in bathrooms.

Instances of both wet and dry rot have been seen many times on the BRE site inspections. During rehabilitation work, adequate procedures to prevent recurrence were not always carried out. For example, where bearings of joists had been carried directly in pockets in the external wall instead of on a sleeper wall laid parallel, the new joists or replacement sections were fitted into the old sockets without the benefit of a DPC with the risk of rot recurring. Cases also occurred where outbreaks of dry rot were inadequately dealt with; for example untreated material was allowed to remain closer that 300 mm to the original outbreak.

Also seen were cases of bare earth solums beneath suspended timber ground floors with no cover whatsoever, and, in some cases where the dwellings were more than 120 years old, the joists actually touched the bare earth. Since there was totally inadequate ventilation, the risk of rot could be expected to be high, and had indeed occurred in these cases. For durability, ventilation is a prime requirement (Figure 2.57).

Mild steel sheet components used in repair work, such as chairs or stirrups, will need surface protection (eg hot dip galvanising). The life of a zinc coating is dependent on the environment to which it is exposed and is proportional to the thickness of the zinc. The thickness of the zinc coating required to give protection depends upon many factors, but, generally, small components are formed from pregalvanised sheet.

Joist hangers made to BS 6178: Part 1 give a choice of protective treatments:

- zinc coating not less than 600 g/m^2
- zinc coating not less than 275 g/m^2 plus a paint finish

So far as the durability of chipboard decks is concerned, impact tests have been carried out by BRE[131] and are considered further in Chapter 8.5.

BRE Defect Action Sheets 22[132], 35[133] and 36[134], and BRE Digest 245[135], detail methods for determining the presence of rising damp in walls. Advice on the specification and detailing of DPCs and DPMs is given in BRE Digests 54[38], 364[37] and 380[136]. BRE Information Paper IP 19/88[137] summarises an approach to house inspection for dampness in timber. Guidance on the recognition of wood rotting fungi is available in *Recognising wood rot and insect damage in buildings*[108] and on methods of assessment and treatment in *Remedial treatment of wood rot and insect attack in buildings*; BRE Digests 299[138] and 345[139] are also relevant.

Finally, the following should be noted:

- improved levels of draughtproofing in a building may increase the risk of condensation in floors

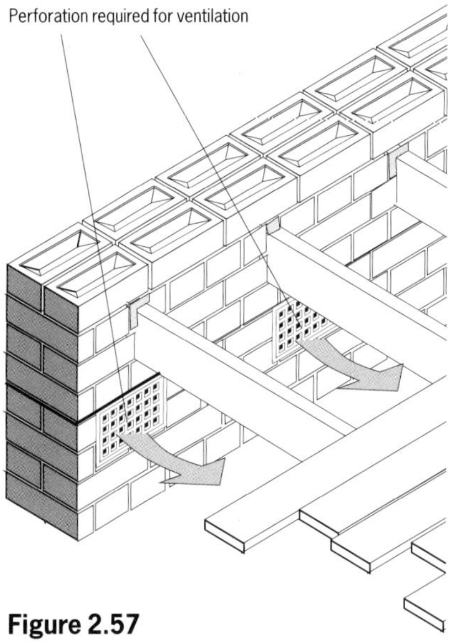

Figure 2.57
Ventilation is vital for durability of ground floor timber joists

Figure 2.58
This ceiling has been removed to help with drying out the floor above, soaked by driving rain entering through unglazed windows

- the addition of impervious floor coverings over timber floors may induce moisture accumulation in the floor deck
- changes to external ground levels can lead to reduced ventilation or even flooding of the underfloor void

Upper floors
Timber upper floors might well be expected to have a long service life and to require little maintenance. However, as seen in the introductory chapter, nearly all the timber upper floors inspected in the BRE site inspections or covered by the *English House Condition Survey 1991* had faults, albeit about half of them being minor in character (Figure 2.58). The main problems seemed to be wood borer infestation, rot in joist ends, and shrunken and squeaking boards.

Decking materials which are not resistant to moisture may be weakened if subjected to repeated water spillages as is likely in bathrooms, kitchens and certain light industrial premises. These floors should be protected with a continuous waterproof membrane. PVC sheet flooring, turned up to form skirtings at walls, and welded at seams, is often recommended for this purpose.

Maintenance
Repairs to floor joists are normally feasible, especially where bearings sited on external walls have deteriorated due to dampness and subsequent rot. One suggested method has been given earlier in the section on strength and stability, but other methods may be appropriate, including splicing in new sections and holding them with epoxy glued dowels.

Where services are laid within the void of a chipboard decked floor, and access may be necessary from time to time, it may be worthwhile considering the provision of access traps in the decking rather than destroying whole sheets. A convenient method of cutting such traps is by using an annular plane (Figure 2.59 and Building Research Station Information Sheet 3/76[140]).

Cleaning out and unblocking the airbricks make a big difference to underfloor ventilation rates. Where too few airbricks have been provided, it is recommended that new ones are inserted to bring the free area up to the current building regulations requirements of 1500 mm^2 per metre run, or 500 mm^2 per square metre of floor area, whichever is the greater. It can be difficult to install more vents into buildings which have thick or rubble filled walls. In these cases the removal of old fired clay airbricks and their replacement by modern plastics louvred vents which have a greater free area can provide a solution.

Work on site

Storage and handling of materials
It has become customary to store structural timber in the open, and not to worry about moisture contents. This is not a practice that is recommended, even though there is evidence that wet timber installed in a building rarely is attacked by rot while drying down to its in-service moisture content. However, it may twist and distort in drying.

Manmade boards are, however, another matter. They should be kept in dry storage before use.

Restrictions due to weather conditions
Floorboards, particularly chipboard floors, should not be laid if the building is not weathertight. Even completed buildings sometimes show evidence of previous wetting of flooring. Many cases have been seen where decking has been ruined by adverse weather conditions through being installed too early (Figure 2.60).

Figure 2.59
The BRE annular plane. This invention facilitates the cutting of access traps in order make service connections within boarded floors. The plane cuts a circular opening with profiled shoulders – into which the circular trap locates – eliminating the need for noggings which are often a source of movement

Workmanship

Problems can be encountered in jointing plasterboard ceilings under joists of nominal 38 mm thickness, since there is very little margin to insert two nails and at the same time keep the required distance from the edge of the boards. Existing ceilings have been found where the plasterboard nails were placed too close to the edges of the boards; the nails split the edges of the joists and failed to hold securely so that the ceiling sagged. Extra battens should have been used in these circumstances, nailed alongside the joists.

The moisture contents of timber tongued-and-grooved flooring should preferably not exceed 20% at the time of laying if wide shrinkage gaps in the finished floor are to be avoided. In well heated buildings, moisture contents could easily drop to around 8% leading to shrinkage of the flooring.

Chipboard must not be allowed to cantilever over joists fixed parallel to walls, otherwise they may show fractures. All tongued-and-grooved joints in chipboard should be glued along the whole of their lengths.

Where water services are accommodated within timber joisted floors, boards should be removable at intervals of not more than 2 m and at every change of direction of a service pipe, whichever is the less. In the case of ground floors, pipes will need to be insulated to BS 6700[141].

Failure of replacement floorings on old suspended timber floors can often be attributed to inadequate preparation of the surface of the boarding before laying the new flooring. The panel above describes suitable techniques for reducing the risk of problems occurring.

Figure 2.60
Wet and ruined floorboarding resulting from an unglazed entrance door

Preparing old timber bases to receive new floorings

Bases should be assessed to ensure that they are sound, rigid and dry. As overlaying may affect the drying upwards of a timber floor, those suspended timber floors at ground level should be checked to ensure that the voids are adequately ventilated to the outside air, and that there is a minimum air gap of 150 mm between the joists and unprepared ground or oversite.

Boarded floors having battens or joists secured in clips set in concrete, or dovetailed and set directly into the concrete, should not be covered by impervious floorings unless it is known that there is an effective DPM beneath.

All protruding nail heads should be punched below the surface of the timber and any loose boards fixed.

Most textile floorings can be laid directly over timber strip and boards, although it is common to provide a separating layer of building paper, paper felt or synthetic fibre felt.

For sheet and tile floor coverings, it is usually necessary to provide an underlay of hardboard, chipboard or plywood. Before laying an underlay, uneven timber floors should be levelled by planing, sanding or applying in situ latex underlayments, or combinations of these. The thickness of the underlay and the mechanical fixing centres are important factors when installing these underlays (BS 8203[35]).

Before laying ceramic tiles and other heavy floorings, the floors should be checked to ensure that they are capable of carrying the extra dead and any likely imposed loads. Stiffening of the joists may be needed. For tiling, a stiff floor is required if cracking of tiles is to be avoided. Boarded floors should be covered with plywood sheets of minimum thickness 15 mm and screwed at 300 mm centres. Movement may still occur between sheets and probably lead to cracking in tiling carried over the board joints. If feasible, joints in the tiling should be made directly over joints in the plywood sheets so that, if movement does occur, the resultant cracking at tiling joints can be filled with a soft jointing material. Some proprietary underlays are available for laying tiles over boarded floors.

Inspection

The fixings of underdrawn ceilings need to be inspected to make sure they are adequate. Plasterboard fixing nails should penetrate at least 20 mm into timber, and be spaced at 150 mm maximum centres.

External ground levels and the locations and sizes of airbricks can be checked by external observation. Airbricks can become partially blocked by cobwebs which then collect dust and other detritus, as well as being blocked by soil heaped against external walls. Additions like porches and other extensions can also block existing ventilation routes.

Floorboards will need to be removed to examine the construction below. Obstructions to cross-ventilation, such as areas of adjacent solid floor, should be identified.

In older buildings, wrought counterboarding was sometimes used over an unwrought first layer of boards. The direction of joist spans may not therefore be apparent at first glance.

Boards may have to be lifted to check the condition of joist bearings.

The particular problems to look for are:

Inspection of ordinary timber floors including upper floors

- ◊ timber quality not up to standards assumed in building regulations
- ◊ floor joists not of adequate section for their span and so deflect excessively
- ◊ floors not level, particularly where joists change direction of span within rooms
- ◊ step hazards at doorways where firring pieces inserted to make floors level
- ◊ floors moving because compressible packing inserted at joist bearings
- ◊ heading joints in adjacent floorboards not staggered between joists
- ◊ lateral restraint straps absent or ill fitting, or not nogged or packed
- ◊ floors overloaded during rehabilitation work with demolition debris or replacement unfixed plasterboards (stack height not more than 150 mm)
- ◊ trimmers round stairwells inadequately fixed
- ◊ trimmers to hearths removed in adaptations to floors
- ◊ notches and holes in joists outside permitted zones or excessive in size or number
- ◊ joists twisting where beam filling absent
- ◊ herringbone or other strutting absent in larger spans
- ◊ joist bearings inadequate
- ◊ joists bearing on door linings via studs because wall plates absent
- ◊ wood rot or insect attack, especially at joist bearings on external walls
- ◊ in situ cut ends of joists not treated against wood rot and insect attack
- ◊ flying floors at oriels and ginnels uninsulated
- ◊ building debris in floor voids
- ◊ joist ends in one dwelling abutting those in an adjacent dwelling (should be staggered)
- ◊ traps cut into decks not supported by joists and noggings
- ◊ vapour control layers absent
- ◊ fire stops absent at perimeters of floors
- ◊ gas pipes in unventilated spaces
- ◊ access not provided for water services

Note: items relevant to floorboarding also appear in Chapter 8.2 and to chipboard decking in Chapter 8.5.

Inspection of joist hangers

- ◊ joists not in contact with hangers
- ◊ hangers not in contact with walls
- ◊ joist hangers too wide for joists
- ◊ joist hangers crushing blockwork bearings
- ◊ joist hangers inadequately nailed
- ◊ joist hangers not let into joists so ceilings uneven
- ◊ soft packing directly under joist hangers
- ◊ joist hangers not marked with the compressive stress ratings of the bearings for which they are suitable
- ◊ joist hangers of wrong grades

Inspection of ceilings

- ◊ broken keys on lath-and-plaster ceilings
- ◊ fixings to underdrawn lath-and-plaster ceilings inadequate
- ◊ insulating or asbestos cement boards (instead of gypsum boards) not providing fire resistance
- ◊ ceilings sagging because battens and noggings absent at joints
- ◊ perimeter noggings for ceiling boards not in place
- ◊ joists too narrow to allow plasterboard fixings (nails too close to edges)
- ◊ plasterboards nailed at centres too far apart
- ◊ popped nail heads in plasterboard skim coats
- ◊ plasterboards not laid to break joints
- ◊ condensation and mould growth on ceilings

Additional points for inspection of prefabricated timber floors

- ◊ staples overdriven into deck
- ◊ assembly tolerances exceeded
- ◊ staples absent from panel joists
- ◊ bolts connecting adjacent panels not tightened
- ◊ lifting holes unfilled leading to interconnecting cavities
- ◊ joist clips not nailed to head binders in timber frame floors (no lateral restraint)

Additional points for inspection of timber separating and compartment floors

- ◊ nailing inadequate for holding two layers of plasterboard
- ◊ gaps round service penetrations
- ◊ insufficient mass for acceptable levels of sound transmission
- ◊ gaps allowing airborne sound transmission
- ◊ isolation or resilience against impact sound inadequate
- ◊ insulation quilt absent or damaged
- ◊ new ceilings suspended from old floors (should be completely independently supported at perimeters)
- ◊ sealing of air paths inadequate for reducing sound transmission
- ◊ cavity barriers absent at perimeters of floors

Additional points for inspection of timber ground floors

- ◊ rot in wall plates over sleeper walls
- ◊ sleeper walls insufficiently well founded
- ◊ ventilation holes or airbricks blocked or absent
- ◊ DPCs absent where joists abut groundbearing slabs
- ◊ thermal insulation under decks inadequately supported
- ◊ joists not preservative treated when required
- ◊ impervious coverings over floor surfaces at risk of damp
- ◊ oversite concrete or other material inadequate
- ◊ insulation for water service pipes inadequate

Chapter 2.2 In situ suspended concrete slabs

This chapter deals with suspended in situ concrete floors used at both ground level and upper levels, whether they are solid (Figure 2.61) or are composed of in situ beams and pots.

Concrete, of course, has existed for some time, dating even from Roman times. However, its use in floors did not become widespread until the advent of Portland cement in the 1820s. Since then developments have been rapid.

Figure 2.62
A special floor under construction. Such floors are used in highly serviced buildings including hospitals. The photograph shows the sinusoidal lower plane in position

Figure 2.61
An in situ suspended floor slab. This one is congested with reinforcement and concrete placing would need good quality control

The variety of designs of concrete suspended floors is numerous, and, in practice, floors can be found which are a combination of the techniques to be described in the next two chapters. For example, precast beams may have been used which have projections from the top surface, designed to provide for composite action with in situ structural concrete toppings.

Characteristic details

Basic structure

It is important, in any inspection, that the existing structural system of the floor is identified, including how it is supported and the direction of span. Often the continuing structural adequacy of an existing in situ floor will be assumed on the basis of its freedom from apparent defects such as significant cracking, deflection or vibration. This assumption may be satisfactory where no changes are envisaged affecting loads on floors or on the supporting structure. In situations where some alteration, conversion or adaptation affecting loading is likely, a detailed engineering assessment will be necessary. An assessment may be necessary in any event if deterioration has occurred.

Most floors serve also as a diaphragm and assist in the redistribution of horizontal (wind) loads on walls in one direction to walls running in an opposite direction. Connections between walls and floors are therefore important. Since many in situ

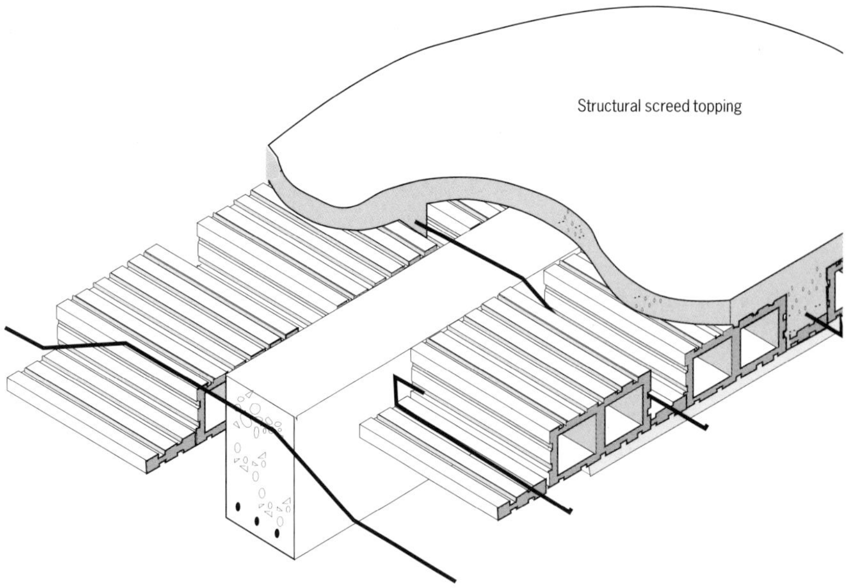

Figure 2.63
Hollow ceramic blocks or pots in a reinforced concrete floor. The concrete topping is integral with the beams

concrete floors will be found to span in two directions, lateral restraint can be taken care of by the bearings, and further strapping will not usually be present.

Surveyors may also encounter floors in which a space is formed between two concrete slabs, an upper and a lower, with a plenum chamber in between for carrying heating (Figure 2.62 on page 95). This space may also be used for carrying services, particularly in highly serviced buildings such as hospitals.

In building types of other than domestic scale, slabs may be found with flat or ribbed soffit designs. Flat soffit floors may be either of solid concrete throughout or be ribbed with pot infilling (Figure 2.63). The ribs either have soffits left from the shuttering or be lined with slips leaving a similar surface to the underside of the pot and giving a consistent surface material for any directly applied ceiling plaster.

Slabs in this category can be found at ground floor level (Figure 2.64) and at upper levels. Those slabs at ground floor level were often cast on permanent shuttering which has been left in place. Such shuttering might consist of woodwool slabs or corrugated sheets of various materials.

Alternatively, slabs may have been cast over hardcore but still be supported by perimeter walls. However, there are two distinctly different types of this design: some with full reinforcement at the bottom of the slab, where the floor is assumed to span with no support from below, and some of an unusual design with nominal reinforcement near the top of the slab designed to receive support from the ground as well as from the perimeter walls. In this latter case the reinforcement is there to minimise the development of cracks at the bearings on the walls and the slab is essentially groundbearing.

Flat plate slabs in upper floors have been used mainly where ceilings are to be formed directly on the soffit. Coffered soffits of in situ tee beams can be found mainly where suspended ceilings are used, and more rarely where the soffit is left exposed as cast (Figure 2.65).

The in situ hollow pot floor operates effectively as a tee beam floor with the pots laid on shuttering, reinforcement placed in position, and the concrete poured to appropriate depth over.

In larger buildings, a technique which was sometimes used was the lift slab principle. This involved two or more simple two-way span slabs which were cast in situ, one on top of the other on a separating membrane, with the lower cast slab acting as shuttering for the one above it (Figure 2.66). On curing, the slabs were then jacked up steel columns and supported on brackets welded in situ to the columns. The slabs naturally have flat soffits. Expert advice may be needed if modifications to the slabs are to be undertaken.

In domestic scale buildings, in situ concrete upper floors have been confined mainly to sections of floors forming hearths in otherwise timber floors (Figure 2.67).

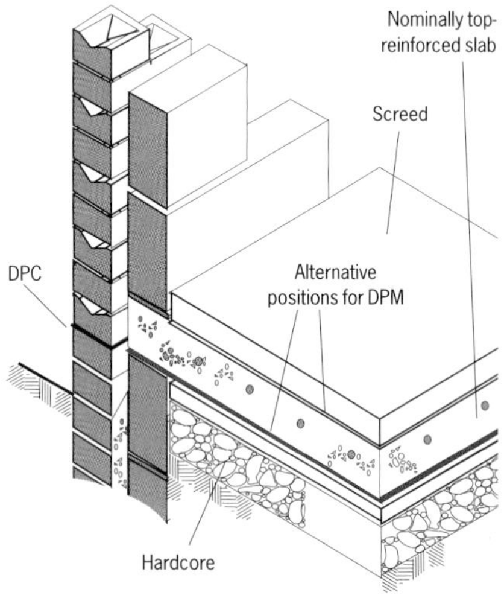

Figure 2.64
A slab which has been cast on the inner leaf of the external wall as well as over the hardcore can provide better stability over deep fill, but it may cause disruption in the inner leaf or DPM or crack at the perimeter

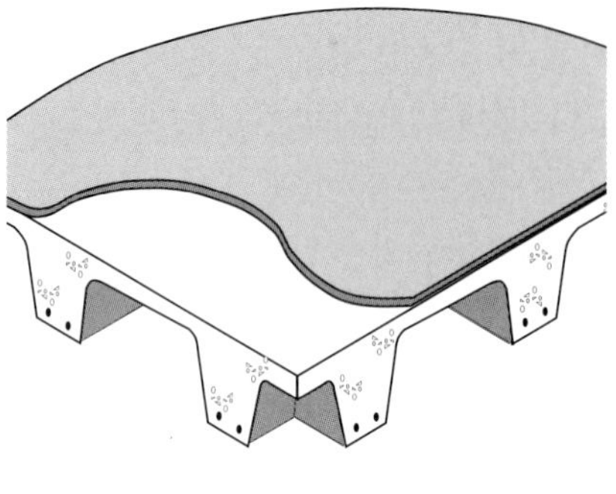

Figure 2.65
An alternative form of floor, a waffle floor, formed by in situ concrete beams integral with the slab. Coffers are formed on the soffit by formers which are later removed

Concrete slab suspended upper floors in individual houses are most likely to be found in system built constructions. One case in point is the BRS Type 4 house where the first floor and roof are of cast-in-situ reinforced concrete construction (*BRS Type 4 houses*[(142)]). Frequently, though, the floors in other concrete systems were of traditional suspended timber (*Fidler houses*[(143)], *Incast houses*[(144)] and *Cast rendered no-fines houses*[(145)]).

The in situ hollow pot floor (mentioned on page 96) is still used, but almost universally in houses in certain areas of continental Europe where it is preferred to the suspended timber first floor. They have not generally found favour in houses in the UK.

Other alternatives occasionally met, although not in domestic scale buildings, include downstand beams, sometimes curved in plan and placed diagonally. This type of construction has mostly been used in roofing, but occasionally for floors over large spans.

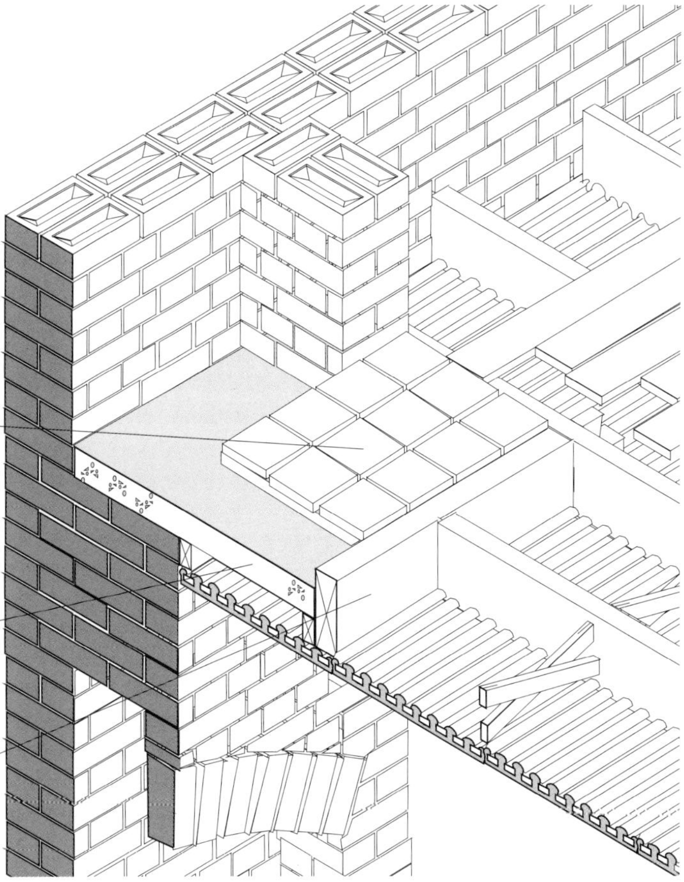

Figure 2.67
A concrete hearth slab in a timber first floor

Figure 2.66
A lift slab operation in progress in the 1960s

Bearing details

Many slabs will have been designed to span in both plan dimensions, with the main reinforcement running in both plan dimensions.

Mushroom heads on columns reduce potential bearing problems of flat slabs where downstand beams need to be avoided. There are no particular points to watch for.

Minimum bearing dimensions for reinforced concrete beams required by BS 8110:Part 1[(146)] are:

- 100 mm on Grade 30 concrete
- 110 mm on Grade 25 concrete
- 105 mm on masonry
- 75 mm on steel

Extra steel may need to be welded onto steel I beams to obtain the necessary bearing.

Where buildings are situated in close proximity to sources of vibration (eg railway lines), the main beams carrying the floors are sometimes mounted on resilient bearings to absorb some of the vibrations. These bearings may need to be replaced from time to time if they are prone to fatigue. The supporting column head and the underside of the beam at the bearing might have to be jacked apart to enable this to be done.

Non-loadbearing abutments

Floors in this category will often be required to provide lateral restraint to the walls which carry them. One-way spanning floors, since they do not bear on walls in the non-span direction, may still be required to provide lateral restraint. Strapping may have been used in such cases. For two-way spans, non-loadbearing abutments will be confined to smaller apertures containing interrupting walls including those enclosing ducts and lift shafts.

Protection against radon and methane

An outline of the problem of radon was given in Chapter 1.5. Some parts of Chapter 2.1 are also relevant.

Services

In most cases, the horizontal distribution of major services will be carried within the void of a suspended ceiling (Figure 2.68). However, in some cases, voids within the floor depth formed by hollow infill units or cast-in ducts may be found to contain water pipes and electricity cables. Water pipes in new construction should all be accessible, but some earlier installations will not be. Gas distribution pipes should only be carried in ventilated voids.

See also Chapter 4.6 for more detailed information on ducts and their covers.

Main performance requirements and defects

Choice of materials for structure

Concrete mix design can be complex, with many permutations of cement type, fine and coarse aggregates, plasticisers, retarders, pigments, air entraining agents and waterproofers. It is next to impossible to determine exactly what mixes have been used in existing slabs, though some routine, if not very simple, laboratory tests can reveal such matters as cement:aggregate ratios. It should be noted that an excess of cement in a mix, as well as insufficient cement, can lead to failure.

In situ testing for strength may be needed if there is evidence of deterioration in the construction.

So far as reinforcement goes, steel fabric should comply with BS 4483[147] and steel bars with BS 4449[148]. Fabric is normally supplied in sheet form or in roll form 2.4 m wide. Fabrics designated A, B and C are normally used in slabs, whereas the D mesh is commonly used in screeds.

Other forms of reinforcement, such as steel fibres and polypropylene fibres, may occasionally be seen incorporated into the wet mix. Their presence does not necessarily signify that they have been incorporated as reinforcement as they may be used to control cracking.

Several cases of non-standard or poor quality concrete mixes have been found in the BRE site inspections of older properties including examples of coke breeze aggregates and mixes with a large percentage of voids (absence of fines). These floor slabs were invariably extensively cracked and uneven, and were candidates for replacement. Floors cast using modern methods of placing and compacting the concrete are unlikely to suffer such deterioration (Figure 2.69).

Figure 2.69
Poker vibrators in use when casting a floor slab. Concrete floors laid with the help of this equipment are less likely to suffer from inadvertent voids

Figure 2.68
Services installed under a pot floor. The suspended ceiling has been removed for access

Strength and stability

Principles of modern building[6] mentions the characteristic types of structural failure of reinforced concrete slabs:

- tensile reinforcement reaching its yield strength before the concrete in compression has developed its full strength
- crushing of concrete occurring before the tensile steel reaches its yield strength
- yielding of shear reinforcement accompanied by diagonal cracking
- bond failure where the bars slide within the concrete

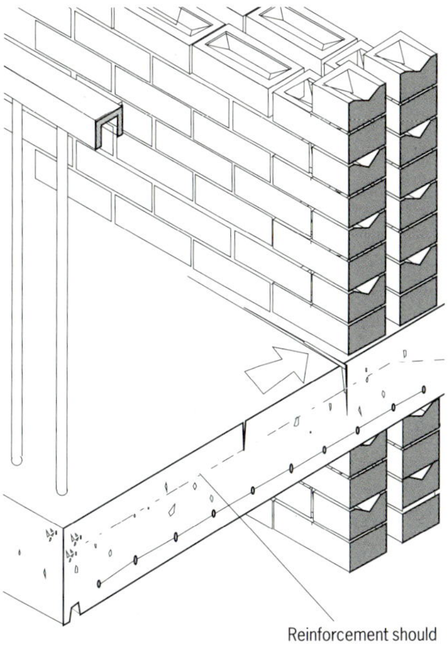

Figure 2.70
Misplaced steel reinforcement is common in cantilevered balconies

All these modes of failure may show as cracks in the concrete and any signs of displacement should be vigorously investigated (see the next section).

Misplaced steel reinforcement is common in the cantilevers of balconies. The first signs of this are the formation of cracks at the bearings (Figure 2.70).

Dimensional stability, deflections etc

Coefficient of linear thermal expansion per °C: 7 to 14×10^{-6}.

Reversible moisture movement: 0.01–0.07%.

The dimensional stability of reinforced concrete floor decks may sometimes be a cause of concern. Dimensional changes in the decks can cause damage to the supporting structure and, also, in some circumstances, give rise to distortion in the deck itself. The most important causes of excessive deformation are:

- thermal expansion and contraction
- drying shrinkage
- elastic deformation due to self weight and imposed loading
- creep due to prolonged loading

Floors are exposed to solar radiation through windows and roof lights, but not to anything approaching the effects on roof slabs. Floors in buildings which are heated intermittently will also be affected to some extent by dimensional changes. Domestic scale buildings are not known to be significantly affected by these changes.

Cracking

Many in situ reinforced concrete floors have been designed to span continuously over intermediate beams rather than each span designed and built as a simply supported structure. Provided the reinforcement has been installed correctly, it will be relatively unusual to see cracks in the structure of the floor in line with beams and columns. Cracks at bearings, particularly where ground floor slabs have been cast on the inner leaves of external walls over deep fill, may occur; they are not easy to see, though, where hidden behind skirtings. Where concern exists over the continuing serviceability of concrete floors, load testing may provide some information.

If the restraints are insufficient to resist the size changes (usually shrinkage) of the structure, cracking, for example of reinforced concrete decks, will result. In most cases the defects caused by thermal movements can be readily recognised. They almost always show on plaster finishes to ceilings. Most of the movements which induce cracking will be shrinkage rather than expansion since floors are normally not subjected to massive changes in temperature such as occur sometimes with inadequately insulated roof slabs.

Cracking can take slightly different forms in framed and unframed buildings. Where the slab is supported off masonry, a slip plane sometimes occurs at the bearing, and any shrinkage of the slab will then be accommodated by sliding over the bearing. The movement will tend to show as a simple horizontal displacement at the wall head.

If the slab has been cast into the frogs or perforations in the masonry,

Case study

Drying shrinkage of large slab in a residential home

Cracking of internal blockwork was discovered in various parts of a two storey old people's home. Much of it was clearly due to drying shrinkage in the longer stretches of walling which was subjected to considerable restraint. However, some conspicuous cracking in the transverse block wall was also noted, concentrated near each end of the ground floor of the building. In these areas, drying shrinkage of the lengthy one-piece reinforced concrete upper floor slab was concluded to be the major cause of the cracking. Nevertheless, structural safety neither had been nor could conceivably be impaired as a result. Further small movements were to be expected, so concealment rather than repair was the recommended remedial action.

any shrinkage may well take the top course with it, and in this case a horizontal crack forms just below the bearing (Figure 2.71).

Where the slab is supported off a reinforced concrete frame, particularly where it is cast integrally with the frame, shrinkage is usually accommodated by the smaller cracks dispersed through the structure, but this depends to a large extent on the disposition of the reinforcement (Figure 2.72).

Where ground floor slabs have been cast on permanent shuttering, there is sometimes a risk that the bearings of the permanent shuttering may have deteriorated, or hardcore may have been insufficiently compacted, with a consequent risk of settlement at these points. Settlement would show as a gap opening up at the floor/wall interface which may or may not be covered by a skirting (Figure 2.73).

Excessive deflection of suspended concrete floor slabs will usually be accompanied by cracking unless the original design provided for transverse reinforcement to control such cracking. The mechanisms of crack development are described in *Cracking in buildings*[149]. They are explained also in *Principles of modern building*.

Exclusion of damp

Dampness is a potential problem with this type of floor at ground level, including situations where permanent shuttering which has been left in place within the subfloor void. In a heated building, though, a suspended concrete ground floor is warmer than the ground and, therefore, conditions discourage the transfer of water vapour from the ground, via the void, to the underside of the floor where it could condense. This is even more the case where the subfloor space is ventilated to the outside. Even so, transient conditions can exist where moisture transfer does occur. It is therefore prudent to provide these floors with an integral DPM, particularly where moisture sensitive floorings are to be applied. The provision of a DPM beneath a screed also cuts down the drying time as construction water in the concrete slab does not need to be taken into account. Only the screed, in such a case, needs to be dried.

Timber flooring supported by battens fixed into, or onto, an in situ ground floor cast onto permanent shuttering with an unventilated space beneath are considered very vulnerable to rising damp. If groundwater problems also exist, the

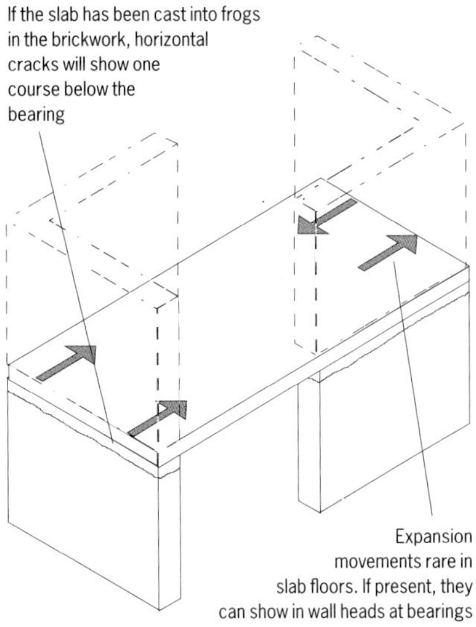

Figure 2.71
Cracking in an unframed building

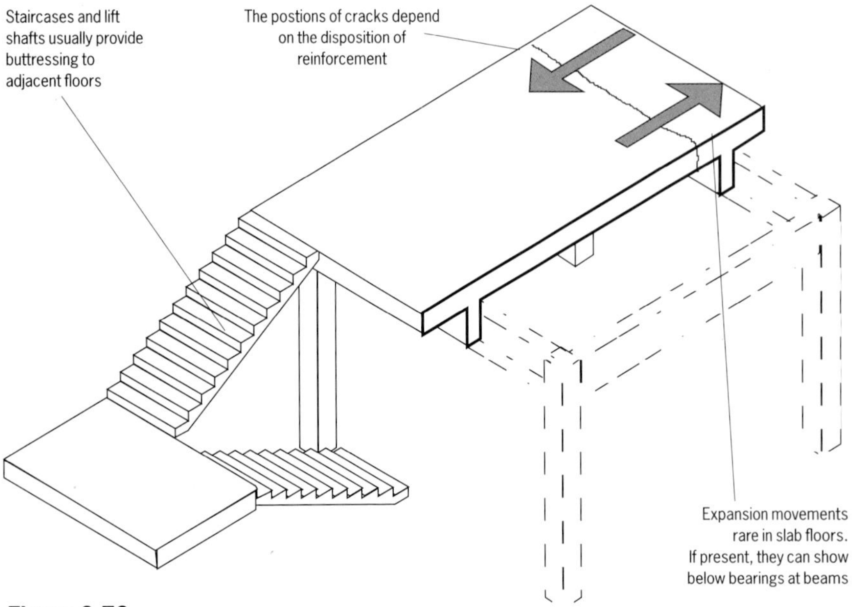

Figure 2.72
Cracking in a framed building

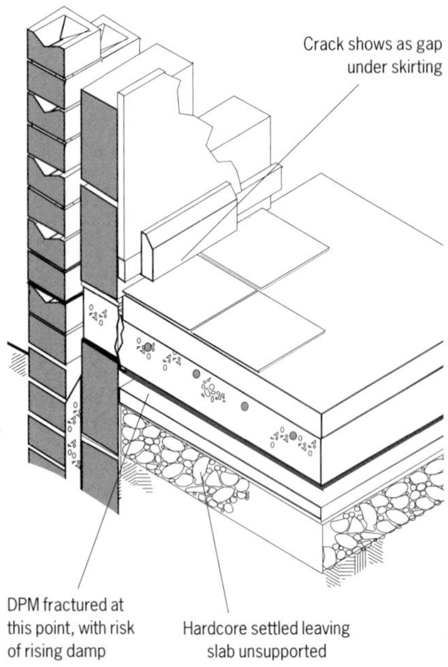

Figure 2.73
Settlement can show as a gap opening up at the floor/wall interface. The gap may be concealed by a skirting or floor covering

risk is further increased. These floors are likely to date from wartime shortages of timber and many will by now have been replaced (Figure 2.74).

Where services run in ducts within the floor, it is essential that DPMs are continuous under the ducts. Water and gas mains and electricity cabling may need to be sleeved so that the integrity of the DPM is maintained.

Thermal properties, ventilation and condensation

There are two schools of thought as to whether voids under suspended concrete ground floors need to be ventilated. Under the Building Regulations 1991 Approved Document C (1985)[150], voids in these situations did not need to be ventilated, although a DPM was required if they were not. In Clause C4 of the revised Approved Document C (1991)[42], ventilation is required only where there is a risk of accumulation of gas which might lead to an explosion. In this case, the actual open area provided by airbricks should be at least equivalent to 1500 mm^2 for each metre run of wall. In Scotland the figure is also 1500 mm^2 per metre.

BS 5250[126] recommends ventilation, but gives no figure. National House Building Council standards suggest a figure of 600 mm^2 per metre run of wall. The higher figure will certainly be needed if there is no DPM in the ground cover.

The BRE view is that the risk of water vapour condensing on the underside of the slab, and travelling upwards, justifies the provision of a DPM within the construction. This type of floor is quite often laid at an early stage in the construction schedule and can be subjected to large amounts of rainfall. A DPM on top of the structural slab and beneath any screed controls this moisture.

There is usually no requirement to provide thermal insulation for intermediate floors within a heated building unless the floor forms a division between heated and unheated spaces. In flats and maisonettes it has often been assumed in the past that all occupiers will heat to a similar standard and that no heat transfer will occur. This is not the case in practice, and significant heat transfer may occur vertically between well heated and poorly heated rooms.

Perimeters of floors are especially susceptible to condensation and mould growth if they extend to the outer face of the outside wall, forming a cold bridge. The effect can be amplified where a solid floor extends to form a balcony on an upstand or downstand beam; in some situations the only solution may be to provide insulation internally both above and below the slab.

Where the cavity continues past the downstand beam, there is the opportunity to continue the insulation past the beam.

Where the balcony is continuous with the floor slab, the soffit of the slab, and the upper surface wherever possible, should be insulated with a material offering at least 0.6 m^{2}K/W thermal resistance for a distance of at least 300 mm from the outer surface of the construction.

Where a concrete intermediate floor is built into an external cavity wall containing thermal insulation within the cavity, there is a risk that the insulation may be damaged or missing, perhaps leading, in turn, to a cold bridge and condensation.

Exposed floors which have been insulated may still give rise to cold bridges, especially at supports.

Case study

Condensation on the soffits of a flying floor in a block of flats

A block of flats suffered persistent condensation and dampness problems. The area most affected by condensation was the reinforced concrete suspended floor to the lowest level flats over an open parking area. Condensation and mould growth was found in a number of flats. Remedial work carried out some two years prior to examination of the problem consisted of the installation of a floating chipboard and polystyrene foam floor in the lowest flats, but this measure had not been altogether successful. The new recommendations were to further upgrade the thermal insulation of the floors.

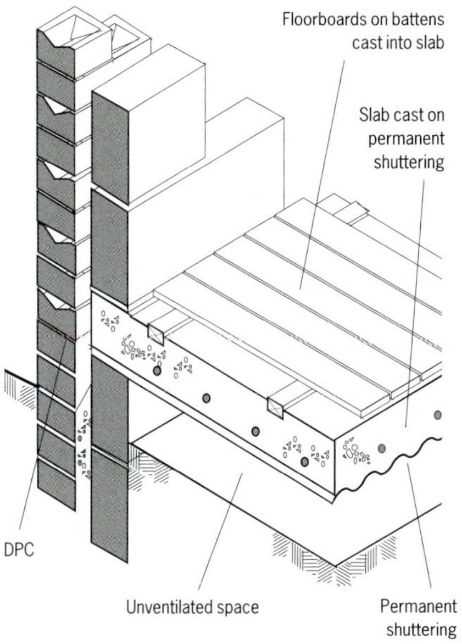

Figure 2.74
An in situ solid ground floor cast onto permanent shuttering with an unventilated space beneath

Case study

Condensation on the soffits of a flying floor in an office block

A block of offices had been built on a suspended floor slab over a car parking area. A floor which suffered from condensation had been insulated with sprayed mineral fibre, but downstand beams had not been treated. The condensation, which had collected as moisture on the surface of the floor adjacent to the external walls, had resulted in severe wetting to carpets. Condensation had also been seen on the inside surfaces of external walls. Remedial measures suggested to the building owner included providing extract fans for the kitchens, placing cork insulation on the floors and improving the thermal insulation to the external walls, particularly at wall/floor junctions.

Fire

Solid concrete floors generally have good fire resistance, provided reinforcement is adequately covered. Problems are more likely to arise where localised perforations occur in the floor, for example where pipes pass through a floor or where cracks exist between the floor edge and a wall. Assessment of the fire resistance of existing construction can often only be made by close examination and comparison with published performance data for similar constructions. Floor constructions which contain a continuous cavity may require cavity barriers. The contribution of any existing suspended ceiling to fire resistance should be taken into account.

Some floors have been built with permanent shuttering of woodwool or woodchip slabs which were left in place. Although, under test, these floors perform well, in practice a lot depends on the quality of the concrete in the slabs themselves. It will therefore be a matter of judgement as to their likely performance in fire (Figures 2.75 and 2.76, and BRE Current Paper CP 68/78[151]).

It is important to achieve adequate performance from the joint between slab and wall. In tests carried out to assist in defining appropriate solutions for retrofitting prefabricated concrete housing systems, for example, the in situ concrete or grout between walls and floor slabs played a significant role in achieving adequate performance.

In the site inspections carried out by BRE in connection with the studies of quality in rehabilitated housing, where services penetrated the floor, holes and gaps around service pipes which should have been filled were often left unfilled.

Sound insulation

Solid in situ concrete floors are generally able to provide good insulation against airborne sound, provided they have adequate mass and are built into heavyweight supporting walls. The addition of a floating floor above a slab can improve both the airborne and impact sound insulation, but careful detailing and workmanship is essential to avoid any direct contact between the floating floor and the structural floor (see Chapter 4.2). The ability of an existing floor to accept the additional weight of sound-reducing measures must be checked.

Woodwool permanent shuttering has been found to have a detrimental effect on flanking transmission in separating floors.

Durability

In concrete decks, deterioration could be due to a number of causes such as carbonation of the concrete, corrosion of reinforcement from excessive water used in cleaning operations, where the concrete abuts an external wall and dampproofing is defective, and where there is insufficient cover to the steel.

BRE have had to investigate many cases of failure of the bond between the soffit of a concrete suspended floor and the plaster ceiling, including collapses causing injury to building occupants.

Plastering of in situ concrete soffits and ceilings is one of the most difficult areas in which to obtain a good bond between plaster and

> **Case study**
>
> **Entrapped moisture leading to the breakup of a floor**
>
> The suspended reinforced concrete floor of a meat factory started cracking due to water trapped in the concrete. Following examination on site, recommended remedial action included:
>
> - drilling of escape holes along the line of the cracking to drain trapped construction water
> - removal of loose material around the crack
> - mechanical (not chemical) cleaning of corroded reinforcing steel
> - regrouting of the damaged areas
>
> Deterioration like that discovered in this case could be made worse by carbonation of the concrete, or the use of high alumina cement – a very slight possibility. Since most high alumina cement was used in precast concrete, it is dealt with in Chapter 2.3.

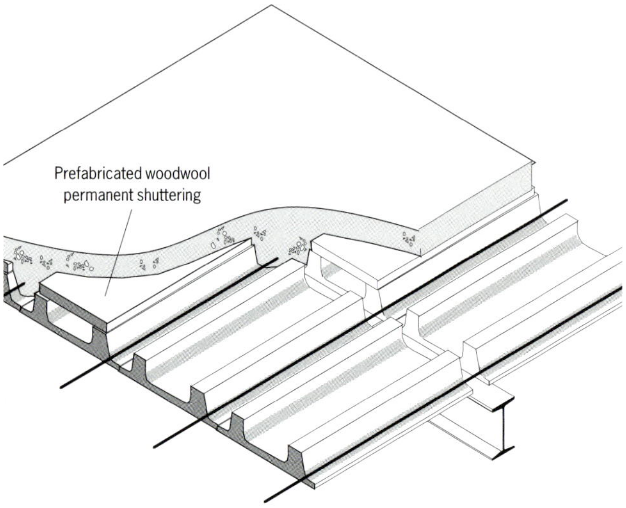

Figure 2.75
Woodwool or woodchip permanent shuttering for a concrete floor formed as two distinct parts

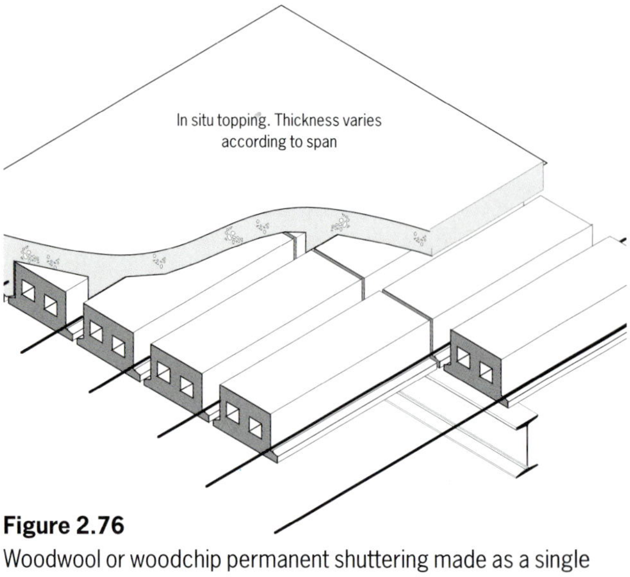

Figure 2.76
Woodwool or woodchip permanent shuttering made as a single hollow beam

Case study

Detachment of plaster from ceilings in a hotel

Detachment of the plaster finish from the concrete soffits of the in situ concrete floors of a hotel was mainly attributed to differential drying shrinkage movement of the plaster and concrete floor slabs thereby imposing shear stresses at their interface. Remedial measures suggested included the removal of all loose plasterwork, slight roughening of the concrete soffit and replastering in accordance with the original specification. These measures were appropriate because the drying shrinkage which caused the problem had by then taken place. Alternatively, for extra confidence, a PVA (polyvinyl acetate) or SBR (styrene-butadiene rubber) bonding coat could have been applied to the concrete soffit prior to replastering.

concrete. In research which was carried out by the Building Research Station in the 1950s, it was found that the type of coarse aggregate used in the concrete was of crucial importance. Concretes made with limestone or crushed brick aggregates were the most difficult to plaster successfully. Suction of the concrete, together with the differential thermal expansion and drying shrinkage between concrete and plaster, was also of considerable importance. In more recent buildings, the use of bonding agents has become more common and, in consequence, the risk of loss of adhesion in modern buildings is probably much lower than for earlier buildings.

Maintenance

Maintenance of the concrete slab is neither practicable nor necessary. However, it is necessary to maintain the integrity of weathertightness at bearings to reduce the risk of corrosion of reinforcement.

Work on site

Storage and handling of materials

Portland cement is the most vulnerable item. It needs to be stored off the ground in covered accommodation.

Restrictions due to weather conditions

Concrete should be protected from frost if it is laid before the shell is weathertight. If frosted, concrete becomes friable and will certainly not develop adequate strength. If it has not been cured properly, for example in very hot weather, the construction will not develop its full strength.

Workmanship

Aspects of workmanship are covered in BS 8000:Part 2:Section 2.1[152] and Section 2.2[153].

Failure of replacement floorings on old concrete bases can often be attributed to inadequate preparation of the surface of the concrete before laying the new flooring. The feature panel on page 104 describes suitable techniques for reducing the risk of problems occurring.

Preparing old concrete bases to receive new floorings

Before starting to apply any new flooring, the existing concrete base should be assessed for strength, cleanliness, cracks and, from the point of view of some new floorings, moisture. Before starting to make this assessment, it is prudent to remove all existing floorings to expose the surface of the base; alternatively to assess the existing floorings to ascertain whether they can be overlaid. The amount and type of preparation will depend on the flooring to be applied.

Strength

Most existing groundbearing concrete bases will have sufficient strength to support new flooring (see Chapter 3.1). Suspended floors may need to be assessed by a structural engineer if the weight of the proposed new flooring greatly exceeds that being replaced.

Where a good bond between the base and the new flooring is essential (eg with resin floorings and polymer modified cementitious floorings), it may be necessary to test the strength of the concrete. This can be done in situ by using a Schmidt hammer or, better still, by a pull-off test. In the latter test, a value of at least 1.0 N/mm^2 should be achieved. If the concrete is too weak for the intended flooring, it may be possible to enhance the strength of the top of the slab by resin impregnation. Where this is not possible, either the concrete will have to be replaced or it can be overlaid by a new concrete screed or slab, weight permitting.

Figure 2.77
Burning off contamination from a concrete floor

Cleanliness

Clearly all loose debris must be swept or vacuumed up before laying any new flooring.

Much adhered surface dirt, which might affect the bond strength of any new layer applied to it, can usually be removed by shotblasting with integral vacuum cleaning. More severe contamination might be removed by scabbling or planing the concrete surface.

In industrial situations it will often be found that the slab has been impregnated to a significant depth by oils, greases and other chemicals which may affect the bond of a new layer (eg a resin flooring), or cause swelling (eg in rubber and PVC floorings) or softening (eg with mastic asphalt). A liberal application of a degreasing agent or detergent followed by mechanical scrubbing and washing with clean water may remove a lot of the contamination from the top part of the concrete but not from further down in the slab. This operation will wet the slab, which may interfere with the application of some floorings (eg sheet and tile), and residual oil may slowly bleed back up through the slab to affect the new flooring at a later date.

Another technique sometimes used to remove organic material after shotblasting is to burn out the material from the surface using a special gas burner (Figure 2.77), and then to apply a penetrating resin seal to the top of the concrete while the latter is still warm and before any residual oil has time to migrate back to the surface. This resin coating bonds well to the surface and effectively seals in the oil to provide a clean surface on which to apply the new layer, whether it is a resin flooring or a latex underlayment.

It is not recommended that new sheet and tile floorings are laid over existing similar materials as various interactions can take place. Removing the existing floor coverings often leaves adhesive residues which are very difficult to remove. Some of the bitumen adhesives used to fix thermoplastic and vinyl asbestos tiles can be particularly obstinate. As many of these adhesives are soft or slightly rubbery, they are not easily removed by shotblasting. Some can be solubilised by organic solvents, but this is not usually a wise procedure as it is difficult to clear up the resulting mess: the solvent penetrates the slab and may affect the new material. Good results have been obtained by mechanically removing as much of the old adhesive as possible by scraping, sanding etc, and then applying a high quality latex underlayment at least 3 mm thick.

Cracks

Cracks should be assessed in an attempt to know why they appeared in the first place and whether they are still widening or are reducing. Most random cracks will have been caused by drying shrinkage, but cases of settlement and overloading may be found. Straight line cracks may be day joints, construction joints or fully formed movement joints. Except for the latter, most joints will be stable and can be filled with hard incompressible material. Joints should be raked out, preferably to a depth of 50 mm, before filling. Suitable filling materials are cementitious grouts or underlayments. The best results will be found with low viscosity thermosetting resins (eg epoxy resins) which will give excellent penetration of fine cracks, adhere well to the sides and are virtually incompressible.

Dampproofing

In any examination of an existing floor, it will be often found that it does not contain an effective DPM. Also, many thermoplastic and vinyl asbestos tile floors have been laid without DPMs because they were moderately tolerant of damp conditions. Wherever, though, a moisture sensitive flooring is to be applied, a DPM will be required. This can be done in one of the following ways.

- A DPM is laid and covered with a sand and cement or concrete screed at least 50 mm thick.
- A mastic asphalt screed 18 mm thick is laid which will act both as the DPM and the screed.
- Where the thickness of the floor cannot be increased, a surface DPM based on a proprietary epoxy resin may be used. Some floorings can be laid directly to the resin surface; others require the application of a 3 mm layer of a latex underlayment.

Inspection

Floor surfaces and soffits should be inspected for any visible cracking, and supporting walls examined for signs of crushing at bearings. Surface cracking may indicate only a problem with the screed or the finish. Should any signs of deterioration be found, or any alterations envisaged, an assessment may need to be made by a qualified structural engineer which includes locating reinforcement and opening up to check construction, particularly at bearings.

When examining reinforced concrete structures for performance over time, and, in particular, finding apparent evidence of deflections and bow, it is as well to remember that the elements of the structure may have been manufactured and erected with such deviations. Permissible dimensional deviations given in BS 8110:Part 1 for new construction allow 6 mm in members up to 3 m long, 9 mm in 3–6 m, 12 mm in 6–12 m and a further 6 mm for every 6 m above 12 m.

Existing reinforced concrete floors may show defects on routine visual inspections which call for further testing. The visual evidence will include cracking, corrosion and spalling, disruption of anchors in post-tensioned structures and any evidence of water penetration at external wall bearings. Simple tests, for example checking the position of reinforcing rods using an electromagnetic cover meter in the manner prescribed in BS 1881:Part 204[154], may be appropriate for relatively unskilled people to use. However, this technique will not detect the presence of voids. More sophisticated tests, in accordance with other Parts of BS 1881, should normally be left for experienced structural engineers to specify and interpret; for example those involving cutting cores for testing compressive strengths, and the use of gamma radiography for detecting the presence of voids. Inspection by specialists using radar may also be useful for detecting voids and misplaced reinforcement. Load tests on completed structures or parts of structures may also be called for by the structural engineer in certain circumstances, and assessments carried out on cracking and the recovery of a structure from deflection.

Internal fracture tests and ultrasonic pulse velocity tests can be useful in identifying areas of very low strength, but are not likely to give the necessary degree of assurance for a basis of appraisal.

Visual inspection of the soffits of floor and beam components offers a viable method of monitoring the general structural condition of components where the protection to the reinforcement has been lost. Where cracking near bearings is found, the beams should be propped and the circumstances assessed by a structural engineer.

With fully bonded screeds on concrete floors, it is probably safe to assume that any cracks in the structural floor will reach the surface, but cracks in screeds will not necessarily be replicated in the structure. It must be remembered also that unbonded or floating screeds may hide structural distress in the floor below, so inspecting the soffit may be important.

The problems to look for are:

- ◊ DPCs absent under bearings of beams
- ◊ floors cracked and uneven
- ◊ concrete mixes with a large percentage of voids (absence of fines)
- ◊ internal walls showing rising damp
- ◊ supporting structures showing damage caused by dimensional changes in decks
- ◊ cold bridges at perimeters of floors
- ◊ cracking at bearings in slabs over deep fill
- ◊ cracking at the tops of walls and partitions
- ◊ moisture problems caused by insufficient drying times
- ◊ strengths of fixings into soffits for suspended ceilings and services inadequate
- ◊ ventilation areas available are less than 600 mm^2 per metre run of wall, or 1500 m^2 where there are no DPMs or where there are risks of flammable gases accumulating
- ◊ holes or gaps in above-ground floors at junctions between elements of structures and where services pass through floors
- ◊ sound insulation in compartment floors defective
- ◊ floating floors in contact with subfloors
- ◊ woodwool permanent shuttering creating flanking sound problems
- ◊ bonds between plastered soffits and decks above not satisfactory
- ◊ cracking, corrosion, spalling and disruption of anchors in post-tensioned structures
- ◊ unbonded or floating screeds hiding structural distress in floors below (inspecting soffits is important)

Chapter 2.3 Precast concrete beam and block, slab and plank floors

This chapter deals with suspended precast concrete floors used at both ground level and upper levels.

Precast concrete beam and block floors have been widely used in housing and other domestic scale construction since the early 1980s (Figure 2.78). They have also been used successfully to replace suspended timber floors in older properties which have had to be replaced because of rot.

Reinforced concrete plank floors have tended to be used in building types other than housing, and may be found made either from dense concrete or from autoclaved aerated concrete. Many precast beams and planks are prestressed. In many cases the planks are hollow.

Since the early 1990s, it has increasingly become evident that the use of the term pot, hitherto reserved for ceramic infill units, has also been used for highly perforated concrete blocks. The reason for this change in terminology is somewhat obscure and could lead to misunderstandings.

Figure 2.78
Precast concrete beam and block floors for housing have increased considerably in popularity since the early 1980s

Characteristic details

Basic structure

Beam and block floors

Beam and solid block floors, sometimes called house floors, consist of inverted tee beams set at a distance that will allow them to accommodate precast infill blocks (Figure 2.79). The latter are frequently standard walling blocks to BS 6073:Part 1[155] and Part 2[156], or proprietary blocks, but are usually solid and 100 mm deep. Blocks made

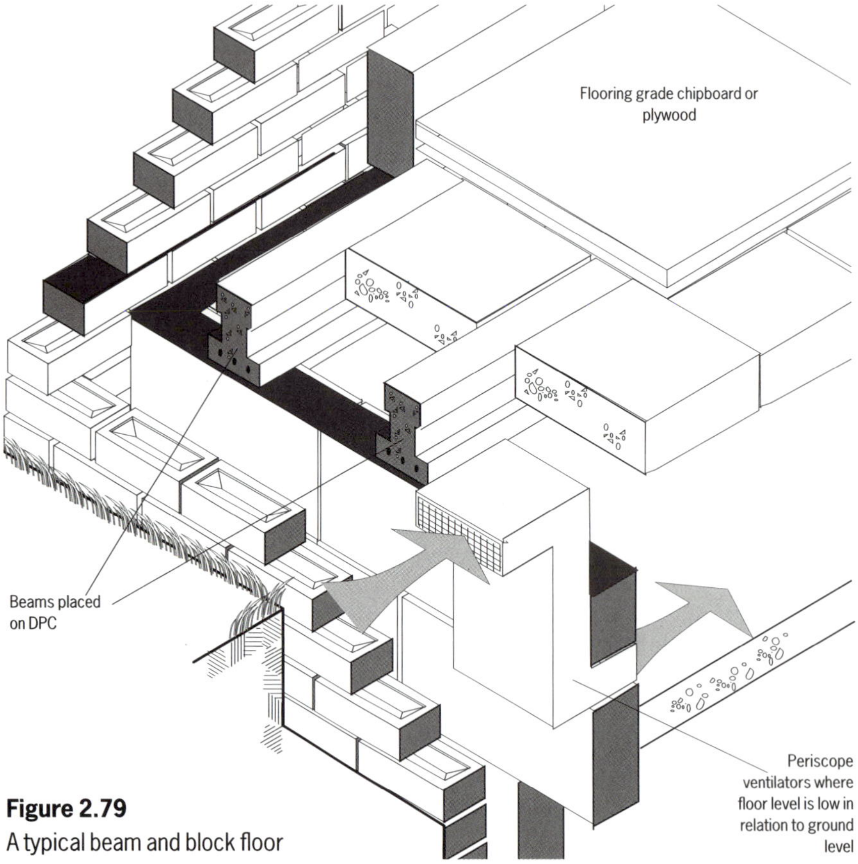

Figure 2.79
A typical beam and block floor

from dense or lightweight aggregate, aerated concrete and foamed polystyrene have also been used. Variation in the size of the blocks means that the spacings of the tee beams can vary. In positions where increased loads on a floor are expected, and therefore extra local strength is required in the floor (eg under internal partitions built off the floor), twin or triple beams adjacent to each other may be found.

In new construction the floor is assembled when the walls are at DPC level. The beams are set on the inner leaf of the external wall and should not protrude into the cavity. They must be placed onto a DPC to prevent both moisture and soluble salts moving from the brickwork and ground into the beams. It is especially important to prevent the ingress of salts which can increase the rate of corrosion of the reinforcement.

Once in position, the top of the floor is treated with a cement based grout to lock the beams and blocks together. This grout is sometimes also referred to as a regulating layer implying that it can cover minor imperfections in the surface but cannot replace the levelling function carried out by levelling underlayments (see also Chapter 4.3). When the building is watertight the floor is usually finished by laying a sand and cement screed or a floating chipboard floor. However, to meet thermal insulation requirements introduced in 1991, it has become common to place a layer of insulation over the beam and block before laying a screed or chipboard.

Beam and block floors have become a very popular form of constructing house floors since the 1980s. It is a particularly useful form of floor construction where significant depth of fill would be required for a ground supported base or where methane or radon protection measures are required.

Because of possible settlement of ground supported slabs, the National House Building Council have insisted on a suspended floor where the depth of fill exceeds 600 mm.

Many designs of precast floor require some additional in situ work, either to provide continuity of reinforcement between bays or to provide additional compression concrete on the upper portion of the floor deck.

In the direction of span, lateral restraint requirements for the adjoining wall can be met by the amount of bearing provided for the floor. However, in the non-span direction, strapping may have been be needed where the ground-to-floor dimension is considerable (Figure 2.80). Strapping will be especially needed in buildings over five storeys.

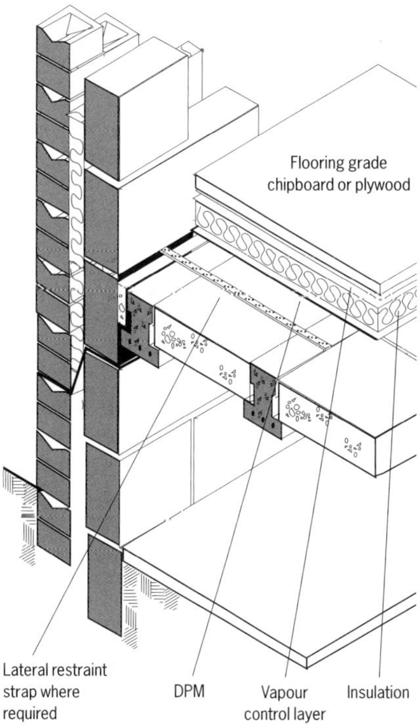

Figure 2.80
Strapping has been needed sometimes over a precast floor to provide lateral restraint in the non-span direction

Other kinds of precast concrete floors

There is an enormous variety of proprietary designs of precast concrete floor units for use in floors other than for housing. They have, in the past, included plain reinforced or prestressed versions of the following:

- inverted tee beam carrying hollow blocks – a more robust version of the domestic floor
- single or multiple tee beams (Figure 2.81)
- solid or hollow slabs or planks (Figure 2.82)

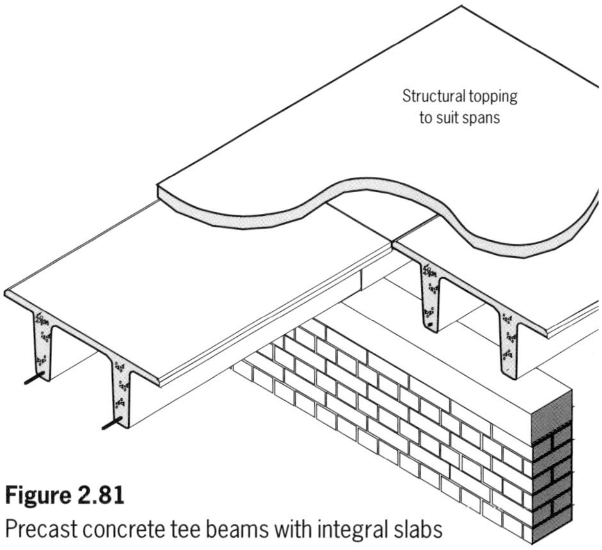

Figure 2.81
Precast concrete tee beams with integral slabs

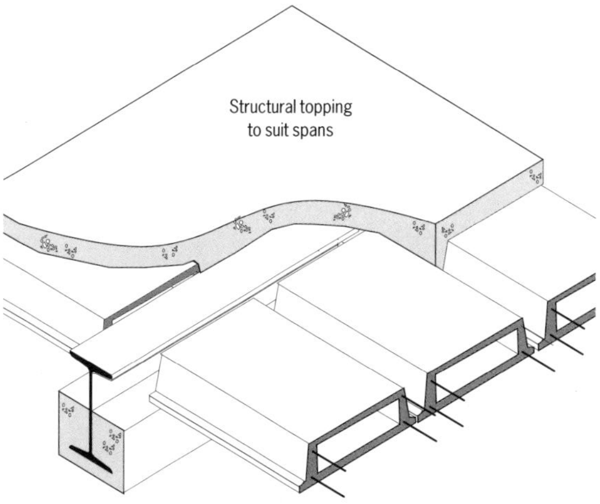

Figure 2.82
A hollow plank floor

- permanent shuttering in the form of profiled coffers or troughs carried on precast solid I section or lattice beams
- solid prestressed planks
- inverted channel beams

Many of these designs rely on in situ concrete toppings to provide continuity between units and to complete the floor. Many designs had protruding stirrup or stud shear reinforcement around which the topping was cast. Because of the narrowness of some of the spaces between beams or planks, it is possible that the topping was not always bedded well into them (Figure 2.83).

Wherever a change of use is contemplated for buildings having these floors, and in the absence of detailed drawings and specifications, a thorough investigation involving opening up the floor will be necessary to find out exactly what has been provided. Even where removal of the topping or screed reveals precast beams or slabs, there is no guarantee that they are not hollow or prestressed. Imposed floor loads allowed for in original designs can vary by a factor of seven or eight in spans of the order of 3 m, while some of the lighter imposed loads allowed spans of the order of 10 m or even more. There is a need for a complete reassessment of the floor construction in terms of flexure and shear, and the required resistance in terms of the anticipated future magnitude and distribution of loadings.

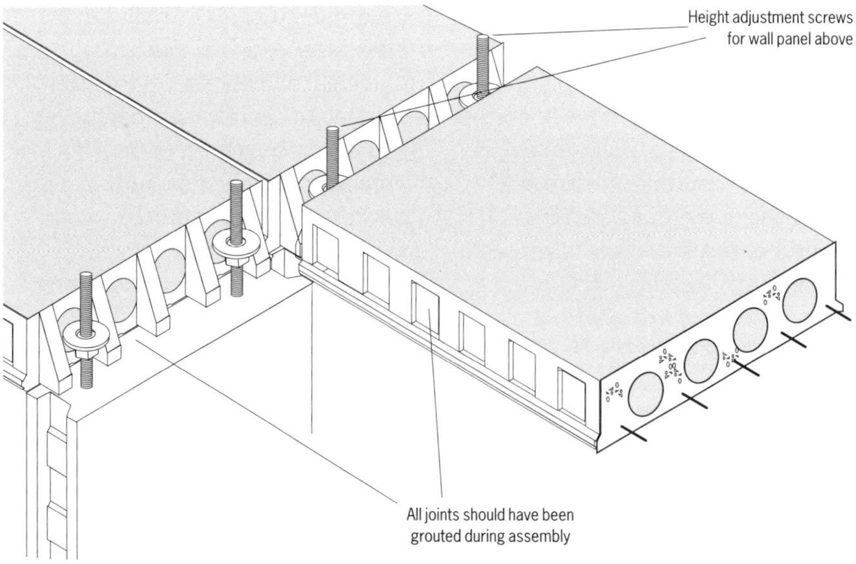

Figure 2.84
A typical precast concrete floor in an industrialised building

Bearing details

Buildings constructed to design criteria dating from the early 1950s may not have met minimum bearing requirements – some have showed less than 50 mm bearings for prestressed floor planks. Industrialised buildings with precast concrete floors can have bearings which vary considerably (Figure 2.84).

Bearing dimensions have largely been dealt with in Chapter 2.2.

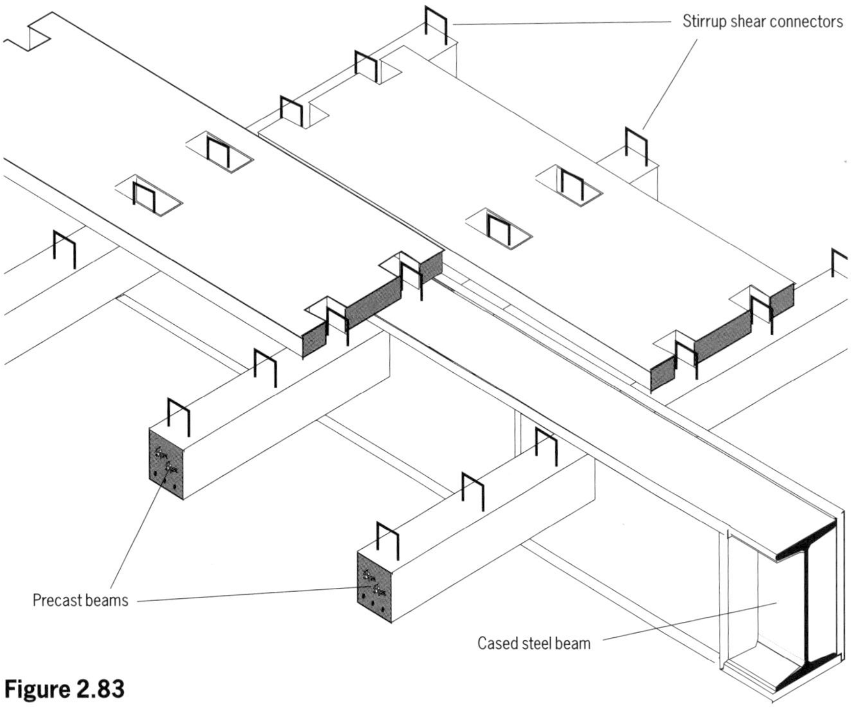

Figure 2.83
Protruding stirrups give fixity when the topping is placed

Non-loadbearing abutments

Accuracy in the forming of suspended floor slab edges, whether of prefabricated or in situ construction, can also vary considerably. As one example, the deviations measured on the perimeters of industrialised buildings in one programme of BRE investigations showed overall deviations of 75 mm in any one storey, and 125 mm when all the floors were included. Appropriate allowances for inaccuracy will need to be made therefore in any refurbishment of such buildings.

Protection against radon and methane

Beam and block floors covered with a continuous gasproof membrane, and with the underfloor space ventilated via airbricks to give the same open areas as for a timber suspended floor (see Chapter 2.1), have proved to be very effective against radon and methane. However, if this system proves in practice not to be completely satisfactory, connecting an electrically powered fan in place of one of the airbricks to provide increased ventilation rates to the subfloor void is a simple alternative. In the case of radon, the fan may either blow into or extract from the void. Expert advice should be sought when choosing fans.

Services

Water supply pipes below suspended floors can be damaged by frost; therefore it is essential to take the following precautions.

- DPMs and vapour control layers are sealed.
- Supply pipes are insulated throughout their length within the void (BS 6700[141]).
- A gastight seal is provided at the points of entry.

Main performance requirements and defects

Choice of materials for structure

There is a very wide variety of proprietary concrete and composite concrete floors available, and an even greater variety has been used in the past. For dense and lightweight slabs, the main materials have included Portland cement; high alumina and rapid hardening cements, some of which may not have been of UK origin; aggregates of various kinds of minerals, some of which may have been of marine origin; fillers (eg pulverised fuel ash) either used as a powder or pelleted and sintered; and additives of various kinds (eg foaming agents for lightweight concrete, and chloride accelerators). Reinforcement may consist of deformed bars as well as plain, and there has been a variety of protection including resins and latexes. In any investigation of defects, it will be important to discover as much as possible about the original specifications before examination on site is undertaken.

Strength and stability

The behaviour of beam and block floors under static loads will have been considered in design. Precast prestressed beams or slabs are vulnerable to failure due to inadequate bearing widths, often resulting from inaccurate setting out and from poor connections to walls parallel to span. Beam-and-pot floors may have inadequate grouting to 'lock' the floor components together. If, by chance, the grouting is missing, and there is no binding action from the screed or other floor finish, no load distribution occurs and the beams can be regarded as a series of separate units. However, even a floating screed does produce some load distribution[157].

Where any doubts exist on the performance of system built, precast unit floor assemblies, it is possible to load test them in situ. Approximate methods of analysis using linear theory and heat conduction – leading to assessment of load distribution, thermal deflections and load corrections – have been developed by BRE for use in load testing of structures, though these procedures will normally be used by the structural engineer.

Sample individual beams which make up precast floors can be tested in the laboratory to verify the design assumptions and quality control procedures.

Poor lateral restraint is commonly found at bearings of beam-and-slab floors with external walls. In the non-span direction, strapping may have been needed. Analysis of movement will indicate whether the strapping is either missing or ineffective (Figure 2.80).

Dimensional stability, deflections etc

Coefficient of linear thermal expansion per °C: 7 to 14×10^{-6}.

Reversible moisture movement: 0.01–0.07%.

Grouting the joints between the beams and blocks in a beam and block floor is required in order to ensure that the floor acts as a composite whole, and that the individual beams and blocks do not deflect separately when under load. Grouting is not always carried out satisfactorily; indeed, evidence from Housing Association Property

Case study

Deflection of insulation caused by inaccurate laying of a beam and block floor

Chipboard had been laid over foamed polystyrene on beam and block floors in a number of houses. Some of the floors had been found to be exceptionally springy and a number did not meet the normally accepted limits on tolerance for levelness. Where flooring is laid directly over foamed polystyrene, there will always be some springiness because of the resilience of the foam. However, in this case, it was excessive.

The BRE investigator concluded that the excessive springiness had been caused by some of the beams not having been installed level, either in their span or in relation to adjacent beams and blocks; hence the polystyrene could deflect between the high points. Differences in level were as much as 15 mm with many cases of around 8 mm.

Taking the intensity of loads on the flooring as those given in BS 6399:Part 1[19], it was shown that the intensity of loads to be carried by the polystyrene over the beams would vary between 8 kN/m^2 and 4 kN/m^2 if the bearings were level – well within the capacity of the polystyrene strength quoted as 70 kN/m^2 at 10% compression.

As the excess springiness had been caused by gaps between the flooring and the structural base, it was recommended that, to avoid re-laying the floor, grout be introduced into the gaps by drilling holes in the chipboard and polystyrene. Suitable grouts containing polymers and superplasticisers are widely available. Although there was some concern about injecting a water based grout between the DPM and the insulation because of entrapment of water, the quantities are relatively small. In the event there were no problems. Where a greater degree of confidence in the solution is required, even injection of thermosetting resins might be less expensive than taking up the whole floor.

Mutual (HAPM) indicates that as many as 1 in 10 of the sites they inspected did not adequately address this need. The evidence of inadequate grouting will tend to show most in finished floors where the spans of beams vary, such as at trimmed areas.

Some plank floors may show cracking on soffits, transversely to the direction of span. This will be due to deflections in service exceeding those allowed for in the design. Where any plank floors have deflected more than $^1/_{150}$ of the span, the advice of a structural engineer should be sought. There is no point in rescreeding such floors to level since the extra weight of the screed material will simply exacerbate the deflection problem.

The surfaces of beam and block floors are often uneven. This has led to unstable floors when finished with preformed insulation boards and chipboard. The preformed materials touch (or sit on) only the high spots. When trafficked (walked on), the insulation or chipboard is pushed down into the valleys so that it undulates dynamically, springing up and down and causing furniture to rock. With new construction this problem can be prevented by levelling the beam and block surface with a thin layer of mortar. A 1:6 cement:sand mortar has adequate strength, and the normal thickness requirements for a levelling screed do not apply. Dynamic rocking of the floor might be prevented if the surface regularity of the base does not exceed 5 mm under a 3 m straight edge before insulation is laid.

Where an existing floor is subject to such instability, injection of a water based grout or thermosetting resin into the voids has been found to be successful; if, though, any doubts exist about finding such voids with the flooring still in place, it will require removing and re-laying to the above criteria.

Floors which have a structural topping of in situ concrete are unlikely to suffer from these irregularities provided the bearings of the precast units remain stable.

Cases have been seen by BRE staff where prestressed beams carrying infilling have cambered to different amounts, leading to cracking of the flooring laid over them (Figure 2.85 and see further note in the workmanship section of this chapter).

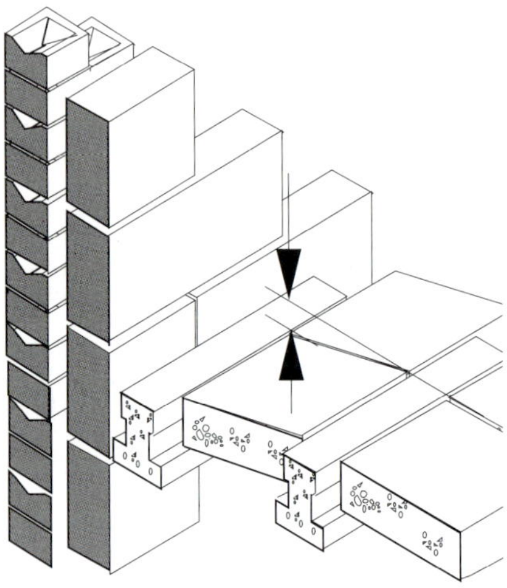

Figure 2.85
Differences in camber between adjacent beams leads to significant variations in screed thickness and, possibly, difficulties with floorings

Case study

Differences in camber and deflections in adjoining beams in a block of flats
Cracks appeared in the ceilings of some flats early in the life of the building. The ceilings were treated with a proprietary finish but the cracks continued to appear through the finish. The cracking was confined to ceilings of ground and first floor flats where the structural floor above was of prestressed beam construction. Ceilings elsewhere in the building, where supporting construction was plasterboard on timber joists, were said not to have been affected by cracking.

With a floor of this type the beams can function independently. Slight differences in deflection of the prestressed beams, which occurred because of shrinkage and creep effects during their early life, were the most likely cause of the cracks developing. Unless the flooring was to be removed and the beams locked together with a suitable grout, differential movements, and hence cracking, would continue to occur.

Vibration

An important consideration in the design procedure for precast floors including replacement floors, especially those shallow floor structures constructed from precast concrete slabs supported on steel beams, is to assess their performance under vibration. It is also important to ascertain their fundamental frequencies and to examine their response to human actions[158]. (Further information on vibration characteristics of floors is available from BRE.)

Exclusion of damp

The various national building regulations require a DPM to be provided under a precast concrete floor if the ground level beneath the floor is below the lowest level of the surrounding ground and the ground beneath the floor will not be effectively drained. These ground floors are normally constructed

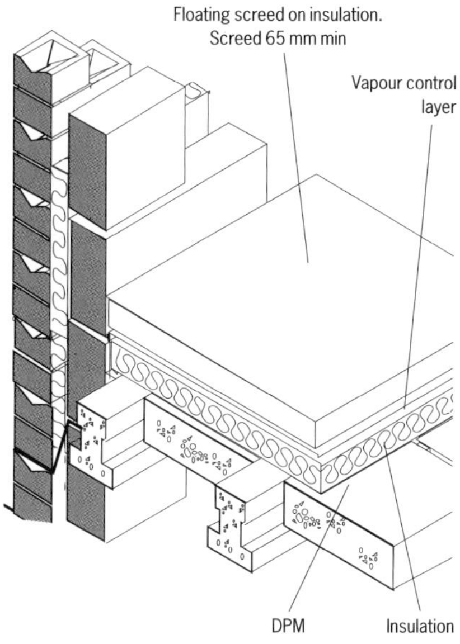

Figure 2.86
Insulation as a separate layer between the structural floor and the floor covering

when the walls are at DPC level and before the building is made watertight. It is possible, therefore, for rain to saturate the blocks or other components of a floor. Subsequently this leads to a long drying time before moisture sensitive flooring can be installed; it often holds up completion and occupation, or leads to premature floor failure. It is therefore good practice, in BRE's view, that a DPM or vapour control layer should always be placed above the structural floor and below the screed or timber panel product to control the upwards transmission of construction moisture. It is also vital to provide a turn-up to the DPM, otherwise moisture from accumulated rainwater can migrate up the inner leaf of the external wall (Figure 3.11).

A DPC must be placed below any beams or planks, at their bearings on supporting walls, to prevent moisture rising up the walls. This DPC also serves, rather more importantly, to prevent soluble salts reaching the components of the floor; salts will induce corrosion of the reinforcement.

BRE site inspections have revealed cases of precast beams projecting into and across external wall cavities due to inaccuracies in site work. A projection may provide a route for penetrating damp, more especially if it collects mortar droppings which bridge the cavity. Linkage of wall DPCs with floor DPMs is also necessary. In just over 1 in 10 cases examined by HAPM where precast concrete floors had been used, linkages were unsatisfactory.

Thermal properties and ventilation

The void beneath a suspended concrete ground floor may be ventilated or unventilated. In BRE's view some ventilation is advantageous as it will help to keep the subfloor void at a lower humidity than it would otherwise be, reducing the possibility of moisture transferring across the void. The free area provided by ventilators or airbricks should be not less than 600 mm^2 per metre run of wall. In areas where there is a risk of methane or radon rising from the ground, or where there is no DPM over the ground, the underfloor space certainly should be ventilated; this ventilation should be provided at not less than 1500 mm^2 per metre run. If ventilation is provided, it may be appropriate to include some thermal insulation in or under the floor. Where the finished floor level is to be set close to the original ground level, a cranked ventilator is often used (shown in Figure 2.79). These ventilators do impede the free flow of air, but only to a negligible extent unless obstructed by other parts of the construction.

The original ground floor may of course have been provided with some thermal insulation, particularly if built since the 1980s. Insulation may be found:

- above the structural floor in a separate layer under the flooring (Figure 2.86)
- as a lightweight screed (Figure 2.87)
- within the structural blocks infilling the beams (Figure 2.88)

Control of dampness and condensation

Condensation occurring at thermal bridges has been observed many times by BRE with this kind of floor

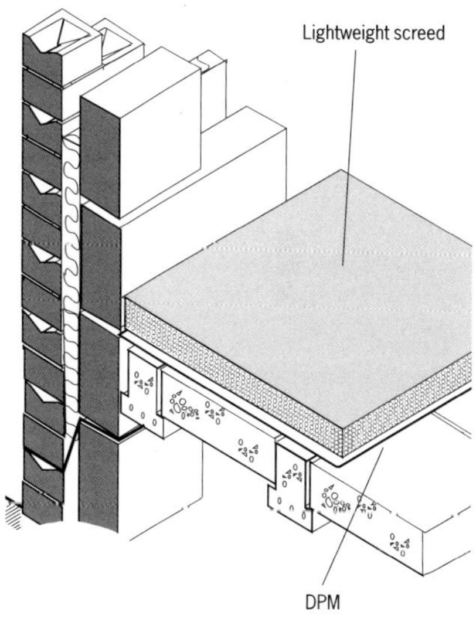

Figure 2.87
Insulation given by a lightweight screed

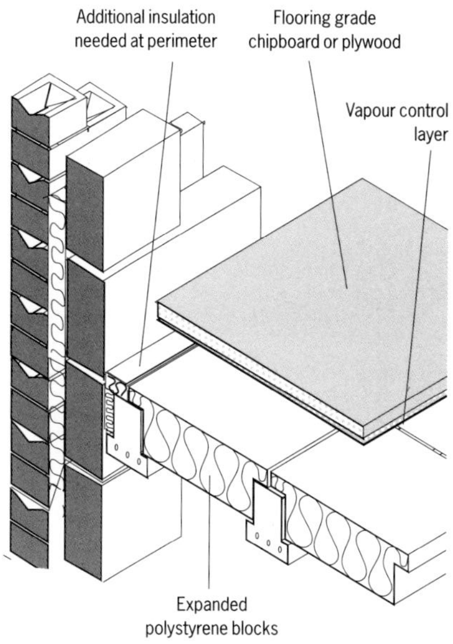

Figure 2.88
Insulation within the structural blocks infilling the beams

used as a ground floor, especially where the thermal insulation in the cavity of an external wall has not been taken sufficiently low to overlap that within the floor.

Fire

Most beam and block floors, depending on the materials used, can be expected to achieve at least a half hour fire resistance. The main exception to this are those which have used plastics infill blocks, particularly in ground floors where there is no requirement for fire resistance. The fire resistance of floors which have employed solid concrete infill blocks may have been enhanced by applying a plastered finish to the soffits. If the floor construction has been given a satisfactory test rating, its composition is unimportant from the fire point of view.

Most precast concrete floors with at least 50 mm in situ concrete toppings can be expected to achieve one hour fire resistance, although some designs can achieve more than this with no additional protection (Figure 2.89). Additional performance may come from replacing the ceiling finish. There may, though, be insufficient cover to the steel reinforcement to permit upgrading beyond two hours.

Holes or gaps in a floor are the main problem for fire resistance, especially at junctions between elements of structure; for example where a floor abuts a wall or beam (Figure 2.90).

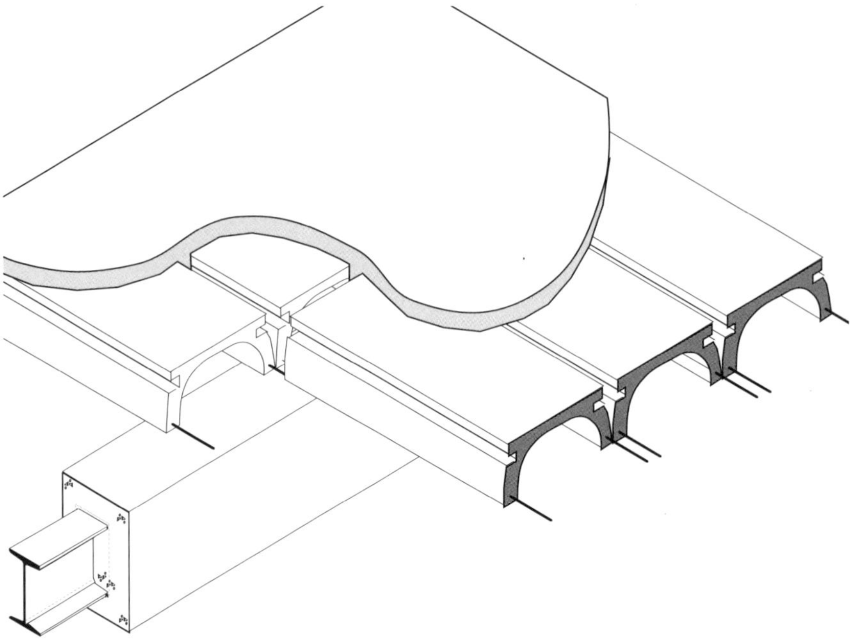

Figure 2.90
Inverted trough floor panels will sometimes be found to have been filled in at bearings. Where filling is absent, voids may need to be filled in for fire reasons

Sound insulation

Airborne sound

There have been cases of insufficient grouting of beams and blocks where timber instead of tiled screeds provided the flooring. This has led to defective sound insulation in separating floors. There is also the possibility that these floors lack the mass necessary to achieve adequate insulation against airborne sound.

HAPM have found that nearly 1 in 5 of beam and block floors they inspected, where used as compartment floors, did not meet the suggested Building Regulations 1991 Approved Document E(86) mass criterion.

Impact sound

Adequate performance of impact sound insulation can be achieved by a resilient layer placed beneath a floating screed or timber deck, or by a resilient flooring. Around 1 in 5 cases fail to satisfy current Approved Document criteria for impact sound insulation. Tests would normally be expected to be called for when unusual construction is proposed, but there is little evidence that these are routinely carried out.

Flanking sound

The sound insulation of separating floors can be compromised by flanking sound paths; for example through lightweight blocks used as thermal insulation in external walls. There has often been a dilemma for the designer since control of flanking sound paths can operate against thermal insulation requirements for the external wall. There is evidence that around 1 in 8 new dwellings do not conform with current requirements for the control of flanking sound transmission.

Figure 2.89
The Frazzi Monolithic Fire Resisting Floor with 4-inch pots and 2 inches of cover concrete. From an old advertisement

Durability

Durability of beam and block floors can be adversely affected by damage occurring during transport and

Figure 2.91
Damage to floor beams. This could have implications for the serviceability of a floor if it occurred on a significant scale, though any topping will tend to reduce the effects of damage

laying, reducing or eliminating cover to the reinforcement (Figure 2.91).

Some industrialised systems have had composite floors. For example, in the Keyhouse Unibuilt system, precast concrete slabs at first floor level were laid on timber bearers (Figure 2.92). (Further information on most of the industrialised systems is available from BRE in a series of publications which have not been referenced here.)

If deterioration of prestressed concrete beams has occurred, the possibility that they were made from high alumina cement concrete (HACC) may need to be considered. There is also the possibility that HACC was used in buildings built between about 1930 and the mid-1970s when the risks associated with this material were brought to the attention of the industry following the collapse of three roofs.

In essence, the mechanism of deterioration in HACC is as follows. Hydrated high alumina cement undergoes a change in its mineralogical composition with time through a process known as conversion. Conversion shows as a reduction in strength and an increase in porosity. Strength loss is greater in mixes with high water:cement ratios; increased porosity brings greater vulnerability to attack by alkalis. In time, the loss of strength could be as much as half the original value. Cover to reinforcement in most prestressed HACC beams is less than 20 mm, so inducing corrosion cracking.

The rate of conversion in HACC increases with rise in temperature, both during initial hardening and subsequently in service. Wet service conditions, including high humidities, also favour conversion. Moreover, the strength of wet HACC is slightly lower than that of dry HACC. Deterioration in suspended floors will, therefore, tend to be concentrated at bearings on external walls, where deficiencies in dampproofing or cold bridges may be evident. High rates of conversion produce a disproportionately greater vulnerability to sulfate attack.

Chemical attack may produce further loss in strength where the components are subjected to persistent dampness; in some cases the effects on strength have been very severe. Where this possibility exists, the question of durability should be addressed by a specialist.

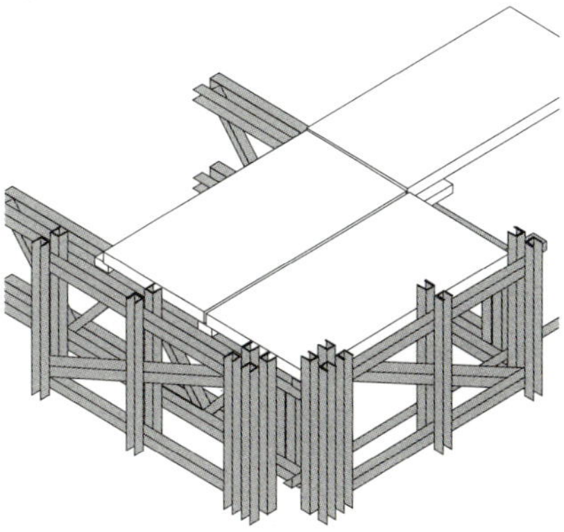

Figure 2.92
The Keyhouse Unibuilt housing system has concrete floor panels carried on timber bearers on steel beams

Since the early 1990s BRE has carried out surveys on 14 buildings between 20 and 35 years old to assess the performance of precast HACC components in service. The single most important finding from these investigations is that the majority of floor (and roof) components of HACC are carbonated to the depth of the steel. Subsequent reinforcement corrosion has not caused significant disruption to date, but the future risk from corrosion in these beams could be significant unless they are kept dry. Chemical attack, which can greatly affect the structural performance of HACC components, has not been identified as a great problem in the 14 buildings investigated although a few isolated instances of major disruption have been found in some of the components exposed to persistent water penetration. In the case of floors, this would normally be confined to the bearings on external walls.

Inspection for HACC deterioration is covered separately in the inspection section at the end of the chapter.

There is a rather more remote possibility that deterioration may be caused by alkali silica reaction (ASR) in Portland cement based components in the presence of water (BRE Digest 330[159]).

Short span concrete deck floors usually give visual warning of deterioration before they collapse, but longer spans may be more problematical, especially where corrosion can proceed unseen. Problems may also arise where the structure is post-tensioned, and where ducts may not be fully grouted and tendons may be at risk of failure. Identification of the kind of construction is therefore vital to proper monitoring of the performance of longer span floors in concrete.

Sometimes it is possible to obtain information on longterm performance by simulated or accelerated testing of prototypes (Figure 2.93), though this is not, of course, possible with existing structures.

Maintenance

Maintenance of reinforced concrete is neither practicable nor necessary, though maintenance of the weatherproofing of buildings at their perimeters is vital to continuing serviceability.

Work on site

Storage and handling of materials

Infill blocks are normally supplied prepackaged and wrapped. If the covering has been damaged and the blocks have become wet, there may be shrinkage cracking which becomes apparent after occupation of the building.

For Portland cement, see Chapter 2.2.

Restrictions due to weather conditions

See the same section in Chapter 2.2.

Workmanship

The ±15 mm suggested by BS 8204: Part 1[160] for levelness of concrete floor slabs may not have been easy to achieve with prestressed cambered beams, particularly where adjacent spans are of unequal dimensions, and particularly also where partitions are built off such beams (see also Figure 2.85). Inaccuracies greater than ± 15 mm may therefore need to be tolerated, affecting the choice of screeds and floorings.

In the BRE site inspections, several cases were seen where ceramic infill blocks had been shattered by clumsy attempts to make holes for the passage of services.

Care should have been taken in original construction not to use unsuitable packs at bearings – cases of compressible packs were encountered on BRE site inspections which would lead to settlement and subsequent cracking of floor finishes.

Failure of replacement floorings on old concrete bases can often be attributed to inadequate preparation of the surface of the concrete before laying the new flooring. The feature panel in the same section of Chapter 2.2 describes suitable techniques for reducing the risk of problems occurring.

Further aspects of workmanship are covered in BS 8000:Part 2:Section 2.1[152] and Part 2:Section 2.2[153].

Figure 2.93
Concrete I-beams being lowered into artificial ageing tanks at BRE

Inspection

The problems to look for are:

- ◊ DPCs under bearings of beams absent
- ◊ beams damaged during transit and installation
- ◊ internal walls showing rising damp
- ◊ cold bridges at perimeters of floors
- ◊ grouting to 'lock' floor components together inadequate
- ◊ signs of movement at floor/wall joints
- ◊ springing up and down of preformed insulation boards and chipboard causing furniture to rock
- ◊ strengths of fixings into soffits for suspended ceilings and services inadequate
- ◊ ventilation areas beneath ground floors less than 600 mm^2 per metre run of wall
- ◊ holes or gaps in floors at junctions with walls
- ◊ sound insulation in separating floors defective

Diagnosis and testing of HACC conversion requires specialist techniques going beyond simple visual examination. However, four general points can be made that may assist surveyors.

- Since conversion rates are greater at high temperatures and humidities, surveyors should be alert, when surveying a building in which these conditions exist, to the possibility that HACC may have been used in its construction.
- The probability that HACC has been used structurally in the UK is low in buildings constructed after about the mid-1970s when the potential severity of the problems arising from conversion became widely known.
- The possibility that HACC has been used in buildings constructed before about 1930 is not great. (The first British Standard specification for high alumina cement was published in 1940.)
- Virtually all the HACC produced in the UK went into the manufacture of prestressed X or I beams.

Inspections of all existing structures in which HACC has been used should follow the recommendations of BRE Digest 392[(161)]. In the absence of chemical attack, the strength assessment guidance issued by the Department of the Environment[(162)] in 1975 and 1976 has been shown to be appropriate for prestressed construction, but the risk of corrosion to reinforcement is an increasingly important consideration in the assessment of HACC components. An assessment should be carried out by specialists with experience of HACC structures. It is likely to be both disruptive and expensive, but the following outline gives an idea of what is involved:

- confirmation of the use of high alumina cement according to the above guidance
- establishing section profiles and numbers of reinforcement wires
- comparing sections with allowable capacities given in guidance issued by the Department of the Environment
- if sections differ from the official guidelines, calling in specialist petrographers to examine lump samples by special test methods
- assessing any shortfall in capacity and deciding whether to replace or strengthen affected structures

Chapter 2.4 Steel sheet, and steel and cast iron beams with brick or concrete infill

This chapter deals with floors of differing materials in situations where the primary method of support is ferrous metal (Figure 2.94). It includes floors where concrete is cast onto profiled steel sheets used as permanent shuttering in a composite steel and concrete floor, as well as concrete or brick barrel or waggon vault construction spanning between steel or cast iron beams. These barrels are usually found to be limited to spans of 2 m or just over. Vaults of greater span without the assistance of beams are dealt with in outline in Chapter 2.5.

Characteristic details

Basic structure

Preformed galvanised steel sheet

This material has been used frequently as permanent shuttering. Floors of this character have been used, depending on imposed loads and intermediate propping, for spans of up to 6 m. The sheet steel, which has been used in various thicknesses and widths, acts as part of the reinforcement in the floor. It may act also in conjunction with steel mesh fabric in the upper part of the in situ slab cast onto the sheets. Profiling or dovetailing of the sheets allows the concrete to interlock with the steel (Figure 2.95).

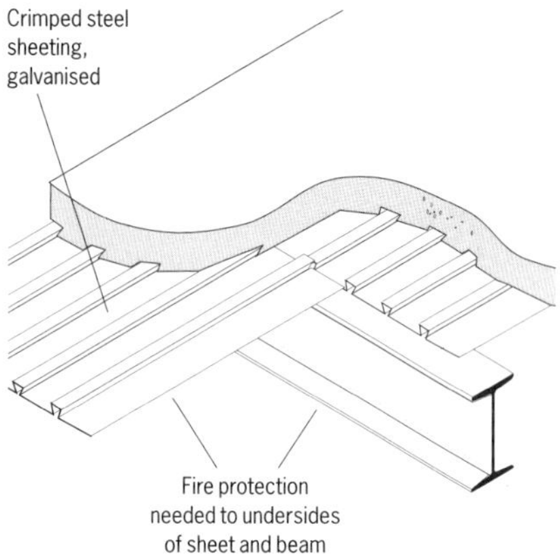

Figure 2.95
A dovetail section steel sheet floor, with concrete topping, carried on steel beams

Sheets which have relatively flat dovetailed or otherwise keyed soffits can be plastered direct, with the strength of the key depending on the ability of the plasterer to force the plaster up into the dovetailed slots or keys and the design of the key. Suspended ceilings are used frequently to hide both floor structures and services.

Floor joists formed from sheet steel may also be found. They are usually galvanised and were used sometimes in place of timber joists in lightly loaded floors of up to 6 m span, mainly in system built houses. They will be either of C section or Zed section carrying plywood decks.

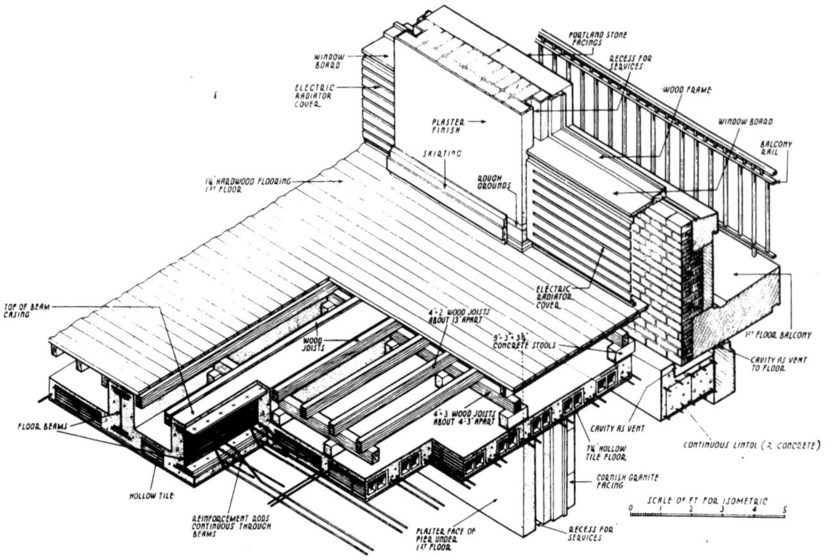

Figure 2.94
This drawing – probably from the 1930s – shows a highly unusual quadruple composite floor consisting of a timber boarded deck supported by timber bridging joists. These are supported, in turn, off an situ concrete structural floor with pot infilling which is supported off steel joists. The steelwork makes it all possible – but hardly economic, even for the time it was built. BRE investigators have learnt never to be surprised at what they might find during site visits

Preformed profiled steel sheet decking with timber floorings

Some of the industrialised housing systems of the 1950s and 1960s had composite floors which consisted of steel decks with plywood floorings laid over. One such example is the

Roften system† which had a floor construction as shown in Figure 2.96. The decking and the structural frame carrying it received a coat of red oxide paint and a coat of bitumen paint; they need to be inspected regularly for condition.

Space frame floors

Space frame floors (Figures 2.97 and 2.98) will only be seen comparatively rarely in older buildings – such construction was mostly used for roofs. However, space frame floors may have been overloaded in service, and a useful precaution would be to remove sections of the suspended ceiling beneath these floors to check the condition of the ends of struts and ties at the nodes. Load testing may be another possibility (Figure 2.99 on page 118).

Cast iron, wrought iron and steel girders

Some of the earliest historic warehouses have cast iron girders. These were later superseded by wrought iron, and later still by built-up steel sections which were used in latc Victorian times for carrying heavy floor loads; for example in warehouses (Figure 2.100 on page 118). A great variety of decking was used. The in situ concrete portions of floors of this kind have been dealt with in Chapter 2.2, and the timber in Chapter 2.1, but masonry infilling is dealt with in this chapter in the following section.

So far as the main structure is concerned, it can be reasonably straightforward to distinguish between constant section rolled steel or wrought iron beams (which are usually riveted) and cast iron beams (which have non-standard sections usually bearing directly onto padstones), but the possibility of mixed types of materials should always be considered. The date of the building can also offer a clue to identification since cast iron beams (but not columns) began to go out of favour in the 1880s[163]. Wrought iron had been largely superseded by steel girders by the early years of the twentieth century.

Masonry arches on steel or cast iron beams

This type of floor is sometimes called a jack-arch floor. The infilling may be either of stone (Figure 2.101) or of brick (Figure 2.102, both of these drawings on page 118). The construction was carried out with the aid of centring, and mortar effectively carried out the function of voussoirs to create the shallow barrel shape. These floors are usually levelled off with a lightweight aggregate such as cinders, then topped with a wearing surface of stone flags.

Figure 2.98
Steel space frame floor decks in a Nenk system building

† A range of BRE publications describes the construction of Roften and other systems.

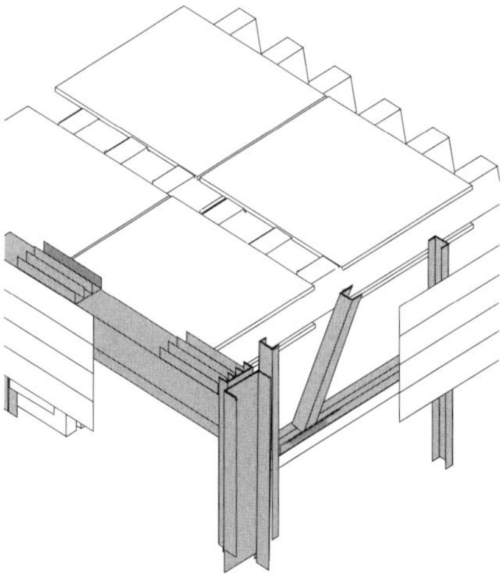

Figure 2.96
The first floor construction in a Roften steel framed house

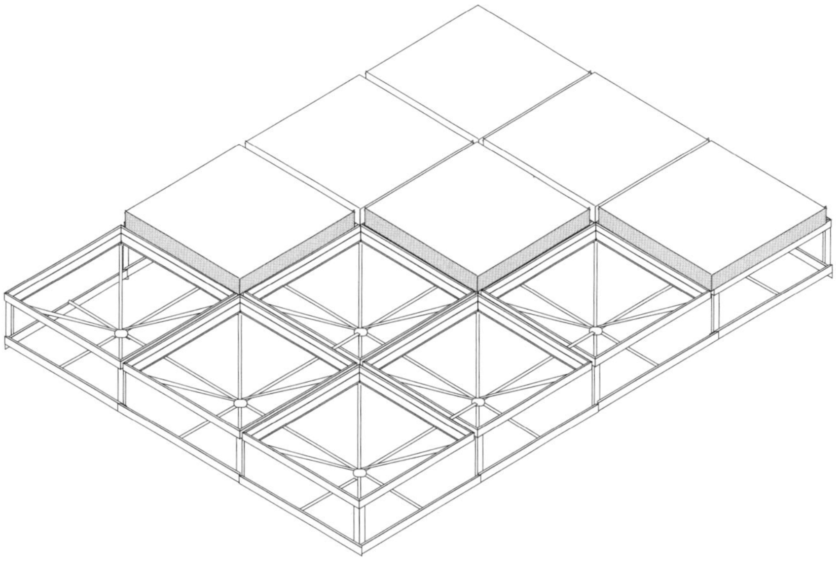

Figure 2.97
A steel space frame floor carrying concrete deck panels

Figure 2.99
Load testing a steel space deck floor in a Nenk system building

Figure 2.103
Load testing a cast iron floor beam from the Palace of Westminster. The floors had shown some tendency to deflect and vibrate under foot traffic

Figure 2.100
The massive riveted steel floor beams carried on cast iron columns at the former Summerstown Goods Depot in London

Figure 2.104
A cast iron beam, in a fort built at the time of the Napoleonic wars, forming the primary support to a triple floor (Photograph by permission of B T Harrison)

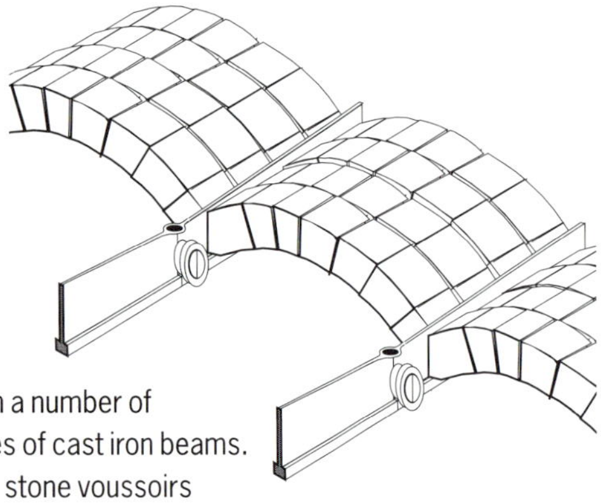

Figure 2.101
There have been a number of different varieties of cast iron beams. This one carries stone voussoirs

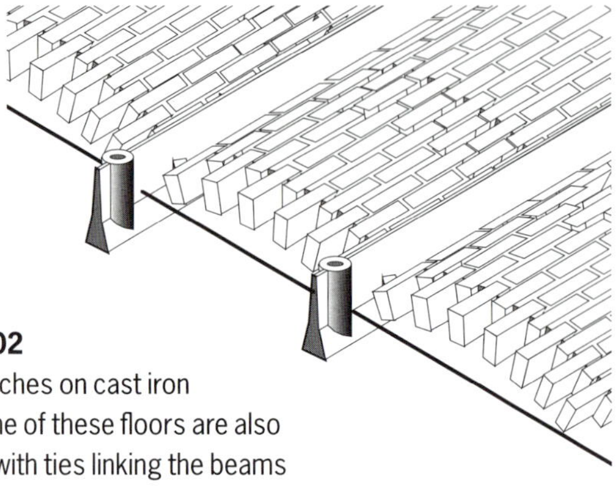

Figure 2.102
Brick jack-arches on cast iron beams. Some of these floors are also to be found with ties linking the beams

In addition to brick arches, concrete – including gypsum concrete – was used to form shallow arches between wrought iron beams. Hamilton (Building Research Station National Building Studies Special Report No 27[(65)]) cites their use by Scott in the Foreign, Home and Colonial Offices, and by Waterhouse in the Town Hall, Manchester. The aggregate used was a mixture of broken brick, stone debris or furnace dross. Portland cements should not be used in the repair of these floors without expert advice because of the risk of formation of ettringite.

Bearing details

Cast iron or steel joists usually indicate heavy floor loads. The combination of these joists and heavy floor loadings usually necessitate special padstones to carry the ends of the joists so that the crushing resistance of the wall units is not exceeded.

Main performance requirements and defects

Choice of materials for structure

The quality of some of the earliest cast iron beams may not be completely satisfactory, but wrought iron and steel may be more consistent in quality.

Although many of the bricks used in jack-arches were produced without the benefit of current standards of quality control, they are very unlikely to have suffered deterioration in the protected environment of a suspended floor.

Paint systems or other protective coatings on sheet steel also may not be to current standards, but may be sufficient for the purpose of protection.

Strength and stability

Masonry barrels

The masonry barrel carried on the lower flanges of wrought or cast iron beams is unlikely to develop cracks in the intermediate barrels where the thrust of the barrel is counterbalanced. However, at abutments of the last spans, movements might have occurred which caused the barrel to spread, and to have developed hinge cracks which will be apparent on the soffits.

Later cast iron beams were usually manufactured with radiused or feathered arrises because, from the castings of earlier times, it was known that cracks developed first from right angled arrises (Figure 2.104). Right angled arrises,

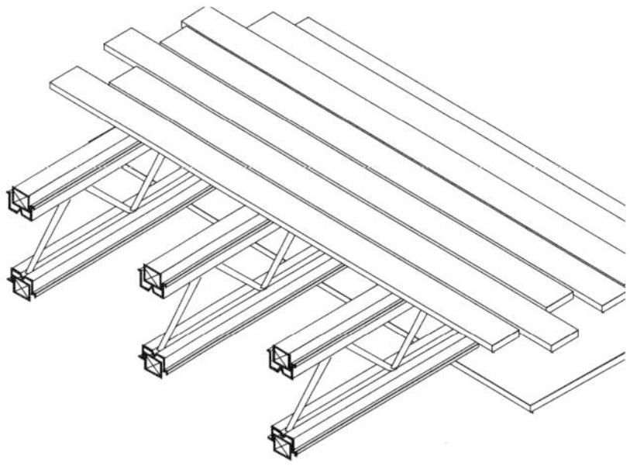

Figure 2.106
A fabricated steel joisted floor. Although the joists in these floors were not always galvanised, significant corrosion in ungalvanised units has been rare

Figure 2.105
Cast iron beams after testing to destruction at BRE. There are considerable differences in the cross-sections and different forms of web stiffening. Casting defects are evident at the junctions between lower flanges and webs; these defects indicate potentially significant weak points in the material and jeopardise the durability of floors. The main problems with cast iron beam floors are their brittle nature and questionable acceptability under current codes of practice

Case study

Effects of finishings on vibration characteristics

A series of vibration tests were undertaken by BRE in a building to evaluate the effects of finishings on the dynamic characteristics of floors. The building tested was a three storey steel frame construction with composite floors consisting of a profiled steel base and a reinforced concrete finish. The floor area tested comprised 42 nominally similar 9 × 9 m bays. The frequency and damping of the fundamental mode of each bay was measured when the bare floor had just been cast, and then when the building had been completed and finishings and furnishings had been added.

Comparing the two tests showed the variation in characteristics which could be attributed to the finishes. On average, the natural frequencies were observed to be higher, especially for areas which had been stiffened due to structural alterations, although for a few of these areas the frequencies were found to decrease resulting from extra supported mass. The damping was found to increase in all furnished areas – by a considerable amount for some.

therefore, are the parts of these beams which should be examined first for signs of cracking.

Beams usually rest upon flanges cast into the heads of columns and are often linked to each other by spigots or pins around which an iron ring is forced. Both pins and rings are at risk of fracture leading to discontinuities in the structure.

Steel joisted floors

There is not much that needs to be said about ordinary hot-rolled steel joisted floor beams and fabricated steel floors (Figure 2.106 on page 119). Main beams can be found that are spaced apart at greater distances than are economically viable for the spans of the floor decks; consequently, in some situations, secondary beams that provide additional support may be seen. However, whatever the disposition of the steel, strength and stability are rarely compromised, and loss of section due to corrosion will be rare. Where corrosion does occur in a floor, it will be confined for the most part to the building's perimeter.

When examining buildings built before the 1939–45 war, it is as well to remember that some steel structures might have been slightly affected by bomb damage (Figure 2.107), but retained because of adequate residual strength. Wartime expediencies often meant that the evidence was hidden and therefore forgotten, to be exposed again only when refurbishment was undertaken.

Dimensional stability, deflections etc

Coefficient of linear thermal expansion per °C:

- cast iron 10×10^{-6}
- mild steel 12×10^{-6}
- aluminium 24×10^{-6}

Reversible moisture movement: 0%.

Floors of the kinds of construction dealt with in this chapter are very unlikely to suffer from noticeable vibration in use, though longer spans will naturally be more susceptible than short. As shown in the case study aside, finishings and furnishings ameliorate the situation.

See also the same section in Chapter 1.4.

Thermal properties

Thermal properties rarely play a major role in the performance of suspended floors above ground level, although flying floors will need to have sufficient thermal insulation both to control heat losses and to avoid condensation forming on sections liable to corrosion.

See also the section on exposed soffit floors and balconies in Chapter 1.3.

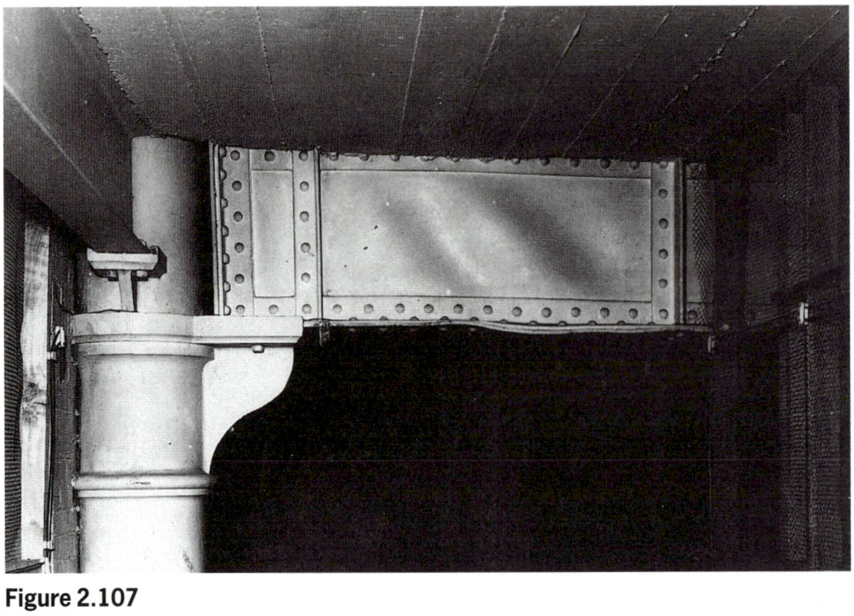

Figure 2.107

Bomb damage to the web and lower flange of a riveted steel floor girder. If this had been a cast iron beam, it would most likely have fractured. The photograph is dated October 1943

Control of dampness and condensation

Unlike most suspended concrete floors which allow drying of excess construction moisture both upwards and downwards, floors cast onto permanent metal shuttering can only dry upwards (Figure 2.108). This is often not taken into account at the planning stage of construction, and delays the application of moisture sensitive floorings. It can take more than two or three years for moisture in the floor and the atmosphere to reach equilibrium, depending on the thickness of the toppings.

Fire

Cast iron beams were not, indeed could not be, protected with masonry on their lower flanges to achieve additional fire resistance, though tests have shown that such unprotected structures remain serviceable for at least 30 minutes and perhaps longer. The danger with such designs in fire, as with other metal structures, is through thermal movements causing disruption to the remainder of the structure. Inspection of steelwork in fire damaged buildings often reveals encastred steel beams punching through external walls.

Intumescent coatings are available to increase the protection of steelwork to two hours where this becomes necessary; for example on changes of use of the building. This protection may need to be renewed periodically.

Sound insulation

As with other forms of construction, insulation depends largely on the mass of the floor deck rather than any relevant properties of the supporting structure.

See also Chapter 1.6.

Durability

Sheet steel floors next to the ground

Where steel decking is used as permanent shuttering next to the ground, no further dampproofing arrangements are normally required. Neither is ventilation of the underfloor void normally provided. In consequence, the underside of the deck may be subjected to longterm corrosion risk from the relatively high humidities if it is inadequately protected. Table 3 in Section 2 of Part 6 of BS 5493[164] refers to steelwork in dry interiors, and Part 7 of BS 5493 to wet or damp conditions †. Ventilation of the void will reduce the risk of accumulation of moisture, but will not remove it altogether.

Concrete on expanded metal lath

One industrialised housing system widely used before the 1939-45 war, Dorlonco, had first floors constructed from profiled expanded metal lath acting as permanent shuttering to in situ clinker aggregate concrete (Figure 2.109). In this system, where the expanded metal lath was used near the external walls and was prone to condensation, the expanded metal lath in most circumstances has rusted badly, raising concern about the condition of the bearings to the first floors. Steelwork in industrialised systems needs careful inspection; further guidance is available in *Steel-framed and steel-clad houses: inspection and assessment*[165].

Cast iron beams carrying masonry barrels

Cast iron is very resistant to corrosion and there will rarely be problems from this source. However, the assessment of the condition of this kind of floor as a whole is far from straightforward. Apart from local crushing or spalling of the brick infill voussoirs, and obvious cracking of masonry or of cast iron, there are no simple visual assessment criteria for the continuing serviceability of masonry vaults§.

Where the masonry is disrupted, it may be possible to strip the infill after suitably propping the barrel and replace with a reinforced lightweight concrete slab to lock the whole structure in place. It is

† ISO 12944, Corrosion protection of steel structures by protective paint systems, is currently being drafted. It is intended that it should also become a CEN standard and would therefore replace BS 5493.

§ BRE has at various times in the past carried out tests on cast iron beams to assess their serviceability. Advice can be sought from BRE Heritage Support Service, the heritage departments or specialist consultants, since they cannot be analysed in the same way as modern rolled steel. Further research on the behaviour of jack-arch floors is taking place under the auspices of the Department of the Environment.

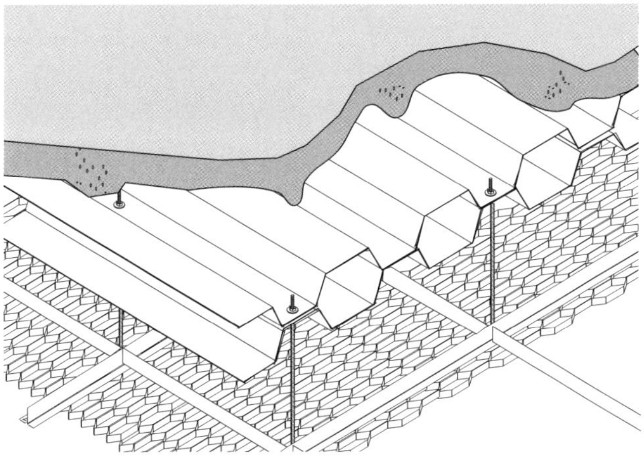

Figure 2.108
Floors cast onto permanent metal shuttering have only one face from which the concrete can dry

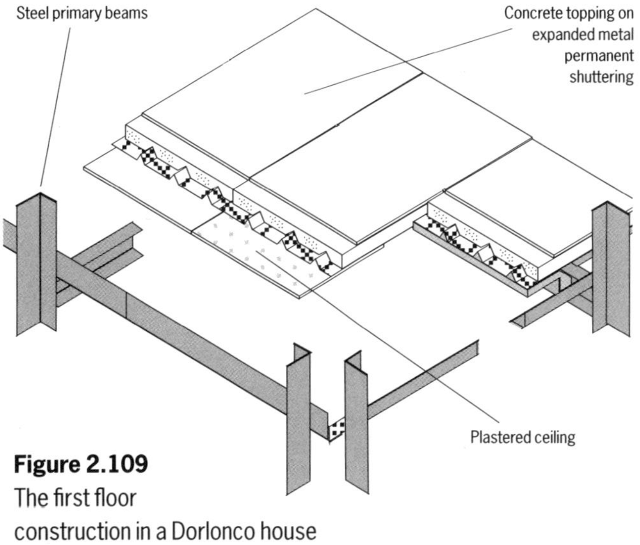

Figure 2.109
The first floor construction in a Dorlonco house

essential to make sure that potential differential movements are taken into account.

Cast iron beams can be stitched but this demands particular expertise. Imperfections in the original castings have not been unknown and may have been cosmetically filled.

Wrought iron and steel beams

Wrought iron is sometimes difficult to distinguish from mild steel and can be much weaker. It has a greater resistance to corrosion than mild steel. Repairs can be bolted on in lieu of the original riveting since welding the material is very difficult.

Plain carbon steel frames will require some protection; the most usual form, in the internal environment of a building, being a paint system (BS 5493).

The steel in steel beams in historic buildings may well have strengths lower than those installed in modern construction, and care is needed in any analysis. The first mild steel beams were rolled in the UK in 1885. Durability of the steel can be compromised by the use of coke breeze in concrete mixes. In writing of Holman's patent filler joist floors, Hamilton remarks: *'The use of fine coke in the aggregate may have seemed at the time a useful way of disposing of a plentiful waste material, but it contained sulphur that slowly oxidised, and many floors of the period later suffered grave damage through the corrosion of the ironwork with which the coke or breeze made contact'*[(65)].

However, irrespective of the date of construction of the building, steel beams will often require careful inspection and, perhaps, need to be exposed. It may also be worth investigating for fluorides in the concrete and carbonation of the encasing concrete as these can affect the durability of steel in moist conditions. During the site inspections carried out in 1984 by BRE staff assessing the quality achieved in rehabilitated housing, the flats seen on one particular site had been built around the 1890s with steel joists carrying concrete landings. The steel joists were exposed to reveal that, in the most heavily corroded case, the webs and most of the flanges had entirely disappeared. However the load which the joist formerly carried had fortunately been redistributed to other parts of the floor construction. Since the joist was situated some distance from the external wall, rain penetration, the first diagnosis considered, was effectively ruled out. Condensation and the occasional stair washing would then have been the only possible sources of water.

Figures 2.110 and 2.111 show parts of the frame of the BRE Materials Laboratory under construction in 1959 in which the deep Vierendeel girders carry precast concrete floor beams, allowing a large void for distribution of services. The steelwork in this case carries a paint system only.

For better assurance of longterm performance – perhaps, for example, where condensation risk is more than normal – the most common protective coating will be hot dip galvanising or an organic coating, or both. The required thickness of zinc will be determined by the required life of the structure; and the size, chemical composition and method of manufacture of the component members of the frame will determine whether that thickness will be possible to achieve. As before, BS 5493 is relevant.

Maintenance

Renewal of protective coatings on steel beams to prevent rust may be required to maintain the integrity of the construction. Building owners should be advised on the need to inspect hidden steel members. It will be largely a matter for judgement on whether opening up the structure for maintenance purposes is appropriate when considering the expected remaining life of the building.

Work on site

Storage and handling of materials

See the same section in Chapter 2.2.

Restrictions due to weather conditions

See the same section in Chapter 2.2.

Figure 2.110
A floor of precast slabs on steel beams

Figure 2.111
The same floor as shown in Figure 2.110 seen from above

Inspection

The problems to look for are:

Steel floors

- ◊ DPCs under bearing of beams absent
- ◊ materials used as packings at bearings unsuitable
- ◊ coatings to steel in aggressive situations deteriorating (eg on the undersides of floors next to the ground)
- ◊ differential dimensional changes in decks causing fracturing of keys between toppings and decks, and between decks and ceilings
- ◊ moisture unable to evaporate from toppings in the time available (up to two or three years should be allowed or a surface DPM be provided)
- ◊ strengths of fixings into soffits for suspended ceilings and services inadequate
- ◊ holes or gaps in floors at junctions between the elements of structures
- ◊ sound insulation in separating floors defective
- ◊ floating floors in contact with subfloors

Cast iron beams carrying masonry vaults

- ◊ movement in abutments
- ◊ cast iron beams cracking, especially those with sharp arrises
- ◊ wrought iron ties rusting
- ◊ hinge cracking or spalling of bricks in vaults
- ◊ mortar in joints eroding
- ◊ pushing out of abutments leading to dropping of the crowns of barrels

Chapter 2.5 Masonry vaults

This chapter is not intended to provide more than outline considerations of masonry vaults carrying floors without the benefit of supporting beams (Figure 2.112). Such vaults can display some of the characteristics of masonry roof vaults, but, in general, they occur over much shorter spans and will often be found in the ground floors over cellars in even quite modest older buildings. These buildings, unlike their vaulted roof counterparts, may not have been investigated by specialist consultants, and this chapter therefore draws attention to points which are amenable to routine visual inspection. It must be emphasised, however, that the care of masonry vaults which show distress is a specialised skill.

Characteristic details

Basic structure

The more pointed the vault, the stronger the arching effect and the smaller the thrusts on abutments (Figure 2.113). However, since Tudor times, partially no doubt to save headroom, shallow barrel vaults and the even flatter multi-centred vaults were introduced over cellars which placed greater importance on the stability of abutments (Figure 2.114).

Indeed, in shallower vaults, and in the absence of undisturbed ground outside the building perimeter to buttress the cellar walls, movements may have led to the insertion of wrought iron ties at or just above springing level to resolve the outwards thrust at the abutments. Such ties are common in factories and small buildings dating from the nineteenth century.

Figure 2.112
A shallow brick vault with the bricks laid in stretcher bond

The webs in ribbed stone vaults may be found in a variety of thicknesses, from 150 mm upwards, though of course the infilling in the spandrels can increase thicknesses considerably.

Bricks in the shallow domes used in floors may be found laid soldier fashion in order to ease the problem of laying them in concentric rings, as in Figure 2.115. In vaults, however, bonds seen may include stretcher and English as well as soldier. Some vaults have joints between the voussoirs packed with slate, presumably to try to counteract the effects of mortar shrinkage.

Hollow pots were used in place of bricks for the structures of some floors in the early years of the nineteenth century. Hamilton[65] cites their use by Nash in the floors of Buckingham Palace, by Wilkins in the National Gallery, and by Barry in the Treasury Buildings. *'The pots were about 8 in high, 4 3/4 in square at the top and rounded towards the bottom; the sides and bottom were scored, and a small hole was made in the round bottom, presumably to act as a key for the plaster. The arches spanned about 6 ft and never more than 7 ft, the rise never exceeding 6 in'.*

Brick vaults may be also found on a smaller scale, in domestic construction, as the structure for hearths carried on timber floors (Figure 2.116).

Bearing and abutment details

As already stated, the stability of voussoir arches and barrels used to support floors depends on the

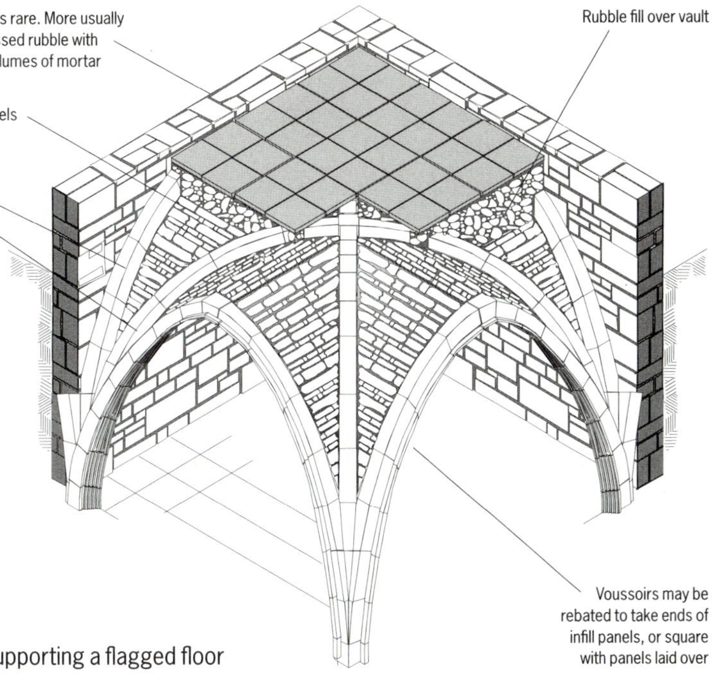

Figure 2.113
A stone vault supporting a flagged floor

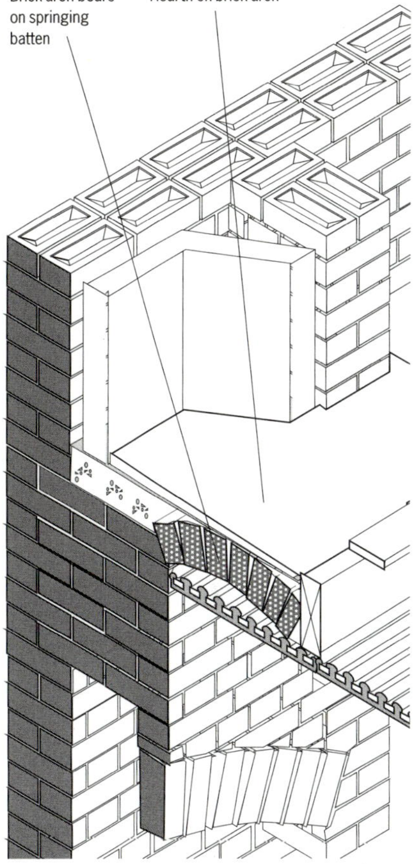

Figure 2.116
Brick arches will occasionally be found supporting hearths in otherwise suspended timber floors

Figure 2.114
This short span stone dome springs from massive stone haunches. The intrados has been whitewashed to enhance interior lighting

Figure 2.117
Part of an eighteenth century brick vault in a former bonded warehouse converted to housing. The bricks in the soffit are laid soldier fashion, but the pointing is less than regular!

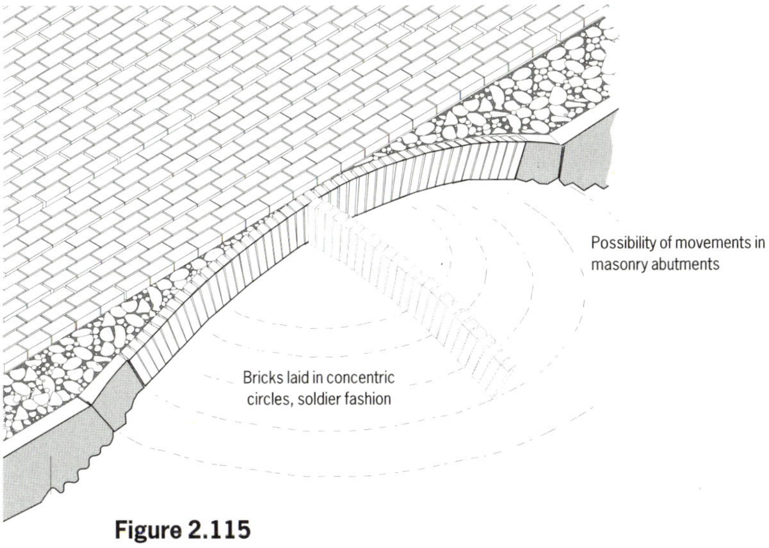

Figure 2.115
A shallow, brick, domed floor

integrity of their abutments. In spite of their size, and the opportunity which exists for buttressing, abutments do move, and it is a matter of judgement when the extent of movement begins to compromise stability. Each case needs to be assessed on its own merits.

Main performance requirements and defects

Choice of materials for structure

It is usually the mortar joints in brick barrels and domes rather than the bricks themselves which provide the necessary wedge or voussoir shapes (Figure 2.117 on page 125). Soft rubber bricks were not normally used in this application, for their crushing strengths, though perfectly adequate for wall arches, might not be adequate for floors. Furthermore, labour costs would have been prohibitive, even in the days when labour was cheap. Stone voussoirs in ribs or groins, on the other hand, will be found carved to form the necessary wedge shape.

Mortars used in vaults will usually be found to be lime mixes, though it is possible that some of the softer mixes may have suffered due to the loads involved with the larger spans. However, since a vault will not normally be exposed to the weather and so suffer erosion, it is a mistake to specify repointing in too strong a cement mortar. Some floors, more particularly those using hollow ceramics, may have been formed using gypsum as a cementing agent. Gypsum used as mortar generally compares very favourably with lime mortars in its properties. Its major disadvantage is the ease with which it will dissolve in water and, therefore, lose strength. It is vital then that floors built with gypsum mortars are kept dry. It is also vital that Portland cement is not used in any repair work on such floors since the combination of materials could lead to the formation of ettringite, a process which is expansive and destructive. See the section on durability in Chapter 3.1.

Strength and stability

The engineering assessment of the strength of voussoir arches is rather complex, and a brief historical review will be found in the National Building Studies Research Paper No 11[12]. The paper points out that methods of analysis are normally based on the assumption that the abutments are rigidly fixed. In practice, though, this assumption is doubtful and absolute immovability of the abutments cannot be relied upon. Not only will the thrust of an arch cause slight spreading of the supports but settlements or other earth movements under the foundations are possibilities which may cause changes after the arch is built. Rise or fall in temperature will also change the assumed conditions. A prolonged spell of either unusually hot or cold weather may be sufficient to cause gaps between the voussoirs.

The single masonry barrel vault will normally be sufficiently stable when carried on massive haunches, assuming that the foundations of the supporting walls or arcades do not move either outwards or downwards. Multiple barrels provide mutual support to sideways thrust excepting end barrels, which depend for stability on abutments.

Since the spandrels of the vaults need to be filled (usually with masonry rubble), to carry the flooring, extra mass is available to bear down on the vault, improving the stability of the vault but tending to increase sideways thrust.

The shallow masonry dome was extensively used, both in Renaissance times and subsequently, for flooring over large spans. Lightness was at a premium, and hollow pots, as described in Chapter 1.6, were frequently used. Ring beams (or ring chains) were rarely provided to resolve the thrusts of floors as they were for roof domes, but relatively small hollow squares to contain the thrusts have been seen in pegged timbers.

Dimensional stability, deflections etc

Coefficient of linear thermal expansion per °C:

- sandstone 7 to 12×10^{-6}
- limestone 3 to 4×10^{-6}
- slate 9 to 11×10^{-6}
- brick 4 to 6×10^{-6}

Reversible moisture movement:

- sandstone 0.07%
- limestone 0.01%
- slate, negligible
- brick, negligible

Following firing, there is an irreversible moisture expansion of clay bricks, but this is normally translated into a slight rise in the arch of a vault.

In the case of arch construction, a small crack each side of the span, showing on the intrados of the arch, might indicate symmetrical outwards movement of the abutments. Cracks showing asymmetrically on one side of the arch on the intrados and on the other side of the arch on the extrados, might indicate settlement of one or other abutment.

The presence of a small crack or two, typically showing as the partial opening of a joint without significant sideways or downwards displacement, is probably not a cause for undue concern, particularly in those parts of buildings subject to temperature and, perhaps, moisture variations. Cracking is, after all, the mechanism whereby stresses are relieved. On the other hand, more than one crack per side probably would be worth investigating further.

Displacement of voussoirs is another matter. Any displacement visible to the naked eye should certainly be investigated by experts. Electronic measuring instruments targeted on unobtrusive defined points in the structure can be used to monitor movements over time.

Thermal properties and ventilation

Old structures will have only small thermal insulation values and usually cannot be upgraded without compromising the historical value of the building.

Control of dampness and condensation

Massive stone floors in unheated buildings will be subject to condensation in exactly the same way as concrete floors; for example when a warm front follows a cold spell.

See also the same section in Chapter 3.1.

Fire

So long as the structure of a dome or vault constructed of pots or bricks remains intact, fire resistance is inherently good. However, concern was expressed in the late nineteenth century that floors of this kind might not survive masonry crashing down from above, and there is probably some justification for taking this view, particularly if toppings are very thin at the crown of the rise. Replacing the topping with reinforced lightweight aggregate concrete will probably improve the behaviour of the construction.

Sound insulation

Floors in this category are usually massive, and there will be few problems of either airborne or impact sound insulation, even, it could almost be said, in the absence of resilient flooring. Sand or ash used as filling could percolate through joints and lead to air gaps in the structure which then become a channel for sound transmission.

Durability

Mortars are likely to be a major point of weakness in vaults and will need to be checked, particularly if water ingress is detected.

Maintenance

Maintenance undertaken without specialist advice should be limited to soffit repointing in lime mortars.

Work on site

The repair of vaults is a specialised activity which should be entrusted to experienced firms.

Inspection

The problems to look for are:

- ribs detaching from the other parts of vaults
- movement in abutments
- hinge cracking in intrados
- anchor plates of wrought iron ties detaching
- masonry or mortar being crushed
- crowns sinking
- ribs and webs bellying outwards
- fine material migrating from infilling or pavement beddings through arches

Any signs of these faults should lead to an inspection by engineers experienced in the assessment of vaults.

Chapter 2.6 Platform and other access floors

Figure 2.118
A platform floor under construction

This chapter deals with two kinds of floors which are used in very specific circumstances in industrial and commercial buildings. While the steel grid has been in use for a number of years, for example to provide access to industrial plant, the use of platform floors (Figure 2.118) has paralleled the phenomenal growth in requirements for servicing commercial buildings with telecommunications and computer facilities, all of which imply the need to accommodate cabling.

Characteristic details

Basic structure

There are two kinds of metal decked suspended floors:

- inverted trays of steel or cast iron designed to span directly between steel joists
- grids of riveted or welded open steel diamond pattern mesh, designed to be left open, spanning between steel joists

Gridded metal (usually steel) decking providing access galleries and floors in buildings containing industrial plant have been in common use for a number of years (Figure 2.119). These floors are transparent to sound, fire, liquids and even small objects, though their utility in specific circumstances is undeniable. They can, however, be subjected to onerous requirements in respect of loadings, and, since they are commonly found in industrial atmospheres, to risks of corrosion. Gridded floors have not figured significantly in the BRE workload except for corrosion risk, the main considerations of which have been dealt with in Chapter 2.4. It should be remembered, however, in considering the degree of protection to be afforded to the steel, that very slender surfaces only are available to resist wear. (The surface characteristics of metal plate floors are dealt with in Chapter 7.6.)

In contrast, a platform floor, albeit complete with its supporting deck, is required to provide the full range of performance characteristics in addition to special requirements relating to demountability. These floors normally provide a system of loadbearing fixed or removable floor panels supported by adjustable pedestals or jacks placed at the corners of the panels to provide an underfloor void for the housing and distribution of services. In full access systems, most or all of the panels are removable. Partial access systems have runs of removable

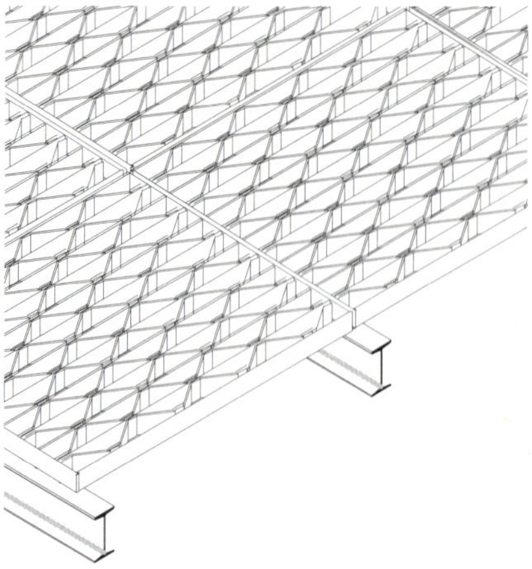

Figure 2.119
A steel grid floor: a type often used in industrial locations

panels or individual traps, or both. Platform floors have become widely used in recent years in offices, telephone exchanges, data processing rooms, electronic control rooms and conference rooms where easy access to underfloor services is required.

Floors within these categories are all secondary; that is to say they require conventional floors beneath them for bearing purposes, and which also will need to possess other performance characteristics such as the required structural strength for supporting the platform, or sound insulation. However, the platform and its supports will be required to posses certain properties over and above those which are normally required of floor finishes, including, for example, structural strength and demountability.

The supporting grid for floors will usually be in the range 500 × 500 mm to 750 × 750 mm. Pedestals are spaced to suit the dead loads actually to be experienced, with a suitable allowance for live loads over the whole area.

Slatted or duckboard floors, commonly fabricated from timber sections on timber battens and laid directly on the flooring, have been widely used in, for example, 'wet' or oil spillage areas in industrial locations (Figure 2.120).

Bearing details

Slatted floors are normally fabricated into panels which are bolted to the perimeter steelwork. They can be shimmed to level, though this is not normally required.

Pedestals for supporting platform floors are usually adjustable for level by screw threads. In most cases the range of adjustment is sufficient to overcome inaccuracies in the base. The size of the pedestal base is a compromise between the area necessary to provide stability against overturning and the possibility of rocking.

Some platform floors have been built using concrete block supports shimmed to level, but many pedestals were glued directly to the base beneath.

Abutments

Most floor systems provide a close fit to perimeter walls, with bracing to prevent sideways sway of the whole support system. Pedestals at the perimeters of platform floors are placed inboard of the edge panels.

Services

The whole purpose of platform floors is to facilitate the installation and maintenance of wired and, possibly, piped services. Power cables are usually laid in conduit or trunking, though all wiring under access floors should be fire retardant and of potential low smoke emission.

Since the suspension system for platform floors is usually in metal, there will normally be requirements for earth bonding in accordance with the requirements of the *Regulations for electrical installations*[74] published by The Institution of Electrical Engineers.

Earth continuity bonding is frequently provided by means of a terminal on the base of the pedestal, but installations have been seen with clips clamped to the threaded shank of the pedestal. It has not been found necessary to bond each and every panel, since continuity is provided by the sheath of the panel, so inspections revealing apparently missing bonding may in fact still prove to be satisfactory. Earth bonding of concrete supported panels is not easy to achieve, and these installations may need to be carefully checked.

Access to the floor void may be provided intermittently instead of making every panel removable. Some designs employ traps cut into the deck, supported by angled shoulders.

Figure 2.120
Duckboarding in an industrial workshop in which vast quantities of water are used and where spillages are inevitable

Main performance requirements and defects

Choice of materials for structure

The results of development work on this type of floor in the UK, undertaken by the former Property Services Agency (PSA), has been published in *Platform floors (raised access floors): performance specification*[166].

The bases and shanks of pedestals are normally of steel while the floor support plates are frequently of die cast alloy. BS 6266[167] requires pedestals not to melt below 600 °C.

Floor panels fitting into the main supporting grid are available in galvanised steel, or aluminium clad chipboard or other timber product, frequently plywood.

Table 2.1
Applications suitable to static floor loads

Floor type	Max uniformly distributed load	Applications
Light	6.7 kN/m^2	Offices without heavy equipment
Medium	8.0 kN/m^2	Offices with heavy equipment and teaching areas
Heavy	12.0 kN/m^2	Public and lightly trafficked industrial areas

Strength and stability

Platform floors have been available to meet three main structural grades, with differing concentrated loads and uniformly distributed loads. Table 2.1 gives examples of uniformly distributed static loads and typical situations of use.

Extra heavy systems are also available for special applications.

Dimensional stability, deflections etc

Coefficient of linear thermal expansion per °C:

- mild steel 12×10^{-6}
- aluminium 24×10^{-6}

Reversible moisture movement: 0%.

Metal casings will govern thermal movements. Moisture movements of boards should be contained within the casings.

Platform floors are normally capable of adjustment to fine tolerances. The PSA specification referred to on page 129 calls for a maximum difference in level between adjacent unloaded panels of 0.75 mm, and maximum differences of ±1.5 mm over any 5 m square or ±6 mm over any complete floor(166).

In a BRE survey of platform floors in use, rocking panels were a regularly occurring fault. The causes appeared to be:

- inaccurate levelling of the subfloor
- duct covers too high (Figure 2.121)
- panel distortion and deterioration under service conditions
- pedestals out of plumb
- pedestal heads not level
- pedestals inaccurately adjusted
- pedestal head caps missing
- detritus below panel seats
- improvised packings

Platform floors may also be susceptible to squeaking, particularly under the action of moving loads, and this is frequently found to be caused by pedestals not being levelled sufficiently accurately during installation.

Electrical resistance

Many platform floors are used in computer rooms where it is important that any electrostatic buildup which can affect the equipment is avoided. This particular aspect will need to be checked for individual applications, though normally the requirement will be in the range 2×10^{10} to 5×10^{5} ohms resistance (*The CFA guide to contract flooring*(168)).

Thermal properties and ventilation

The space beneath the floor surface is sometimes used as a plenum for ventilation systems, and this will need to be borne in mind when any alterations are proposed.

Control of dampness and condensation

With floors next to the ground, the concrete base to which the pedestals are fixed should contain suitable dampproofing arrangements – usually a DPM beneath the slab.

Condensation has been known to occur on the underside of platform floors laid on solid bases at ground level, particularly at the perimeters adjoining external walls. The areas affected are mainly the perimeters of the concrete supporting structure, but may also involve the metal trays if they are in contact with uninsulated walls. If the surface of the concrete is below dewpoint and warm moist air from above comes into contact with it, it will condense on the concrete. Placing insulation in the external wall cavity may be straightforward,

Figure 2.121
This platform floor panel has been removed and inverted to lie over an adjacent panel. The edge of the panel provides a straightedge showing that the top of the duct cover (centre of the photograph), which should have lain below the level of seating of the panel, had been set too high. (The shadows created by the camera flash have distorted the apparent levels of the duct cover and the structural floor.) Consequently the panel did not seat on the pedestals but rocked over the duct. An abortive attempt had been made to compensate by removing part of the underlayer to the chipboard

but remedial action to the floor itself is problematical since the insertion of thermal insulation without a vapour control layer above it will not be effective, and installing a vapour control layer is very difficult, if not impossible, because of the pedestals. Treating the void as a plenum and providing heat to the cold slab in some way may be the most practical solution, though it is not known whether this has been tried in practice.

Fire

Platform floors do not have to be fire resisting to conform with building regulations as they are not elements of structure (Clause 7.4c of the Building Regulations 1991 Approved Document B[59]). However, the structural floor underneath does. Also it is important that any fire within the void is contained, not least so that provision for means of escape is maintained. Cavity barriers may be required, and the lower surface of panels be non-combustible. The PSA specification referred to earlier also calls for Class 1 surface spread of flame within the cavity.

Moreover, there might be a requirement to conform with BS 6266.

Sound insulation

Platform floors add virtually nothing to the sound insulation provided by the structural floors on which they are laid. However, there may be horizontal routes within the plenum for the transmission of sound between adjacent rooms.

Timber platform floors are often found to perform better in the laboratory than in the field and the reasons for this are not well understood. The main parameters that affect the sound insulation of this type of floor have been investigated[169]. They include the density of the resilient layer, the effect of absorbing material, the effect of the joist layer and the effect of flanking transmission.

Durability

The PSA specification calls for a 50 year life for the support system and 25 years for the deck, but the achievement of this, of course, rather depends on the conditions within the building concerned.

A survey by BRE of the use and performance of platform floors, based on a questionnaire sent to building owners and users, and on inspections of the floors in 20 of the buildings, found that the users were generally satisfied with the performance of their platform floors although some problems were found to be common in a number of systems. These included rocking and noisy panels, difficulty in obtaining access to the underfloor void, and faulty or inappropriate floor finishes.

Pedestals often become unstuck from the base. The adhesive frequently, and wrongly, gets the blame for this, whereas the most common cause is that the adhesive has been applied to a concrete surface which has received no preparation whatsoever. In fact the adhesive has been stuck to a weak laitance. Without proper preparation of the concrete surface to which the pedestals are to be stuck, the bond strength will be only as strong as the laitance.

Pedestals and jacks sometimes break because they are insufficiently robust for the service conditions to which they have been subjected (Figure 2.122).

Galvanised steel panels can be scratched and damaged during transport and laying; and in the presence of small amounts of residual water from water based adhesives used to fix floorings, the exposed steel can corrode. The water which does the damage is trapped between the steel and the impervious floor covering.

There is a slight theoretical risk of electrolytic corrosion of parts of the system where dissimilar metals are used – galvanised steel, alloys containing zinc, and copper used for earth bonding tags. However if the floor is in a continuously heated building, the likelihood of corrosion is minimal.

Figure 2.122
A broken pedestal from a platform floor

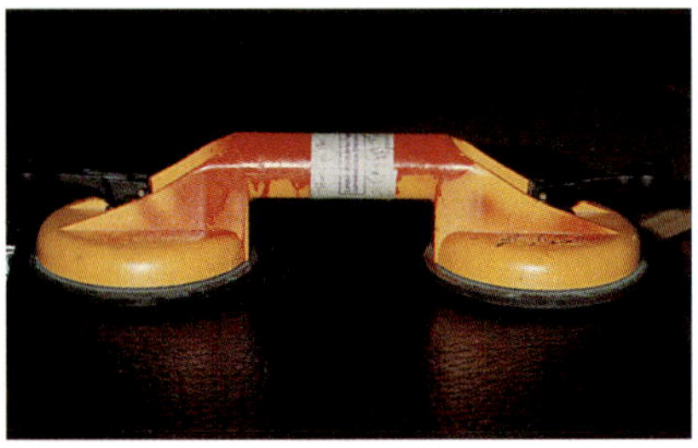

Figure 2.123
A suction device for removing panels from a platform floor

Maintenance

In a survey of platform floors in use, some floors were found where it was difficult or impossible to access the underfloor void and a number of panels showed damaged edges where force had been used.

Moreover, some floor finishes had proved inappropriate to the conditions of use, particularly bubbling of textile flooring and panel edges grinning through the finishes where the joints in the supporting structure did not coincide with flooring joints (the so-called tartan effect). Wheeled vehicles were found to have the most detrimental effect.

Dirt and grit from above finds its way into joints between panels. This effectively locks the panels into position and they cannot then be easily taken up to provide access. In the absence of purpose made lifting devices, robust strong-arm methods have been used to remove panels, perhaps assisted by screwdrivers, wrecking bars and hammers. Unsurprisingly, panels and pedestals can be broken during this process. Distorted panels are put back by further hammering, which leads to further distortion, and this makes it even more difficult to raise the panel the next time.

Work on site

Workmanship

It is absolutely vital that the correct techniques are used to remove panels when maintenance or replacement of services is undertaken (Figure 2.123), otherwise distortion inevitably ensues.

The following points would seem to be crucial to installation or replacement work:

- compatibility between sealants used to seal the surfaces of subfloors and adhesives used to fix pedestals
- holes at the edges of voids to be fire stopped before installation begins
- cleaning voids before installation begins
- removing laitance from concrete surfaces before applying adhesives
- protecting completed work from damage by following trades

Inspection

The problems to look for are:

- ◊ pedestals breaking or becoming unstuck
- ◊ panels rocking and distorting
- ◊ galvanised steel panels rusting under flooring
- ◊ panels jamming leading to further distortion on removal
- ◊ condensation on uninsulated ground floor slabs at the perimeters of floors

Chapter 3 Solid floors

Figure 3.1
Casting a massive reinforced concrete floor in one single pour for BRE's new European testing and research facility at Cardington

By solid floors is meant those floors, which consist for the most part of concrete, receiving continuous support from the ground (Figure 3.1). Although continuously supported beaten earth and grip floors were mentioned in the introductory chapter, they are relatively rare. Chalk floors, compacted using a vibrating plate, are used in certain areas of the UK, mainly for agricultural buildings; when the chalk is dry it possesses adequate strength for light traffic and looks somewhat similar to concrete. None of these more unusual floors are dealt with, other than as a brief mention, in this book.

The opportunity has been taken in this chapter to subdivide solid floors into three types: those which have never had the benefit of additional thermal insulation, those which have been laid over thermal insulation, and those which have been reinforced to act as combined foundations and floor slabs – rafts.

Chapter 3.1

Concrete groundbearing floors: insulated above the structure or uninsulated

In situ concrete groundbearing slabs are the most common base used for floors, and will be found in all types of buildings – houses, offices, factories, hospitals, warehouses etc. The slab may be finished:

- to receive a bonded, unbonded, or floating screed before receiving other floorings (Chapter 4)
- directly as the wearing surface (5.1)
- to receive a granolithic or cementitious polymer topping (5.3)
- to receive thermosetting resins or paints (5.5 and 5.6)
- to receive mastic asphalt flooring (5.7)
- to receive magnesite flooring (5.8)
- to receive, either with or without smoothing compound, sheet and tile floorings (6 and 7)
- to receive ceramic tiles, terrazzo tiles, or slab flooring; either bedded in mortar or fixed with thin bed adhesives (7)
- with prefabricated timber sheets (eg chipboard, plywood etc) either fixed on timber battens or laid directly on thermal insulation (8)
- to receive wood block (8.3)

The quality and thickness of the concrete base will depend on the use to which the floor is put. Plain concrete has been the traditional form of solid floor. This was site mixed, the mix proportions often used being 1:2:4 cement:fine aggregate:coarse aggregate by volume. The strength of the concretes varied greatly depending on the aggregate grading, quantity of mixing water, efficiency of mixing, degree of compaction etc. Consequently, existing concrete bases will be found to vary greatly in their properties. To minimise drying shrinkage cracking, particularly in the absence of reinforcement, it was common to lay the concrete in relatively small areas. The most usual method, the chequerboard fashion, was to lay alternate squares of about 3 m side length, and then to lay the intermediate squares after the initial drying shrinkage had taken place. These mix designs and techniques are no longer used, but they will be found in existing buildings.

Since about 1972, it has been much more common to lay large areas of concrete by the long strip method. This usually entails laying strips between 3 and 4.5 m wide, and several tens of metres long. Alternate strips are laid, and when these are set and hardened, the intermediate strips are cast. To prevent shrinkage cracking, stress contraction joints can be sawn at 6 m intervals, or longer for reinforced floors. Where the concrete is to be covered by certain types of flooring, contraction joints are not provided as it is preferable to allow the shrinkage to form a large number of fine cracks. A fuller description of this technique is to be found in the section on workmanship in this chapter.

Figure 3.2
Blinded hardcore ready for laying the DPM and casting the concrete slab

Characteristic details

Basic structure

Hardcore

The principal uses of hardcore are as a make-up material to provide a level base on which to cast a ground floor slab, to raise levels, to reduce the capillary rise of ground moisture, and to provide a dry firm base on which work can proceed or to carry construction traffic. There are a number of factors that need to be taken into account in the selection of materials for use as hardcore. A variety has been used satisfactorily but difficulties can occur and some of these are discussed in the following pages. Ideally, the materials should be granular, and drain and consolidate readily (Figure 3.2). They should be chemically inert and not affected by water.

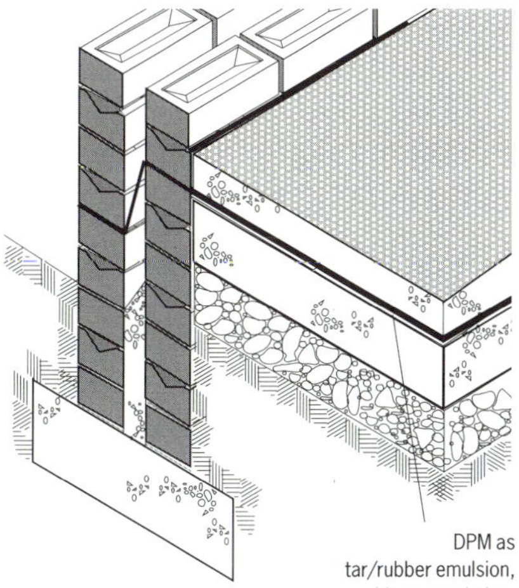

Figure 3.3
A typical groundbearing floor installed in new houses built soon after the 1939–45 war when there was a prohibition on the use of timber. Considerable effort was invested at the time in developing satisfactory sandwich DPMs

Need for DPMs

In all cases a DPM will be required to comply with Clause C4 of the Building Regulations 1991 Approved Document C[42]. This may be positioned above or below the concrete slab and should comply with Section 3 of BS CP 102[40] (see also BRE Digest 54[38]). In addition, thermal insulation in the form of foamed or expanded plastics sheets, glass fibre or foamed glass may be incorporated above the slab. See Chapter 3.2 for illustrations of thermal insulation below the slab.

As well as being required to prevent moisture from the ground reaching the inside of the building, the DPM may be required to fulfil other functions: preventing the interaction of ground contaminants with the concrete, stopping interstitial condensation and retaining constructional water. It may be placed in different positions (Figure 3.3). To prevent interaction with ground contaminants the membrane must be placed below the slab. The time required for the residual constructional water to evaporate should be taken into consideration when deciding on the position of the DPM. A table of drying times is given in Chapter 1.4.

Materials for DPMs

For suitable DPM materials, see the list in Chapter 1.4. For mastic asphalt for waterproofing duty, the appropriate grades are Type R988 or T1097 of BS 6925[170].

Concretes

The concrete for bases to receive other floorings is specified by strength. Grades based on characteristic strength of 30 N/mm^2 or 35 N/mm^2 are recommended; the higher strength may be required for structural reasons. Clause C4 of the Building Regulations 1991 Approved Document C requires the minimum quality of the concrete to be at least mix ST2 of BS 5328:Part 1[171], or, if there is any embedded steel, mix ST4.

Protection against radon and methane

Sealing cracks around the perimeter of solid concrete floors should reduce the seepage of radon from the ground. However, in practice it has been found that this only reduces radon levels by a half to two thirds. This is probably due to the fact that it is difficult to seal **all** the cracks effectively. The major cracks are likely to occur at the perimeters, and, for the most part, may be concealed by skirtings. Sealing does remain a viable option for radon levels up to 400–500 Bq/m^3 if it is properly executed (*Sealing cracks in solid floors: a BRE guide to radon remedial measures in existing dwellings*[172]).

Improvements to the natural ventilation rates of the building above floor level may also be possible and worthwhile. The benefits of such improvements should be measured before resorting to the installation of mechanical means of ventilation. There are too many buildings where radon sumps have been installed unnecessarily.

For solid concrete floors in dwellings where radon levels are above 500 Bq/m^3, a sump is by far the best option. The purpose of a sump is to reverse the air pressure between the ground under the floor and the interior of the building. Essentially it consists of a hole in the ground beneath the floor slab, linked by pipework to the outside. Suction is applied by an electric fan in the pipeline to provide a negative pressure beneath the floor in relation to the interior and to draw out radon laden air (Figure 3.4). The main problem in an existing building is where to site the sump so that the associated pipework can be easily

Figure 3.4
Radon sumps being installed under the ground floor of a new house

Case study

Reduction of high radon levels in housing

BRE were called in to advise on remedial work for a terrace of five houses having high radon levels. The most successful solution for reduction of radon in existing dwellings, which can be used in buildings with either solid floors or suspended floors where the soil below is capped with concrete, involves depressurising the soil immediately beneath the dwelling; this is known as subslab depressurisation or, more commonly, a sump system. It has been found that buildings with quite large floor areas can be remedied using sump systems, with several sumps manifolded to a single fan, often using the same size of fan as would be used for much smaller buildings. This suggested that it might be possible to deal with more than one dwelling with a single sump system provided that administrative and maintenance arrangements could be resolved. In this case, a successful communal system, using subfloor depressurisation from a single sump below the slab, was recommended and installed.

hidden. The results of tests carried out by BRE suggest that the best place for a single sump is close to the centre of a building. However, if a central spine wall has deep footings, a single sump on one side may not be effective and two sumps, one on each side of the spine wall, may be needed. These can be connected by pipework and taken to a single fan.

The type of fill is also crucial to performance. Even where 'clean stone' is used as the fill, that is to say without too much fines, the efficiency is sometimes very poor. A higher permeability fill helps the sump to work more effectively. The width of a shrinkage gap between the slab and wall has negligible effect on the flows in the gap only if it is less than 1 mm wide.

In remedial work, where the soil resistance is considerable compared to the crack resistance, sealing cracks has little or no effect on the rate of soil gas entry into a building unless all of the cracks are sealed.

Services

Penetrations by services through the main area of a slab should have been reduced to a minimum during the construction of the building. Drainage and service ducts, and service entries, will be found mainly to be confined to edge strips to avoid conflict with laying the slabs, especially with large buildings where continuity of operation is important. Innovatory design details, such as the adoption of a solid floor slab in which the access for services to housing was provided by a single service entry box, were shown by BRE to be well worthwhile, and may occasionally be found (Figure 3.5).

Main performance requirements and defects

Choice of materials for structure

Hardcore

Three types of failure can occur to groundbearing slabs from the use of unsuitable hardcore material, or from poor or unsatisfactory techniques in installing it. The three types are:

- inadequate compaction of the hardcore leading to settlement of the overlying slab
- expansion and loss of strength of the concrete base caused by soluble sulfate attack on the concrete, leading to expansion, cracking and heave of the slab
- expansion of the hardcore material leading to uplift and cracking of the slab

These are dealt with in this chapter as follows:

- inadequate compaction – in the section on strength and stability
- sulfate attack – in the section on durability
- expansion of hardcore – under durability

Concretes

A great variety of concrete mixes will have been used in the past. As noted in Chapter 2.2, the constituent materials may have been of poor quality, the concrete may have been subjected to attack by sulfates migrating upwards from below, coke breeze may have been used in the aggregates, and many slabs will have been laid without adequate compaction.

Where it is necessary to remove and replace the slab, and there is no need to incorporate reinforcement, one of the lower strength grades will normally be sufficient for this

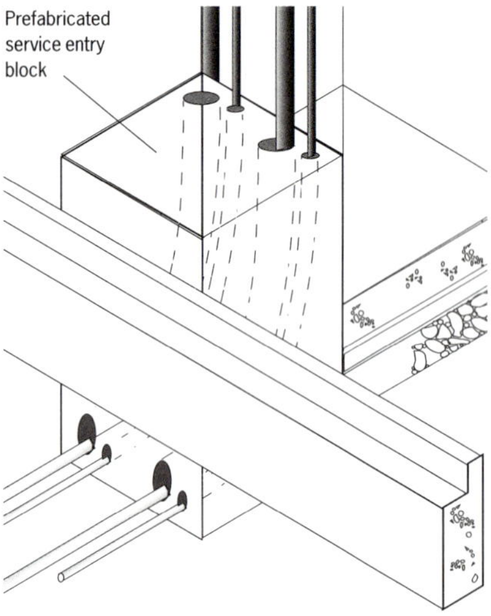

Figure 3.5
A precast concrete block with provision for service entry points

application in groundbearing slabs. BRE Digests 325[173] and 326[174], and BS 8204:Part 1[159] describe the requirements in detail.

Strength and stability

The stability of a solid groundbearing floor depends on the thickness and strength of the concrete (or other flooring such as stone flags), and the support offered by the fill or ground underneath. Problems have sometimes occurred as a result of insufficient compaction of hardcore. The risk of settlement is greatest where large depths of fill (more than 600 mm) have been used under the floor. Settlement can occur also in localised deeper levels of fill over foundation or drain trenches.

However, floors may have been designed to span between walls in these situations. National House Building Council and housing association specifications limit the amount of fill that can be used in such cases to 600 mm. Nevertheless, fill in excess of this depth is still being used in 1 in 20 dwellings according to Housing Association Property Mutual (HAPM).

Fill and hardcore should be compacted in layers not exceeding 225 mm before compaction. BRE Digests 274[175], 275[176] and 276[177] describe the selection and behaviour of fill beneath groundbearing slabs. BRE Digest 251[178] describes the effects of heave and settlement. (Clay heave is discussed later in this chapter.)

Dimensional stability, deflections etc

Coefficient of linear thermal expansion per °C:

- gravel aggregate 12 to 14×10^{-6}
- limestone aggregate 7 to 8×10^{-6}
- lightweight aggregate 8 to 12×10^{-6}

Reversible moisture movement:

- gravel aggregate 0.02–0.06%
- limestone aggregate 0.02–0.03%
- lightweight aggregate 0.03–0.06%

Stability of the concrete slab has been dealt with in Chapter 2.2. However, since in this category of floors the slab rests on hardcore, the hardcore must be adequately compacted to provide the necessary support. In addition, where the floor is built on a clay site, the effects of clay subsoil heave and shrinkage must be taken into account.

Inadequate compaction of hardcore

Settlement of concrete floor slabs directly supported by the ground is a structural defect commonly met with in house floors (Figure 3.6), though less so in the floors of other building types. It is caused by further consolidation of the hardcore after earlier inadequate ramming; or, less frequently, by settlement of the ground beneath after building operations are complete. Hardcore may consolidate if:

- it has been inadequately compacted
- the depth is too great
- materials which can degrade have been included

Loss of support over part or all of a floor area can lead to gaps appearing between skirting boards and the floor, cracks forming in the concrete slab and disruption to partition walls built off the floor. Sites where a considerable depth of hardcore has been used are particularly vulnerable. Where the depth of hardcore varies greatly, it is common for settlement to be uneven. Difficulties can also be encountered at edges adjacent to walls where deep trenches have been dug to construct the foundations, even though the average depth of hardcore is small.

Diagnosis

The first sign that settlement is occurring is that a gap appears between the floor and skirting board. The gap progressively becomes wider, and perhaps uneven. Cracking in the floor slab and partition walls may follow. The amount of settlement can be variable: from a few millimetres up the several tens of millimetres in the worst cases. As the hardcore progressively consolidates, it is usual for the rate of settlement to decrease with time.

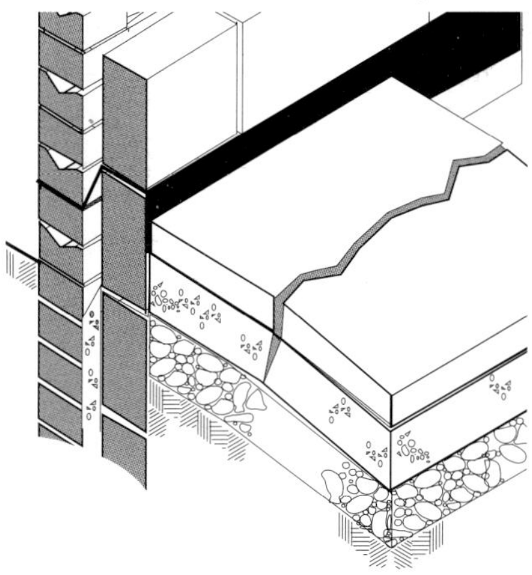

Figure 3.6
Settlement of hardcore under a ground supported slab could lead to cracking of the slab and flooring

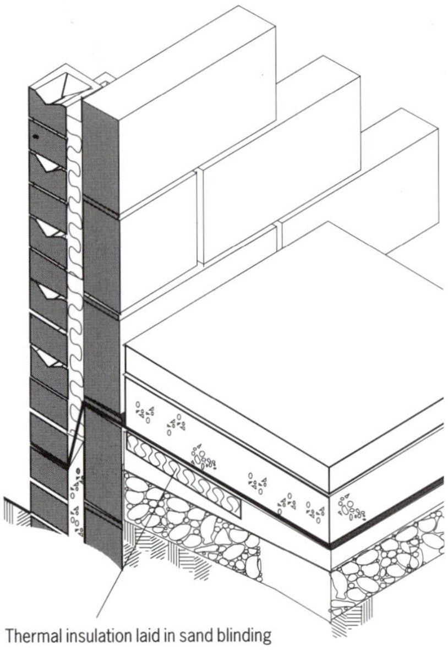

Figure 3.7
Thermal insulation laid in sand blinding

Figure 3.8
The deterioration of this floor indicates that the DPM (if one exists) has completely failed

Repairs and reinstatement

In the worst cases the concrete slab will have to be broken out. The hardcore should then be taken out and examined for degradable material. Provided none exists, the hardcore can be replaced in layers, making sure that each layer is adequately compacted before continuing with the next. Where the depth of hardcore exceeds 600 mm, consideration should be given to replacing the floor with a suspended one.

Some slabs of low rise buildings have been stabilised and repaired using mini-piles (BRE Digest 313[179]). These piles are installed by a variety of contractors, some of whom may not be familiar with the principles of foundation behaviour or the particular problems of pile design and construction.

In cases where settlement has been minor to moderate, and there is no cracking of the slab and partition walls are unaffected, levels can be made up by applying thin screeds or levelling compounds. As settlement decreases with time, this method of repair is particularly relevant for properties more than, say, 15 years old where further movement is likely to be limited.

Effects of clay heave

Heave occurs when clay, which has dried down and shrunk (eg by tree roots drawing water from the ground), returns to its original volume on removal of the cause of the shrinkage (eg the tree). This expansion can be rapid, and perhaps destructive. Information from HAPM indicates that, of sites inspected which were at risk from clay heave, around 2 in 5 did not have adequate precautions.

Two common methods for attempting to provide for movements in the clay subsoil beneath a groundbearing slab are:

- a compressible layer between the slab and the hardcore
- reinforcement in the slab

Neither of these solutions is without risk of failure since rise or fall of the slab will be reflected, for instance, by crushing of finishes and sticking of doors, or by gaps opening up under the skirtings. It is probably better on such sites to select a suspended floor with a suitably sized void underneath.

Apart from the hardcore and the subsoil, the parts of the floor most vulnerable to movements are of course the insulation and decks which are laid over it. The material in common use is chipboard which should comply with BS 5669: Part 2[180] Type C4(M) when laid over a solid floor. This is dealt with further in Chapter 8.5.

Thermal properties and ventilation

Most slabs laid before 1980 will incorporate no thermal insulation whatsoever. Although, in theory, insulation can be retrofitted above a slab which has no screed, in practice there are so many difficulties to overcome such as internal doors, staircases, fixtures and fittings, not to mention sanitary fittings and plumbing, that it is totally uneconomic to do so. Fortunately, as *Principles of modern building*[6] noted, floors of hardcore and concrete laid on the ground do not lose heat in the same way as walls and roofs. The ground acts to some extent as an accumulator of heat, and in fact most of the heat loss occurs within a short distance of the edges adjacent to the external walls. Consequently the average heat flow is dependent on the size and shape of the floor[181]. This topic is described in general terms in Chapter 1.3.

Where the existing slab has a screed, however, and thermal insulation is being considered, it may be possible to remove the screed and fit both thermal insulation and a new floor finish within the available depth. In these circumstances it is almost certain that a levelling layer will be needed over the existing slab. There may be other risks involved in this operation, however, and these are fully discussed in *Thermal insulation: avoiding risks*[31].

In a survey carried out on the buildability of insulated house floors, observations were made of insulation placed above subfloors and beneath chipboard, and above subfloors and beneath a screed. It was concluded that most insulation operations were potentially straightforward, though defects occurred from lack of communication of design intention and inadequate supervision. Defects included missing insulation (varying from 1 to 5%) which created potential cold bridges, and inadequate movement joints for chipboard decks (see Chapter 8.5).

Surveyors may come across floors with edge insulation, built since around 1985. This method can be particularly useful in reducing thermal bridging at the floor edge. It can take two forms:

- vertically inside, outside or within the cavity of the external wall
- horizontally around the perimeter of the floor underneath the edge of the slab adjacent to the wall (Figure 3.7)

The first of these is appropriate to this section and the second to Chapter 3.2

BRE site inspections have shown many cases where edge insulation to an in situ slab has risen as the wet concrete has been poured, creating a cold bridge below the edge insulation.

The U value of a floor with edge insulation is obtained from the U value of the same floor without edge insulation, minus a value for the insulation. The procedure for calculating this U value is given in BRE Information Paper IP 7/93[182].

Where the wall below a solid floor is composed of low density blocks, the wall can provide a useful reduction in heat loss. A method for calculating the U value of ground floors contained by low density block walls is given in *Masonry International*[183]

Control of dampness and condensation

Rising damp

Existing floors without a DPM may perform satisfactorily in combination with a moisture permeable finish and a 'draughty' internal environment (ie the building has a degree of natural ventilation, intentionally or unintentionally). However, reduction of ventilation rates in refurbished dwellings and, particularly, the addition of a moisture sensitive or impervious flooring may result in rising damp becoming more apparent, and therefore demonstrate the need to incorporate a DPM. Floors with slabs of less than 100 mm thickness will be more vulnerable to rising damp, particularly if the concrete is of low quality. Floors where the DPM consists of asphalt laid above the structural floor as a screed are vulnerable to damage by building works and might require specialist repair or alteration. It may be advisable to allow for the protection of asphalt DPMs during rehabilitation work. Where alterations are being considered, how continuity between DPMs and DPCs will be achieved should be assessed.

The following points have often come to the attention of BRE staff on site visits.

- Dampness can penetrate a floor slab where an effective DPM has been omitted (Figure 3.8).
- Localised dampness may result from inadequate laps, punctures in the membrane and discontinuities at service pipes.
- Brush applied membranes may have inadequate thickness or missed areas.
- Dampness at perimeters can indicate inadequate linking of the floor DPM with the wall DPC – this should not be confused with condensation due to cold bridging (Figures 3.9 and 3.10).
- Asphalt screeds may have cracked or be discontinuous where previous works have removed fire hearths or partitions, or provided service routes.

Excess construction moisture

Excess constructional water in concrete bases and screeds must be allowed to dry out before moisture sensitive floorings (carpets, PVC, timber etc) can be laid. A sand and cement screed 50 mm thick laid on

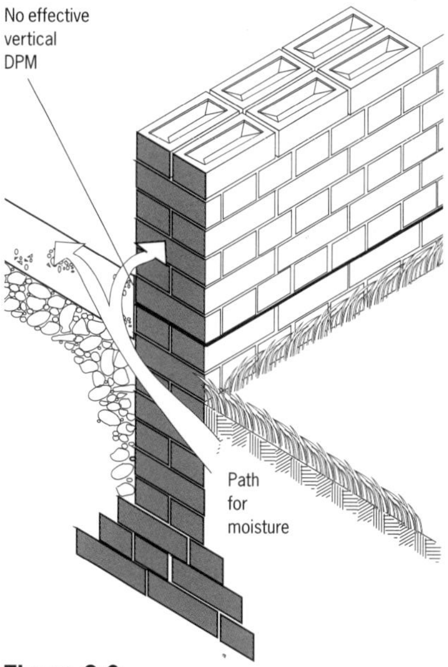

Figure 3.9
An effective vertical DPM is absent at the edge of the slab

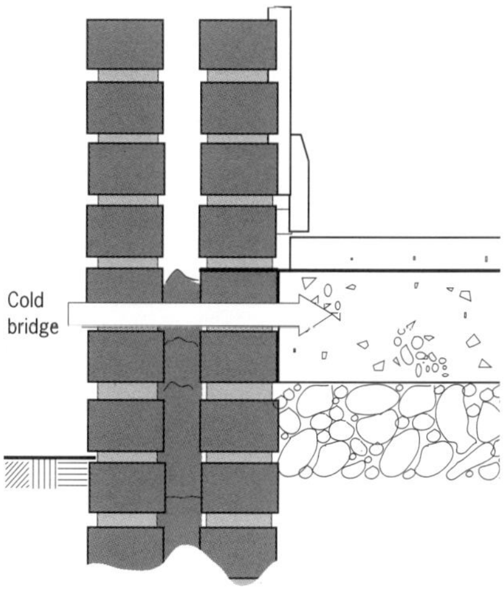

Figure 3.10
A cold bridge at the perimeter of the building

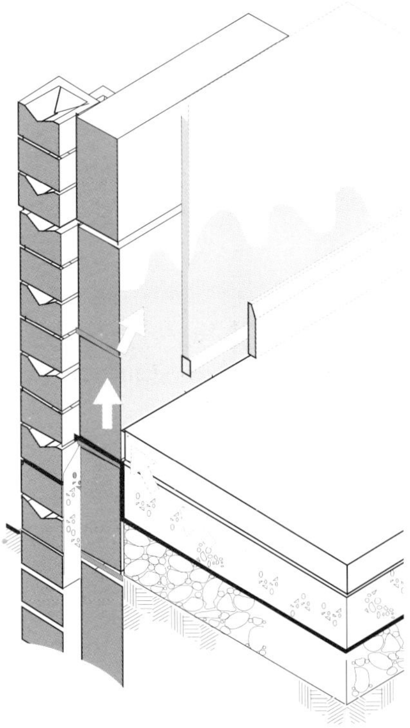

Figure 3.11
Construction water from the slab can seem like rising damp from the ground. Not only can it affect the flooring, it may also affect the wall

a DPM generally needs to be left about six weeks, but thicker slabs require much longer times (Figure 3.11 and Table 1.3 in Chapter 1.4).

Proprietary screed systems are available which dry considerably more quickly than conventional screeds, but these are of little value if laid onto concrete bases which are still wet. Because it is rarely possible to allow sufficient time for concrete bases and screeds to dry to a state which will not induce moisture movement in moisture susceptible decks such as chipboard, a vapour control layer should be placed between the base and the chipboard. Where insulation is interposed between chipboard and the base this vapour control layer is preferably placed on top of the insulation (ie on the warm side). With some prefabricated panels, where the chipboard and insulation are stuck together at the factory, this is not possible and in these cases the vapour control layer should be placed below the panel.

The vapour control layer used should be polyethylene of not less than 125 μm (500 gauge) and preferably of 250 μm (1000 gauge). For bases laid directly on the ground the vapour control layer is in addition to the DPM.

Exclusion of rising damp

In older buildings, many of which will be found without DPCs in external walls, rising damp can occur very easily. Fortunately many of these buildings have suspended floors, only the perimeters of which will be at risk. BS CP 102 was for many years the authoritative source of information on construction standards on this topic, and even now has only been partially replaced by BS 8102[(184)] and BS 8215[(185)].

Existing dwellings being rehabilitated will often have solid floors without DPMs. Many houses built between 1950 and 1966 had floors which were finished with thermoplastic or vinyl asbestos tiles stuck down with a bitumen adhesive. This system was moderately tolerant of damp conditions and as a result it was common not to provide a DPM in the base. When this type of flooring is removed, an assessment should be made of the moisture condition of the base as a DPM may be required before laying moisture sensitive flooring.

Moisture sensitive materials such as chipboard and other timber products, flexible PVC, linoleum and cork tiles, should be laid only on a floor which has a satisfactory DPM. Faults identified during rehabilitation work have included failure to provide a DPM, even in vulnerable situations, and the lack of satisfactory linking of DPMs with DPCs.

Where timber ground floors have been replaced by solid floors, BRE staff have been called in to investigate rising damp showing on the external, usually solid, wall (Figure 3.12).

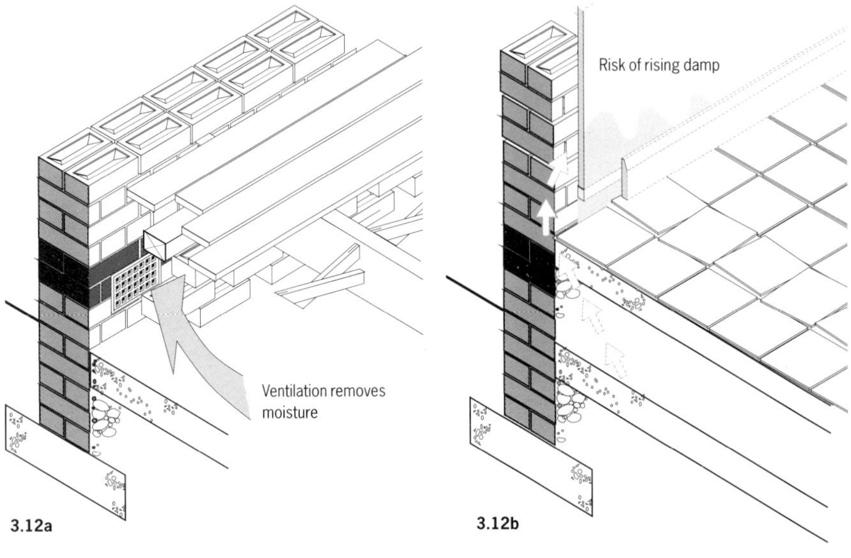

3.12a 3.12b

Figure 3.12
There is a risk of rising damp where a suspended floor (3.12a) is replaced by a solid floor (3.12b), unless an adequate DPM is installed and lapped with the DPC

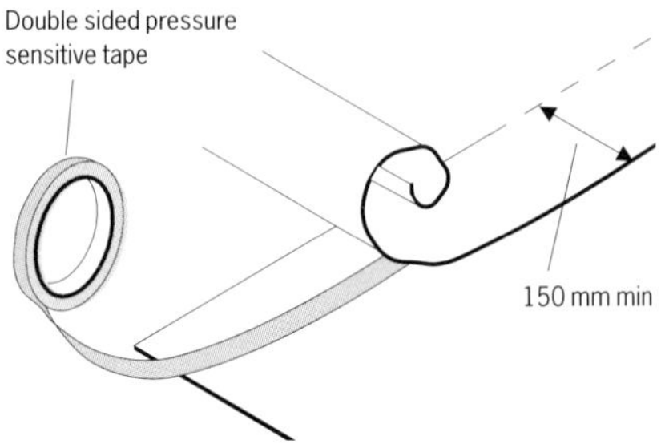

Figure 3.13
Sticking the overlap of a DPM

DPMs must be continuous, whether above or below the concrete base, and be linked to the DPC in the walls. Polyethylene below the base should be at least 300 μm (1200 gauge), or 250 μm (1000 gauge) if the product has a British Board of Agrément certificate or is to the Packaging and Industrial Films Association standard. Where a sheet material DPM is being laid below a replacement floor slab, the joints between sheets must present an impervious barrier to water entering the slab. The preferred method of forming the joint is to overlap the sheets by at least 150 mm and stick the joint with double-sided pressure sensitive tape (Figure 3.13).

If welts are to be used to join sheets, they should be constructed in a three stage operation (Figure 3.14). The welt should be held flat until the slab or screed is placed.

Service entry points are sometimes a source of penetration by moisture (Figure 3.15). The problem, however, occurs mainly in buildings built on ground having high water tables or where the hardcore is persistently damp. When replacing slabs in such locations, it is worth taking trouble with the dampproofing of service entry points (Figure 3.16).

Liquid membranes can be applied to the top of the base by brush or by hot pouring They should provide a minimum thickness when dry of 0.6 mm. Care must be taken in selecting these DPM materials as some of the solvents may not be compatible with foamed plastics insulants. In any event, sufficient time must be allowed for any solvent to evaporate before covering. Also to be considered is the possibility of poor workmanship with brush applied materials. These types of DPM must be covered with a sand and cement screed not less than 50 mm thick.

Yet another alternative is to provide a waterproof flooring such as mastic asphalt to BS 6925. This should be laid in accordance with BS 8204:Part 5[56] and will add around 20 mm to the finished floor level.

The use of sandwich membranes is restricted to situations where the whole floor is being replaced. In these cases it will be beneficial, given adequate headroom, to install thermal insulation below the slab. See also the same section in Chapter 3.2.

Surface DPMs based on proprietary epoxy resin systems have been available since around 1965. They have a good track record but have only gained widespread acceptance since their original high cost fell to more economic levels. They are rarely considered for new work, but are very useful for renovation and change of use where no DPMs were previously present. They can be applied to clean surfaces of concrete or screed and should be covered by a 3 mm minimum layer of latex levelling compound. These surface DPMs are also appropriate for controlling excess constructional water in thick constructions for which it would be uneconomic to wait for them to dry out.

Case study

Rising damp caused by damage to an unprotected DPM

Dampness in the floor of the main hall at a conference centre was found to be associated with water penetration from an area below ground level. Investigation indicated that the DPM was defective. The most likely cause was mechanical damage during construction due to the omission of a protective screed over the DPM. Drying shrinkage of the concrete floor, disrupting the membrane at internal angles, had been a contributory factor.

The solution was to make the hall watertight by applying an internal DPM to the floor and walls. Additional rubble filled soakaways had to be provided externally to prevent flooding of adjacent ground. Weepers were installed in the walls, in the course of blockwork directly above the stepped DPC, to prevent any buildup of moisture in the wall adjacent to the floor.

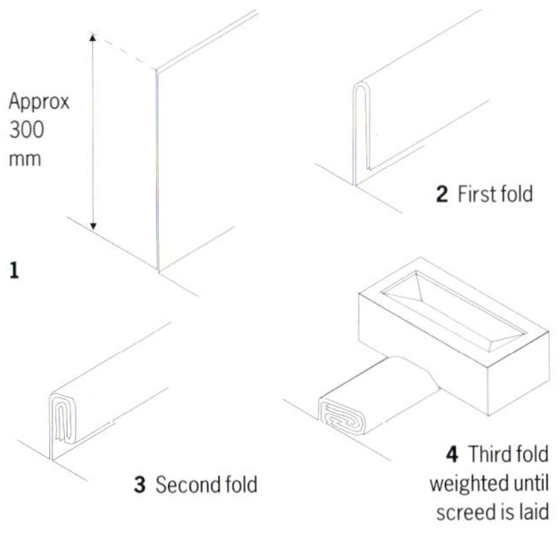

Figure 3.14
Forming a welt in a DPM (in four stages)

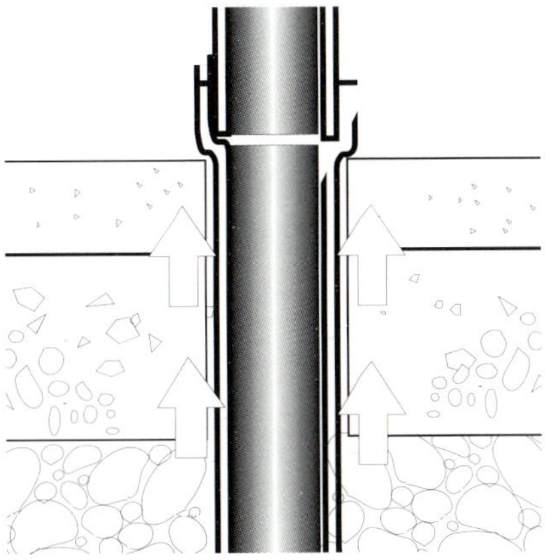

Figure 3.15
Service installations are the most likely points at which rising damp will show

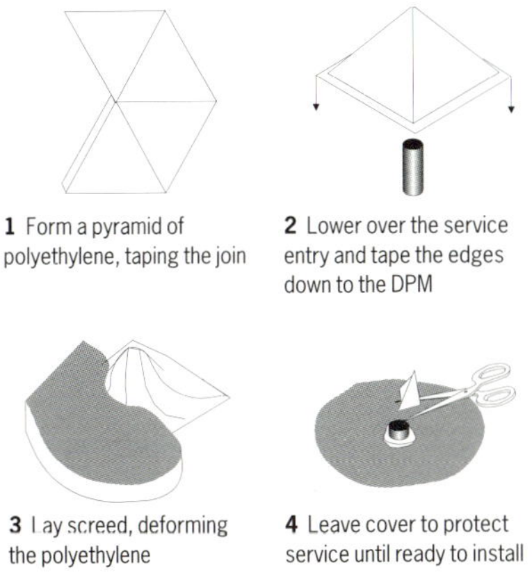

Figure 3.16
One method of protecting a service entry point (in four stages)

Figure 3.17
The floors of large concourses open to the weather often suffer from condensation

Condensation

Solid ground floors are at risk of condensation at their perimeters. Many floors have a cold bridge at the floor edge due to its proximity to low external temperatures and the fact that dense concrete has poor insulating properties. Raft construction is particularly at risk. See also the same section in Chapter 3.3.

Although *Principles of modern building* suggested that the incidence of condensation at the edges of slabs was low, it did point out a need for edge insulation with certain kinds of detailing where the external wall was thin, such as with light cladding or curtain walling. This is still true.

A floor perimeter is likely to suffer surface condensation where the slab edge is exposed to the exterior, abuts a solid exterior wall or abuts the outside leaf in a cavity wall. Depending on the detailing, condensation problems may occur locally below door thresholds, full height windows or infill panels. It can also occur if there is a large accumulation of mortar droppings in a cavity wall which allows cold bridging across the base of the wall cavity.

A number of cases have been reported to BRE where condensation has occurred on a floor slab when a period of cold weather has been followed by a warm front. Users of the main railway stations will be aware of this phenomenon through the appearance of warning notices of slippery surfaces (Figure 3.17). The condensation is caused by moist air in warm fronts coming into contact with the concrete surfaces of station concourses which have not had time to warm up after cold spells. The condition has also been evident adjacent to thresholds in domestic situations where carpets can become quite wet. It is also a common problem in warehouses and other unheated buildings, though it does tend to be transient.

With solid floor construction, unless it is necessary to remove the slab for reasons other than lack thermal insulation and resultant condensation on the slab at the perimeters of the building, the only practical remedy is to add a layer of insulation over the slab. The decision on whether this is worthwhile must be taken in the light of all the other work on skirtings, floor finishes, doors and staircases which might become necessary to make good.

Where the floor finishes and screed can be taken up, however, it may be possible to fit in a thin layer of thermal insulation within the existing depth available using, for example, a board finish instead of a screed. Figures 3.18 and 3.19 indicate acceptable forms of construction.

Alternatively, it may be possible

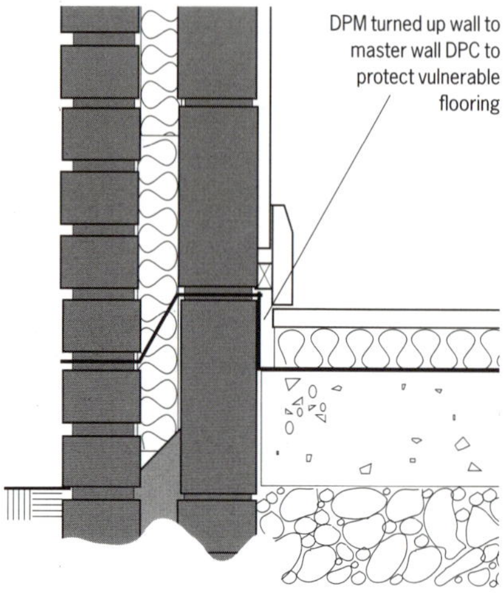

Figure 3.18
Protecting the perimeter of moisture sensitive flooring

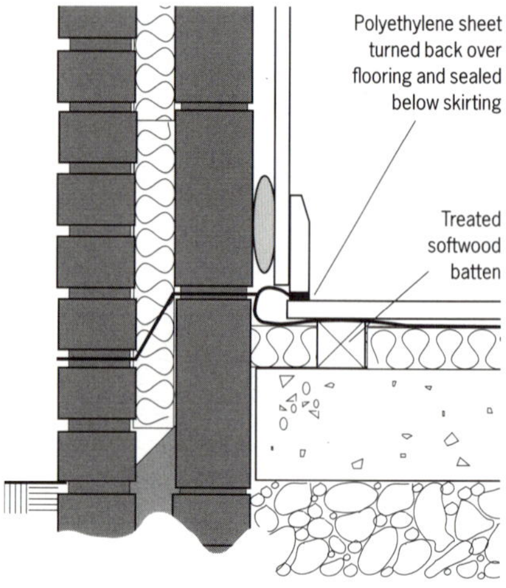

Figure 3.19
An alternative method of protecting moisture sensitive flooring

to replace the screeds with one of the flowing screeds. These are based on gypsum or cement plus additives which can be laid in much thinner layers on insulation than can traditional sand and cement screeds; a thin layer of thermal insulation is then put down before replacing the flooring.

Spillages and leaks

Spillages of water from industrial or other processes, and leaks from embedded water pipes should not be discounted as sources of dampness in slabs – indeed, these sources should be amongst the first to be investigated in such cases. Embedded water pipes should occur relatively infrequently in buildings built in the last quarter of the twentieth century, since Water Byelaws now require pipes to be laid in ducts with removable covers.

Protection from excessive heat and fire

Groundbearing floors are not normally required to possess any specific degree of behaviour in fire conditions. However, high temperatures are often reached under boilers; cracking of the slab (perhaps from drying shrinkage of the concrete), extending radially from the boiler, will often result if the original design and materials specification made no provision for the heat gain.

Boiler heat can also affect the ground underneath the slab, giving rise to drying out and possible shrinkage, and leading to loss of support to the slab. The design of boiler bases is a specialist matter and, in general, requires special protection to both slab and supports.

Concrete which is subjected to high temperatures begins to lose strength at about 300 °C, and will have lost half its strength at 600 °C. In concretes made with common siliceous aggregates, a pink or red colour is evidence of temperatures between 300 and 600 °C. The colour changes to grey above 600 °C and the concrete becomes friable.

Sound insulation

The lowest floor of a building, if carried on the ground, normally does not present problems of sound transmission, although vibrations can be transmitted from one part of the building to another if precautions have not been taken. Auditoria may need to be isolated by means of resilient mountings sandwiched between false floors and groundbearing slabs.

Figure 3.20
Cracking in a concrete floor caused by sulfate attack

Durability

Solid concrete groundbearing floors should not deteriorate if they were originally well constructed with adequate thickness, mix and compaction, and if laid on a prepared base of graded, compacted and inert hardcore. Solid floors of flags or quarries may not be so reliable if laid directly onto prepared ground without a sound concrete bed. Where concrete ground floors are designed to span between walls rather than 'rest' on the ground beneath, the durability and location of the necessary steel reinforcement will be critical to the life of the slab.

Soluble sulfates leading to sulfate attack

Sulfate attack can occur on the underside of a concrete groundbearing slab when fill below the slab contains sulfate salts and the slab is not isolated from the fill by a DPM.

It has long been known that water soluble sulfates from hardcore materials could cause disruption to concrete floors laid directly over them. In damp or wet conditions, sulfates can migrate to the underside of the slab and react with the tricalcium aluminate found in Portland cement to form ettringite. The reaction produces expansion in that part of the slab in contact with the fill. This reaction is expansive within the concrete. The first visible signs of sulfate attack are usually some unevenness in the floor, followed by cracking and possible heave (Figure 3.20). The upper part of the slab is put in tension leading to a map pattern of cracking. Containment of the slab at the abutments forces the slab to distort into a domed shape which, with time, can become quite large (Figure 3.21 on page 144).

In the worst cases the walls bounding the slab can be pushed out. This is visible as walls over-sailing the DPC or as disruption to the masonry (Figure 3.22 on page 144). Where the wall is of cavity construction, the outward movement of the concrete slab can push the inner leaf towards the outer

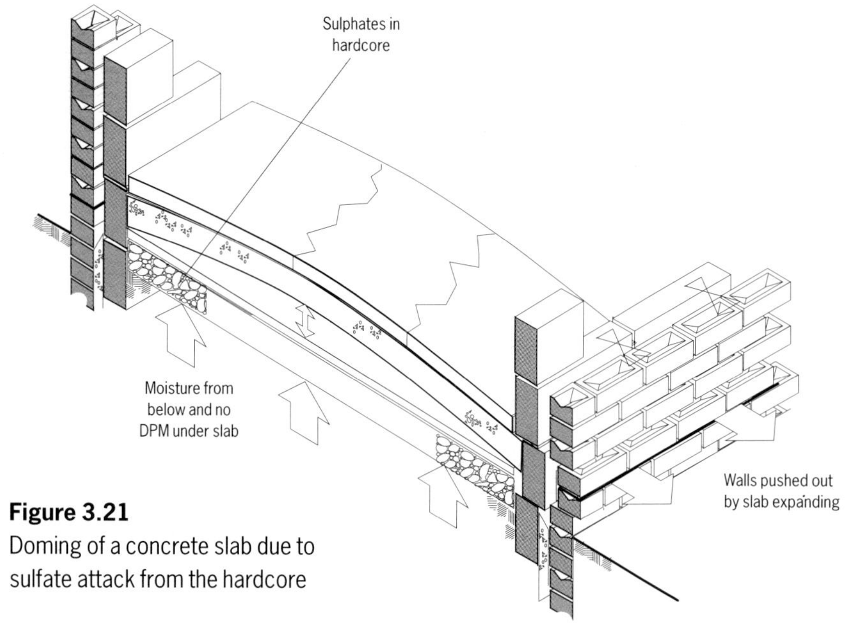

Figure 3.21
Doming of a concrete slab due to sulfate attack from the hardcore

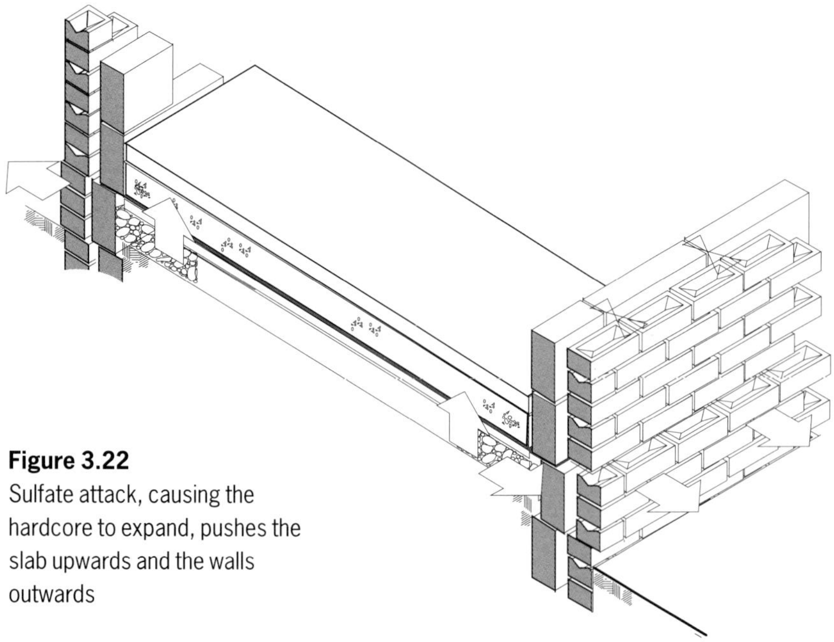

Figure 3.22
Sulfate attack, causing the hardcore to expand, pushes the slab upwards and the walls outwards

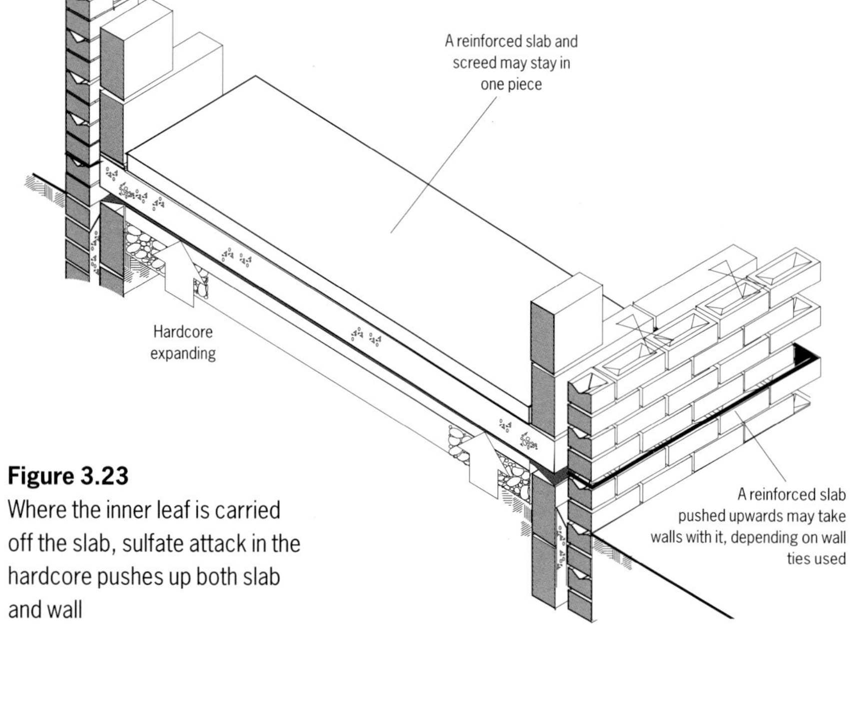

Figure 3.23
Where the inner leaf is carried off the slab, sulfate attack in the hardcore pushes up both slab and wall

leaf without necessarily moving the latter. Alternatively, where the inner leaf is built off the slab, it can be pushed upwards (Figure 3.23).

A further indication of the problem is that an attack may be accompanied by efflorescence on the outer face of a wall which has mortar filled cavities below DPC level. BRE Digest 363[186] provides design guidance where concrete may be subjected to attack by sulfates. Even a thin blinding layer of ash or clinker can contain sufficient impurities to induce problems.

In looking for factors determining the rate of sulfate attack, the following are important:

- the type of sulfate – calcium sulfate is less soluble than magnesium, sodium or potassium sulfate, and will take longer to produce a significant defect
- the quality of the concrete
- the presence of an impervious flooring
- the presence of water

The second and third of these factors will affect not only the rate at which moisture containing sulfate solution can be drawn into the slab by evaporation from the surface, but, in combination with a porous concrete structure, will allow ionic diffusion of sulfates into the concrete where they can react. The existence of a sandwich DPM or an impervious flooring which prevents moisture evaporation from the surface can reduce the rate of sulfate attack since the rate of ingress of sulfate into the concrete is slow.

It is important that the effects of sulfate attack on the slab are not confused with sulfate attack or moisture expansion of the brickwork.

In all cases of sulfate attack, water must be present. Therefore the level of the local water table and the mobility of groundwater can be important (Figure 3.24). A 'dry' site where the depth of the water table is difficult to find in any season is unlikely to give rise to significant chemical attack on concrete placed on it. It is not uncommon for floors in houses at the bottom of a hill to be affected by sulfate attack while those higher up the hill and of similar construction and materials specification remain defect free.

Diagnosis of sulfate attack

Sulfate attack can be suspected if progressive lifting of the floor leads to increasing difficulty in using internal doors caused by the lower edge fouling the floor. Further lifting leads to doming and cracking. In domestic construction, lifting is generally highest in the centre of the room. The presence of doming in the floor, detectable with an adequately long straight edge, is characteristic of sulfate attack and therefore distinguishable from the effects of heave in clay soils or physical expansion of fill material. In domestic cellular construction, the pushing out of walls which contain the slab is also characteristic. A hole broken or, preferably, cored through the slab will often reveal that it has arched out of contact with the hardcore. The lower part of the slab thickness is likely to show white crystalline deposits which are the products of the chemical reaction. If the original concrete was not very strong, there may be sufficient loss of strength at the bottom of the slab for it to be broken with the fingers. In the worst cases it may have deteriorated to a flaky consistency.

If a cracked solid ground floor slab is found in a region where colliery shale fill is likely to have been used, sulfate attack can be suspected. Most of the cases, though, where an attack could occur will have done so by now and will have been remedied already. Sulfate attack may also be caused by the use of pyritic quarry waste or of rubble contaminated by sulfates. Failures resulting from brick rubble containing gypsum plasters, and from brick chimneys where sulfates from coal combustion have accumulated, have also been recorded.

It is possible for soluble sulfates to be contained in the ground beneath hardcore and therefore, in theory, become a problem, but in practice this has rarely been found to be the case with concrete floors.

A floor slab with a pitch mastic or asphalt mastic floor finish is most unlikely to have been provided with a DPM between fill and slab, so that any sulfate-bearing fill would be in direct contact with its underside.

If conclusive proof of sulfate attack is necessary, samples can be taken for examination in a specialist laboratory for the presence of ettringite and thaumasite.

To produce ettringite (which is calcium sulfoaluminate), the reaction requires the presence of all the following three constituents:

- soluble sulfate salts
- cement which is vulnerable to sulfate attack
- persistent wetness

To produce thaumasite(calcium silicate carbonate sulfate hydrate), the following two further conditions must also coexist:

- a source of carbonate (eg from the concrete itself)
- low temperatures of 5 to 15 °C

The reaction producing thaumasite can proceed very rapidly, and the use of sulfate resisting Portland cement or good quality concrete cannot be guaranteed to prevent it. In floor slabs, the mode of failure will be similar to that when ettringite is formed.

Predicting sulfate attack

It is difficult to give precise advice as to whether a building, which is known or suspected of having sulfate-containing hardcore but

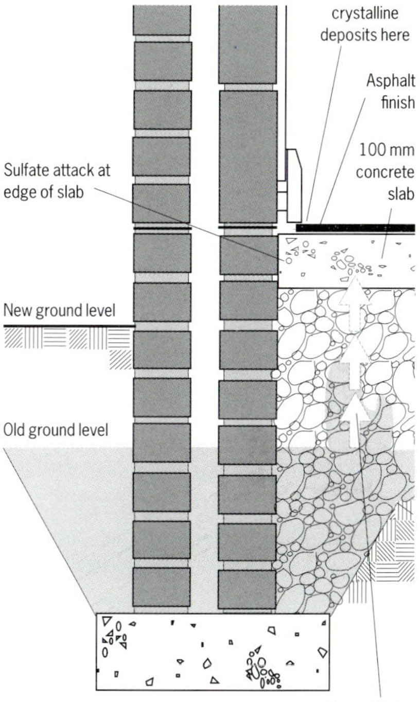

Figure 3.24
Detail of ground floor construction which resulted in sulfate attack

Case study

Sulfate attack in substrates

BRE investigated a number of cases of sulfate attack where solid floors covered with mastic asphalt for 20 to 30 years gave no problems, but, within a few months of the asphalt being removed, the slabs cracked and expanded. The investigators concluded that removing the impermeable asphalt had allowed water to evaporate from the slab, drawing sulfate solution from the hardcore into the base of the concrete.

Case study

Expansion of shale fill under slabs in housing

The cause of floor heave in some houses in the north of England was thought to be a volume increase in the fill associated with oxidation of pyritic shales and the resulting crystallisation of gypsum. Sulfate attack on concrete in contact with the shale might have been a contributing factor. The particular shales causing the trouble were also likely to be intermixed with alum shales. The factors necessary for a shale to be expansive are for it to contain an appreciable percentage of pyrite in a finely divided form and for there to be sufficient calcite present to form gypsum with the sulfuric acid liberated in the oxidation process. The best way of identifying potentially troublesome shales is by chemical analysis combined with an inspection of the source of the shale.

without a membrane between it and the concrete, will be subjected to sulfate attack in the future. For a building more than say 20 years old, where the slab was cast on sulfate-bearing fill without a membrane and which was in fact undergoing sulfate attack, it is not unreasonable to suggest that there would be some visible signs of the attack. Of course, latent sulfate attack can occur on a hitherto dry site if there is a change in the water table or the site potentially becomes more wet.

Repairs following sulfate attack

If damage is sufficient to justify repair, the slab must be broken out and removed; it should not be used as hardcore under the replacement (or any other) slab. The fill material, however, can remain (with further compaction if necessary) since the new slab should be laid on a DPM over the fill and turned up at the perimeter. Therefore the new slab will be isolated from any sulfates remaining in the fill material. However, the risk of attack on the existing foundations should be assessed. Further measures may be needed to prevent attack in the wall below the DPC.

Alkali aggregate reaction

A further form of deterioration of concrete is alkali aggregate reaction (AAR) which takes place between the alkali present in concrete and some aggregates. However, in order to proceed, the reaction requires the presence of water; hence it is unlikely to be found in floor slabs which have been adequately protected. For a short summary of the occurrence and confirmation of AAR, see *Cracking in buildings*[148]; see also BRE Digest 330[159] and BRE Information Paper IP 16/93[187].

Expansion of hardcore

Although sulfate attack is the most common cause of distortion, cracking and upward movement of concrete floors, a small percentage of problems showing these symptoms have been caused by expansion of the hardcore. In these cases of expansion the strength of the concrete remains unaffected, and sideways expansion of the slab does not occur unless it is accompanied by sulfate attack. However, expansion of the hardcore sideways beneath the floor can cause distress to walls and ground beams below floor level.

Figure 3.25
The vertical rule indicates the position of upwards displacement of a concrete slab laid on steel slag. What appears to be a further crack behind the rule is, in fact, the vertical face of a rebate in the slab

Case study

Expansion of slag fill containing magnesia

BRE Advisory Service was called in to advise on defective concrete floors at two factories which showed extensive cracking and upward displacement. The fill material beneath the floor in both cases was identified as steel slag which was well known to cause expansion and heave of floors when used as fill.

Fill material taken from beneath the floors which had heaved was investigated in the laboratory. It was shown to contain a steel slag with a high level of magnesia. Magnesium oxide and magnesium hydroxide have been found to occur extensively in white material associated with the slag and this white material has been shown by autoclave tests to be a principal cause of expansion. In the cases at the two factories, the presence of free unhydrated magnesium oxide showed the slag to be capable of causing more expansion and this was confirmed by the observation of an expansion of 1.4% in seven days produced by a sample of the slag tested in an expansion cell at 80 °C.

In these particular cases, the laboratory examination of samples of the fill showed that its expansive capacity was not yet exhausted, and removal and replacement was therefore the only viable option.

The following hardcore materials have been known to cause problems due to swelling:

- steel slag or old blastfurnace slag (Figure 3.25)
- old broken concrete
- hardcore containing clay
- hardcore containing pyrites

Steel slag, both fresh and from old slag heaps, can contain free lime (calcium oxide, CaO) and magnesia (magnesium oxide, MgO) which, in the presence of water, can hydrate to the hydroxides. The reaction produces a volume expansion. These oxides of calcium and magnesium are the residues of limestone and dolomite fluxes added during steelmaking. On exposure to moisture and air, they convert to hydroxides and carbonates. These reactions, which proceed with an increase in volume, usually take place slowly so even old slag can contain unreacted material. If unhydrated lime or magnesia is contained in the steel slag used as hardcore, subsequent swelling can disrupt the floor.

Some blastfurnace slags from old slag banks have undergone expansive reactions when used as fill. They may contain excessive amounts of sulfate and dicalcium silicate. Under wet conditions the latter can hydrate and subsequently carbonate; the carbonated slag then reacts under cold conditions with sulfates to form the sulfate-bearing reaction product, thaumasite. The formation of thaumasite is expansive and leads to severe heave of concrete floors. Unfortunately, where floors have been built into the inner leaf, heave of the floors also can cause the walls to be lifted. Modern air cooled blastfurnace slags are not susceptible to attack by sulfates by this mechanism.

Broken concrete used as hardcore has been known to disrupt floors when it has been exposed to sulfate solutions, either from the other materials originating in the hardcore or from groundwater.

Materials containing significant quantities of clay can cause heave if they are dry when placed and subsequently become wet.

Hardcore materials containing sulfides, mainly as iron pyrites (FeS_2), can be oxidised to sulfuric acid in the presence of moisture and air, and contribute to the problems of sulfate attack on the concrete. Also, in the presence of soluble calcium compounds, it is possible for fill material to expand, due to the formation of gypsum crystals ($CaSO_4.2H_2O$). Swelling by this mechanism has caused failures in the Tees-side area where a natural Whitbian shale, used as hardcore, contained all the necessary ingredients. Similar problems have been reported from the Cardiff and Glasgow areas, where vulnerable shales form the natural bedrock. Early indications of problems are similar to sulfate attack which may occur simultaneously with shale heave. Cracking, lifting and hogging of floors will be observed together with movement of the external walls[188].

Red shale

The most widespread reported failures to slabs have involved the use of partially burnt colliery shale (red shale). Red shale was frequently used in construction in the 1950s and 1960s as a fill under floors without a membrane being laid between the hardcore and the concrete.

However, red shale also often contained considerable quantities of soluble sulfates. In order to avoid problems, this material should have been overlaid with a DPM, but the use of polyethylene as a membrane material beneath concrete floors did not become common until the mid-1960s. The first specific reference to damage to concrete floors being caused by the use of colliery shale as a fill or hardcore was given in January 1956 (Building Research Station Digest 84, First series[189]) with a further warning being issued in April 1957 (BRS Digest 97, First series[190]). A much fuller description of the problems is given in BRE Current Paper (Design Series) 30[191]. All these documents indicated that moisture expansion of incompletely burnt colliery shale was a significant part of the problem, but the problem was also compounded by spoil containing high levels of soluble sulfates.

It was later concluded that there was no convincing evidence that colliery spoils in themselves had caused disruption by swelling. Clearly the problem is associated with partially or even well burnt spoil containing high levels of soluble sulfates.

Since 1966, it has been a requirement of national building regulations that no hardcore laid under a concrete floor should contain water soluble sulfates or other deleterious matter in such quantities as to be liable to cause damage to the floor. This has effectively precluded the use of red shale beneath concrete floors since that date.

Diagnosis of expansion of hardcore

The number of cases of disruption of floors caused by swelling of hardcore is quite small. Observed effects are often similar to sulfate attack which may occur simultaneously and be a contributing factor. Identification of the hardcore material is crucial for the correct diagnosis to be made and this will usually require expert help. Chemical analysis and X-ray diffraction can identify sulfates, sulfides and soluble calcium compounds. Steel slag which has given problems will often still contain calcium oxide and magnesium oxide, and implies that the expansive reaction is not yet complete. A test is available for assessing the potential expansion of steel slags.

Fixing floorings to power floated bases

Problems have arisen since the late 1960s in fixing floorings directly to finished concrete surfaces. Before this time, screeds were in almost universal use to provide cementitious bases for floorings, with DPMs sandwiched between base and screed. The screeds were usually dry in about six to eight weeks, and had good absorption for water and solvents used in adhesives.

With the advent of polyethylene sheet as a DPM material, it became possible to provide dampproofing arrangements beneath the slab. In the mid-1990s, around 50% of projects were estimated to follow this practice. The use of power floating or power trowelling techniques for smoothing and enhancing the abrasion resistance of concrete floors became more widespread, with the surface apparently suitable for fixing floorings directly to it. Problems ensued, though, with these directly fixed floorings.

- Long drying times were required to eliminate excess constructional water from thick concrete bases. Examples are known where slabs were over two years old and still not dry (see Table 1.3 in Chapter 1.4).
- Suitable methods were not available for measuring the moisture content of thick slabs. Electrical conductance or

Laying a floor in a large building by means of the long strip method

Standard steel road edge forms are set out along the long axis of the building over strips of polyethylene sheeting approximately 1 m wide, all secured in place by steel stakes driven into the hardcore bed (Figure 3.26). The top edge of the forms must be carefully set out to level as they determine the levelness of the finished slab.

In order to provide dampproofing, and to permit easy expansion and contraction of the slab over the hardcore base, a further layer of polyethylene sheeting is laid between the edge forms and over the polyethylene strips under the road forms.

this method is to permit power trowelling to start within as little as one hour after casting. Otherwise, for passive evaporation, several hours may need to elapse, depending on the weather conditions, before the concrete is sufficiently stiff for the finishing process. Vacuum de-watering also improves the abrasion resistance of the concrete.

With large slabs, it may be necessary to finish the top of the slab to avoid the need for, and cost of, secondary finishing layers such as floor screeds. The floating process, to create a smooth level surface, can start as soon as the slab is sufficiently stiff.

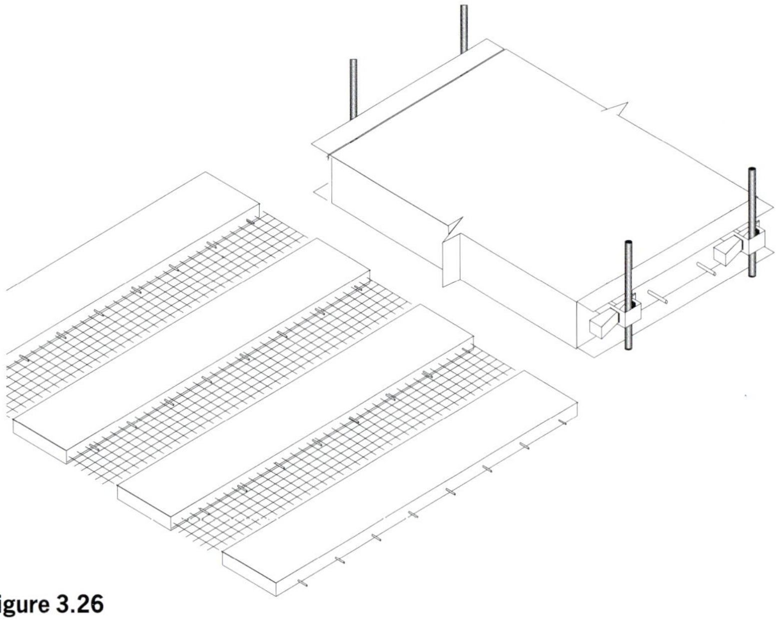

Figure 3.26
Casting a concrete floor slab by the long strip method

As compaction of the long strips requires a vibrating beam screeder, there is the need for working room adjacent to the perimeter walls. Edge strips 900 mm wide are therefore cast first around the perimeter.

The main slabs are cast in alternate strips involving the positioning of crack inducers, longitudinal and transverse continuity reinforcement, and the main reinforcement mesh. Work should be arranged to permit the ready-mix lorry to gain access to that part of the slab furthest from the entrance. If this is difficult to arrange it may be necessary to phase the construction of the external wall so as to permit access for casting the slab. Compaction of the slab is achieved using the vibrating beam screeder, with the addition of poker vibrators at the edges. Usually one strip is laid and finished in a day.

Excess water in the mix can be removed using special de-watering equipment comprising a vacuum pump and fine filters which remove excess water but retain cement in the mix. The main advantage of

Mechanical power floats are usually used. There is no advantage in tamping the surface to a rough ribbed finish since this does not add to bond strength for bonded screeds, and impedes further preparation of the surface where this is required.

Careful curing, by preventing early drying out, is essential to ensure a floor that is hardwearing, and free from cracking, dusting and the effects of drying shrinkage in the surface of the slab. One method which is in common use is to lay plastics sheeting over the slab for seven days. After curing, the edge formwork can be removed, cleaned and reused. Intermediate strips can then be cast.

After the slab has cured, 20–30 mm deep grooves are cut in the top of the slab over the position of the crack inducers in the bottom of the slab. These joints ensure that random cracking does not occur in the slab due to drying shrinkage of the concrete. Should cracks occur, they will form along the line of the grooves. A large disc cutting machine is used to cut the grooves right across the slab.

acetylene bomb methods can give misleading results and are not recommended. The hygrometer method can give reliable results but requires attention to detail and a different technique from that described in BS 8203[35].

- Power trowelling providing a dense impervious surface which hindered absorption of water from adhesives into the concrete.
- The use of spray-on curing membranes resulted in low bond strength of flooring adhesives to the surfaces of concrete floors.
- Movement in joints formed in the concrete base led to defects such as rippling of thin flooring materials.

Many difficulties with thick concrete slabs originating from excess moisture and lack of absorption can be resolved by using suitable surface DPMs, mainly based on solventless epoxy formulations and latex underlays and underlayments. Alternatively, where slabs are dry, the double drop method and pressure sensitive adhesives have been successfully used to overcome poor absorption of very dense surfaces.

Where floorings are to be fixed directly to concrete bases, spray-on curing membranes should not be used. If they have been applied, then it will be necessary to remove them by light shotblasting before laying the flooring. Spray-on membranes also delay drying.

Work on site

Repairs following sulfate attack

If damage to a groundbearing slab from sulfate attack is sufficient to justify repair, the slab must be removed. It is advisable to examine the walls to see if there has been any displacement caused by expansion of the contained slab. However, even if the outer walls have moved slightly (say by up to 10 mm), it is unlikely that stability will have been affected, and usually no action need be taken. More substantial movement will require specialist inspection. Internal partition walls built off the slab may have lifted and cracked, but whether a partition wall needs to be rebuilt depends upon its state.

To prevent a recurrence of the attack and soluble sulfates transferring into the concrete, it is necessary only to provide a polyethylene membrane of 300 μm (1200 gauge) over the hardcore, turning it up at the perimeter. This in theory should prevent ingress of sulfates from the hardcore into the concrete. However, it has become customary to err on the side of caution and to remove the existing hardcore to a depth of 450 mm; then to refill with a sulfate free hardcore before covering with polyethylene and new concrete.

Repairs following other expansive reactions

Like sulfate attack, if damage is sufficient to require repair, the slab will have to be taken out. However, unlike sulfate attack where a sheet of polyethylene can be used to separate the new concrete from the hardcore, the latter will have to be removed also. This procedure will apply to most situations, but there may be rare situations where it can be shown that the expansive process within the hardcore has been exhausted, in which case it can be left in situ.

Storage and handling of materials

See the same section in Chapter 2.2.

Restrictions due to weather conditions

Concrete slabs should not be laid in unheated buildings during the winter unless the concrete can be maintained above freezing. This is normally achieved if the air temperature is above 5 °C for a few days during and after laying. After this time the concrete usually has developed sufficient strength to resist frost. Unless the building is roofed, slabs should not be laid during rain.

See also the same section in Chapter 2.2.

Workmanship

A fuller description of laying a floor in a large building by means of the long strip method is shown in the feature panel on the opposite page.

Accuracy

Standards for flatness and levelness of concrete bases are given in BS 8203[35] and BS 8204:Part 2[192]. The normal standard used for flatness is a maximum of 5 mm deviation from a 3 m long straightedge laid in contact with the finish. This can be reduced to 3 mm where greater accuracy is needed; for example where mechanical stacking in warehouses is to be provided for. There is also a utility standard, where surface flatness is not critical, of ±10 mm. For high racking in warehouses, much more accurate floors are required (The Concrete Society Technical Report No 34[193]).

For normal purposes, for levelness in large floors an overall deviation of ±15 mm is suggested. There are several different methods of specifying and achieving greater accuracy where it is required†.

Preparation for laying new floorings

Failure of replacement floorings on old concrete bases can often be attributed to inadequate preparation of the surface of the concrete before laying the new flooring. The feature panel on page 104 describes suitable techniques for reducing the risk of this problem occurring.

It is important to make sure that any hardcore is not contaminated with old gypsum plasters and that colliery shale is not used.

When re-laying a screed over a layer of thermal insulation, it is good practice to turn up the thermal insulation layer, which is then covered by the skirting (Figure 3.27).

Further aspects of workmanship are covered in BS 8000:Part 2:Section 2.1[152] and Section 2.2[153]. BS 8000:Part 9[194] may also be relevant.

† Specialist advice is available from the British Cement Association or BRE.

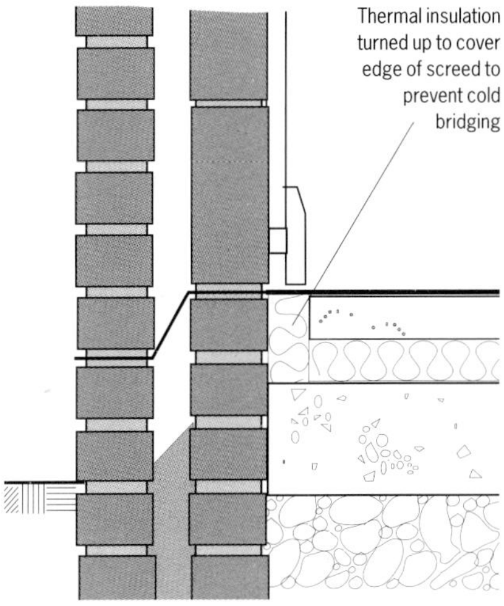

Figure 3.27
Thermal insulation turned up to protect the edge of the screed

Inspection

Bulging, bowing or cracking of slabs and jamming doors could indicate heave, but sulfate attack may also be a possibility. Obvious indications of settlement show as gaps at perimeters of floors, gaps beneath skirtings, and cracked or uneven surfaces. Other signs of settlement include cracking at junctions of loadbearing and slab supported walls. Voids can often be diagnosed by tapping a floor. Confirmation can be obtained by drilling into voids and inspecting with specialist equipment. Floors that appear to be built into walls, apparently spanning the floor area, may not have been reinforced.

Evidence such as puddle staining on the floor, dampness and mould growth beneath floor coverings, and lifting of floor tiles would indicate rising dampness. It is important to differentiate between moisture from condensation (at cold bridging) and rising damp; two problems which require different solutions. Where evidence of moisture is concentrated at perimeters, a check will need to be made for cold bridging. To check the physical presence of a DPM or the quality of the concrete, trial cores may be necessary. The ground around a building will also need to be considered; a poorly drained site will usually increase the risk of rising damp. Signs of dampness resulting from past plumbing leaks and water spillage, and of construction water which has been trapped beneath impervious flooring, sometimes confuse diagnosis.

Surveys may reveal dampness, perhaps mould, in carpets at floor perimeters or where no carpets are present, damp stains on the floor surface. This could indicate condensation, but at doors and windows, rain penetration or condensation run-off from glazing may also be responsible. With timber floor finishes, inspection below the timber could indicate the origin of dampness, but isolated inspection may not be representative. Where evidence of a problem at floor perimeters is found in cavity wall construction it may be necessary to remove a section of the outer leaf to determine the construction and check for cavity bridging. Where there is lifting of floor tiles which extends to the centre of rooms, this is often indicative of rising damp; it may, though, be confused with trapped construction water.

The problems to look for are:

◊ dampness
◊ condensation, especially near the perimeters of floors
◊ rot in timber floor finishes
◊ gaps opening below skirtings
◊ cold bridges at perimeters of floors
◊ concrete friable
◊ doors sticking due to doming of slabs
◊ cracking of slabs
◊ ducts below floors collapsing

Chapter 3.2

Concrete groundbearing floors: insulated below or at the edge of the structure

The main difference between this kind of floor and that described in the previous chapter is that the operational sequence when the building was being built was varied to permit the placing of thermal insulation underneath the slab (Figure 3.28) and, sometimes, vertically at its perimeter.

It is unlikely that these practices were very widespread before the early 1980s except in cold stores. In the absence of working drawings it may be difficult to say with certainty what has been done, although in the worst cases, say of inappropriate compressive strengths of thermal insulation, the fault will become obvious. The presence of edge insulation may sometimes be revealed where it is turned up under skirtings.

Figure 3.28
Thermal insulation overlaid with a polyethylene DPM

Characteristic details

Basic structure

Thermal insulation is placed under or at the edge of the slab as shown in Figures 3.29 and 3.30.

See also the same section in Chapter 3.1.

Bearing details

In this form of construction it is sometimes recommended that the slabs be carried over the perimeter walls to prevent overconsolidation of the thermal insulation layer. To minimise cracking due to differential settlement of hardcore underneath the slab, some reinforcement of the slab is recommended.

Protection against radon and methane

An outline of the problem of radon was given in Chapter 1.5. The same section in Chapter 3.1 is also relevant.

Main performance requirements and defects

Choice of materials for structure

See the same section in Chapter 3.1.

Strength and stability

In a survey of floors insulated beneath the slab, carried out by BRE in the mid-1980s, several cases were found where the insulation, carried over the bearings of the slab, was inadequate in compressive strength. This led to cracking where the fill settled or where the loads of the inner leaf were heavy and concentrated. Incredible though it may seem, in some cases the internal

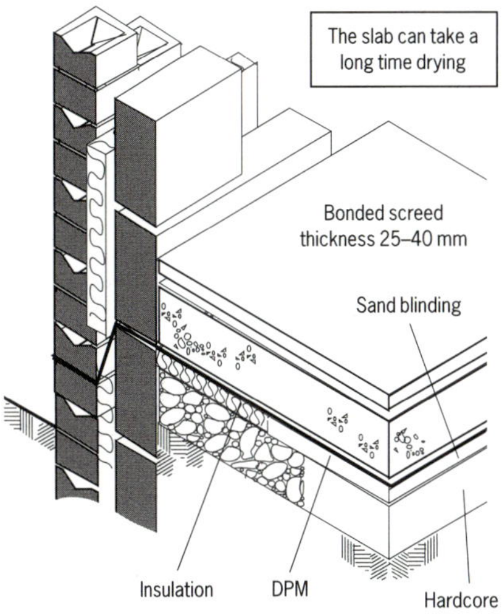

Figure 3.29
A slab insulated below the structure

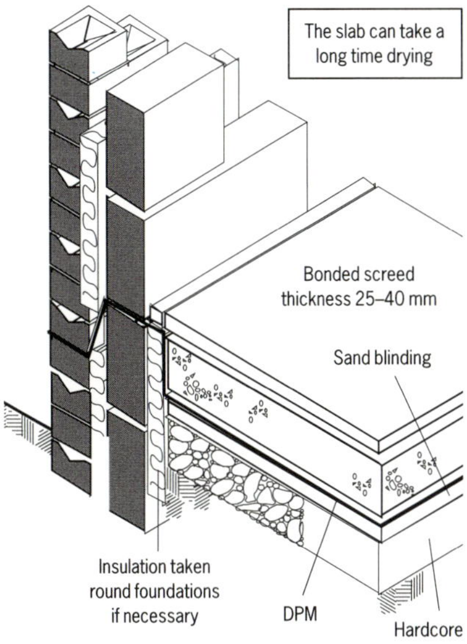

Figure 3.30
A slab insulated at the edge of the structure

leaf had been built off compressible insulation (Figure 3.31).

Deformation of the thermal insulation can be caused either by compression under loads or by gradual deterioration where it is degraded by contaminants in the fill or hardcore. The remedy, of course, is to re-lay the slab using the correct grade of insulation.

Dimensional stability, deflections etc

See the same section in Chapter 3.1.

Thermal properties and ventilation

In a survey carried out on the buildability of insulated house floors, observations were made of insulation placed below concrete floor slabs. It was concluded that most insulation operations were potentially straightforward, though defects resulted from lack of communication of design intention and inadequate supervision. Defects included deformable expanded polystyrene insulation below the concrete and insulation missing from parts of the floor construction (varying from 1 to 5% in particular cases) creating potential cold bridges.

Control of dampness and condensation

Care should be taken on the timing of laying a concrete slab. If it is laid before the roof is in position, as is common with housing, there is a chance that rain will saturate the slab and the insulation underneath it; some types of insulation (eg glass fibre), when saturated, are ineffective as insulation. This may also greatly extend the drying time of the slab. On the other hand, with some building types such as the large single storey shopping complexes, it will be possible to build the roof before laying the ground floor slab.

In housing, where a suspended timber floor has been replaced with a solid groundbearing slab, and the opportunity has been taken to replace or insert a DPC into the external walls, BRE investigators have encountered some problems of linking the dampproofing arrangements (Figure 3.32).

Where the floor has a layer of thermal insulation over the whole or a part of its area, and the wall insulation is of a high standard, condensation might occur at cold bridges where the slab passes into or under an external wall. It is important to avoid the risk of condensation occurring at thermal bridges by making sure that the continuity of insulation is unbroken, especially at the floor/external wall interface and at the interface of the floor with an internal wall occurring within 1 m of an external wall.

Where an existing floor slab is being relaid, the new DPM can be laid either over or under the slab. If laid over, it might need to be

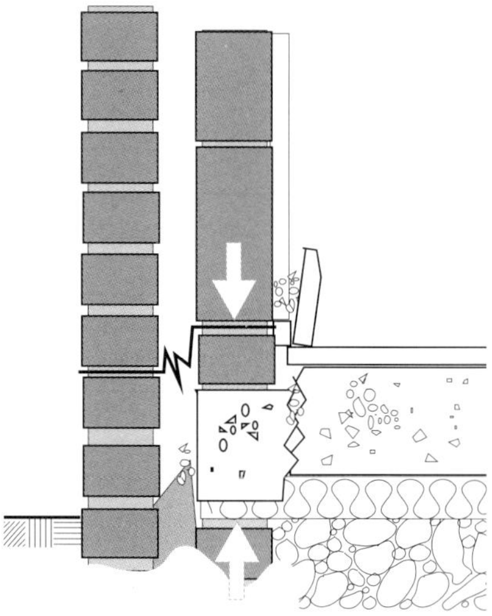

Figure 3.31
Where thermal insulation has been carried through the inner leaf, it will compress causing severe disruption

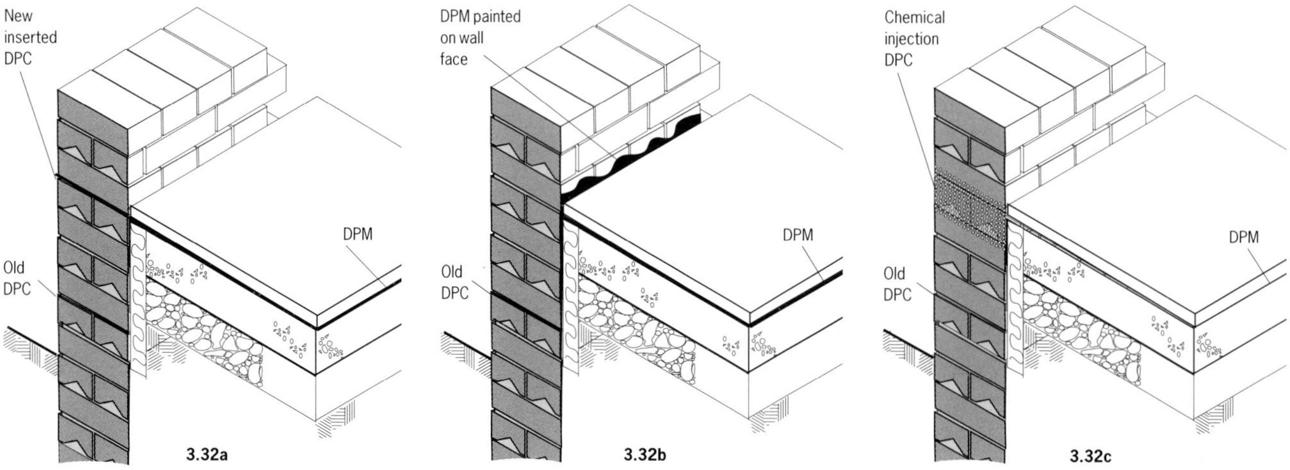

Figure 3.32
Three possibilities for replacing a DPC in a solid wall: a DPC installed to line up with a DPM (3.32a), a DPM painted on the wall face to join an old DPC with a new horizontal DPM (3.32b) and a chemical injection DPC (3.32c)

protected with a screed (Chapter 4). Where the existing slab and hardcore cannot be excavated sufficiently, and the slab needs to be retained at its original level, this will have a significant effect on headroom – as it will call for the removal and replacement of skirtings, doors and staircases, it will rarely be a satisfactory option. Far more likely in these circumstances is the provision of a DPM, such as polyethylene sheeting or a surface membrane, under the slab. The entire procedure is described in Chapter 3.1.

Thermal bridges may also occur where changes of level between dwellings are made as stepped-and-staggered separating walls (Figure 3.33).

Fire
See the same section in Chapter 3.1.

Sound insulation
See the same section in Chapter 3.1.

Durability
See the same section in Chapter 3.1.

Maintenance
See the same section in Chapter 3.1.

Work on site

Storage and handling of materials
See the same section in Chapter 2.2.

Restrictions due to weather conditions
See the same section in Chapter 2.2.

Workmanship
Failure of replacement floorings on old concrete bases can often be attributed to inadequate preparation of the surface of the concrete before laying the new flooring. The feature panel in the same section of Chapter 2.2 describes suitable techniques for reducing the risk of problems occurring.

Several instances of quite severe damage to thermal insulation sheets when casting floor slabs were observed by BRE during site inspections in the mid-1980s. Although the loss of thermal insulation is probably not critical, unless it occurs at the perimeter of the floor, it is worthwhile taking care. It was calculated at the time that some 2% of the insulation at the very edge of the slabs had been lost in this way with a consequent risk of localised cold bridges. Gaps occurring between insulation boards within the body of the floor area were measured to accumulate to a further 4% of the total board area.

The following precautions or measures would seem worthwhile:

- laying thermal insulation immediately before casting the slab
- considering whether or not any necessary DPM can be laid over instead of under the slab
- weighting the boards to prevent dislodgement by wind
- ensuring that care is taken not to puncture the DPM (Figures 3.34 and 3.35)

Further aspects of workmanship are covered in BS 8000:Part 2:Section 2.1[152] and Section 2.2[153].

Inspection

In addition to those listed in the inspection section at the end of Chapter 3.1, the problem to look for is:

◊ settlement of slabs

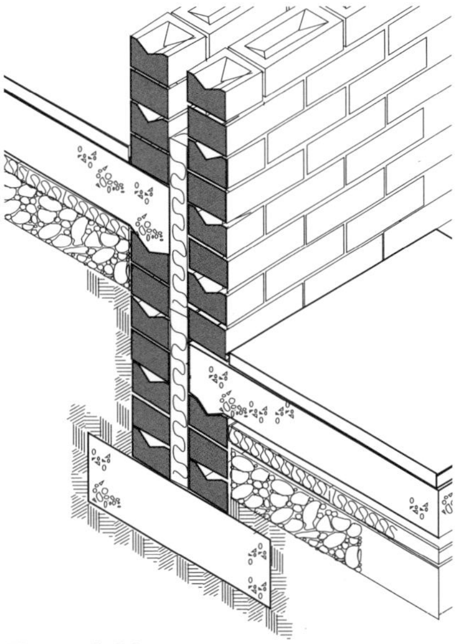

Figure 3.33
At stepped-and-staggered separating walls, the insulation should be carried down to foundation level

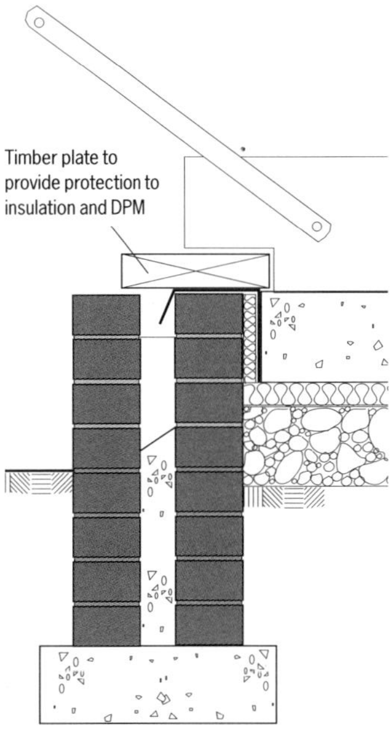

Figure 3.34
Protection to the DPM and insulation when tamping a concrete slab

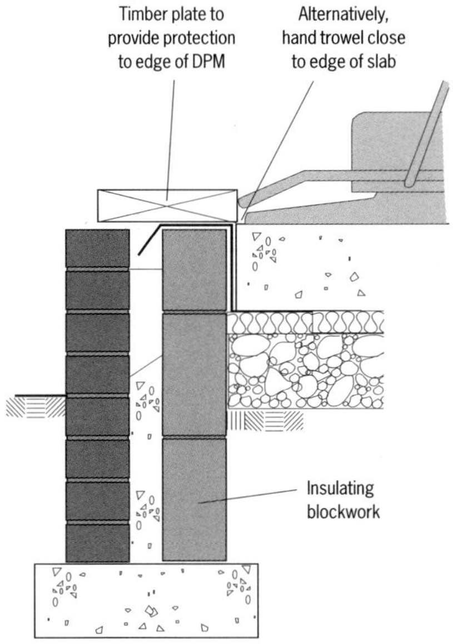

Figure 3.35
Precautions to be taken when power trowelling

Chapter 3.3 Rafts

Raft foundations have been mainly used for small buildings on difficult sites such as made-up ground or over ground liable to subsidence from mining operations. Housing Association Property Mutual (HAPM) investigations indicate that in the early 1990s just under 1 in 10 of the housing association dwellings monitored by them were built on raft foundations. While it is suspected that in recent years the number of sites needing rafts may have increased because of the economic and planning pressures on building land, making difficult ground more acceptable, there are at the same time substantial numbers of houses built on rafts in earlier years.

Characteristic details

Basic structure

The main difference between the raft and the ordinary slab is that the raft is designed to move without significant deflection under the influence of differential ground movement whereas the slab might not do so. In order to move as a whole, the raft is reinforced, often with deep downstand beams (toes) at the perimeter and under loadbearing walls.

See also the same section in Chapter 3.1.

Protection against radon and methane

Since rafts are designed to mitigate the effects of differential movement, it is unlikely that they will crack to the same extent as unreinforced slabs. This implies that rafts will be relatively impermeable to gases from below, though service entry points are the weakest areas. Resealing these with flexible sealing compounds may suffice.

Retrofitting radon sumps under some rafts is practically impossible because of the presence of deep toes and edge reinforcement.

See also the same section in Chapter 3.1.

Services

Services penetrating the raft will need to be flexible and isolated by sleeves from the edges of the concrete, since movements in rafts must be expected to occur.

Figure 3.36
The base being prepared for casting a raft for a multi-storey building. Column bases are already in position

Main performance requirements and defects

Choice of materials for structure

See the same section in Chapter 3.1.

Strength and stability

The design of some rafts has led to problems of settlement or heave. For this reason raft design should always

Figure 3.37
Laying a DPM which also acts as blinding to prevent fines migrating to the hardcore. Taping provides integrity to the seal

be carried out by a structural engineer taking into account the particular conditions to be found on site.

So far as recently built dwellings are concerned, information from HAPM shows that in the early 1990s, the main problems concerning rafts related to lack of adequate structural design, particularly with what has become generally known as semi-raft design. The particular problems pinpointed relate to lack of adequate provision for thickening at perimeters and at positions of internal loadbearing walls.

Naturally, curing the concrete adequately is needed to develop full strength as well as protection from frost in winter conditions, just as with any other kind of slab. Since the slab is often exposed at the perimeter, any weakness in the concrete will soon be apparent.

Dimensional stability, deflections etc

See the same section in Chapter 3.1.

Thermal properties and ventilation

See the same sections in Chapters 3.1 and 3.2.

Control of dampness and condensation

Protection of a raft from rising damp has to be done beneath the concrete, in order to provide continuity. This means that some of the geometry is complicated, particularly at perimeters and thickenings (Figures 3.36 and 3.37). Of course, floorings inside can be protected by applying a surface DPM to the top of the raft and covering it with a screed.

There is a risk of condensation occurring at thermal bridges with this kind of floor. Where the toe of the raft is shallow, there is insufficient depth to allow the insulation in the wall to prevent heat loss through the perimeter of the slab. The construction shown in Figure 3.38 illustrates a method for reducing the loss to a minimum.

Fire

See the same section in Chapter 3.1.

Sound insulation

See the same section in Chapter 3.1.

Durability

Raft foundations are engineered floors in which the condition of the steel and concrete affects the

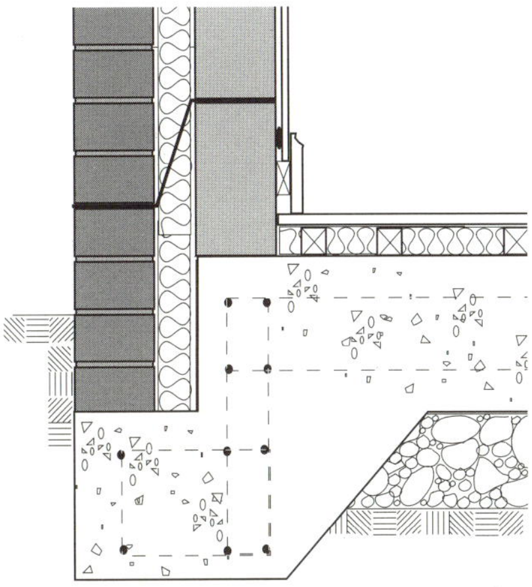

Figure 3.38
Construction should allow for the insulation to be started as low as possible; in other words from a deep toe in the raft

performance of the whole building. Corrosion of steel reinforcement may occur where steel is close to the top surface of the concrete and therefore vulnerable to dampness, and especially at the perimeters of rafts (Figure 3.39).

Maintenance

See the same section in Chapter 3.1.

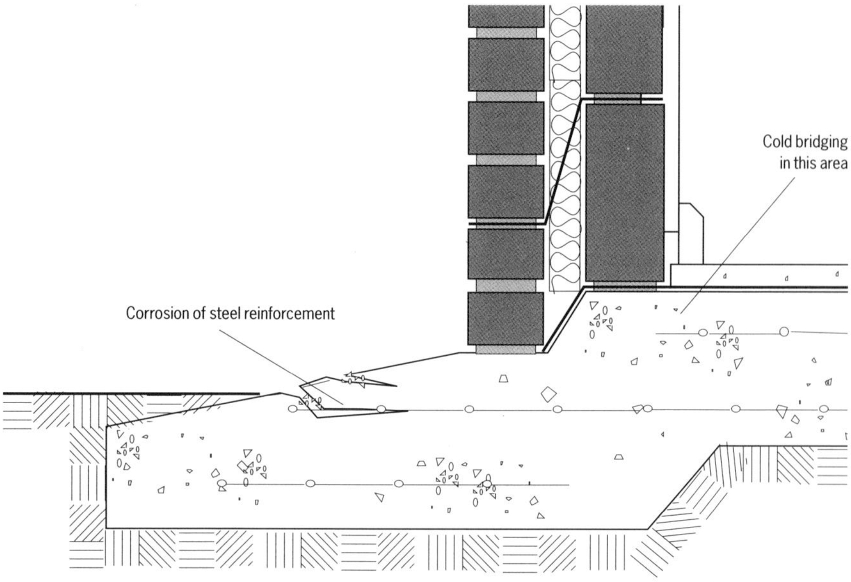

Figure 3.39
The toe of a raft is vulnerable where it extends beyond the perimeter of the building

Work on site

Storage and handling of materials

See the same section in Chapter 2.2.

Restrictions due to weather conditions

See the same section in Chapter 2.2.

Workmanship

Failure of replacement floorings on old concrete bases can often be attributed to inadequate preparation of the surface of the concrete before laying the new flooring. The feature panel in the same section of Chapter 2.2 describes suitable techniques for reducing the risk of problems occurring.

Further aspects of workmanship are covered in BS 8000:Part 2: Section 2.1[152] and Section 2.2[153].

Inspection

In addition to those listed in the inspection section at the end of Chapter 3.1, the problems to look for are:

- ◊ fracturing of services
- ◊ cracking of slabs
- ◊ cold bridges at perimeters of floors

Chapter 4 Screeds, underlays and underlayments

This chapter deals with materials laid over the structural floor to provide suitable level surfaces for the flooring. They will take many forms, although the sand and cement screed will be found to predominate (Figure 4.1).

Some flooring materials, for example mastic asphalt, may be used as an underlay for other kinds of floorings. There can be advantages in using such an underlay as an alternative to waiting for cementitious screeds to dry before laying floorings. Since the materials are identical with those used as the final surface, the descriptions used in Chapter 5.8 are relevant.

Also included in this chapter are sections on matwells, ducts and movement joints since these are often accommodated within screeds.

Figure 4.1
Ruling a sand and cement screed

Chapter 4.1 Dense sand and cement screeds: bonded and unbonded

Figure 4.2
Flooring which has been removed to show a friable screed mix

Dense sand and cement screeds are by far the most common form of providing a smooth and level base for the subsequent laying of floorings.

Investigations of floor screed failures, some involving remedial work of tens of thousands of pounds, have demonstrated that published guidance on design, workmanship and supervision too often goes unheeded (Figure 4.2).

Screeds may be:

- fully bonded to the base
- unbonded (ie not bonded to the base)
- floating (see Chapter 4.2)

Fully bonded and unbonded dense sand and cement screeds are described in detail in this chapter.

Characteristic details

Basic structure

For fully bonded screeds the surface of the concrete base should be scabbled or shotblasted to remove laitance and to expose the aggregate. The optimum thickness of the screed is between 25 mm and 40 mm (Figure 4.3). For thicknesses greater than 40 mm there is an increased risk of the screed debonding from the base.

All screeds laid directly over DPMs (except for epoxy bonding DPMs) must be considered as unbonded. The minimum thickness should be 50 mm but, for certain kinds of floors, the screed may need to be thicker (Figure 4.4).

Proprietary screeds based on cementitious systems or calcium sulfate are available (see Chapters 4.2, 4.4 and 4.5). Some of these are trowel applied, others are self-levelling (see Chapter 4.3). Many proprietary screeds are designed to be laid more thinly than conventional screeds.

The standard appropriate to cementitious mixes described in this chapter is BS 8204:Part 1[159].

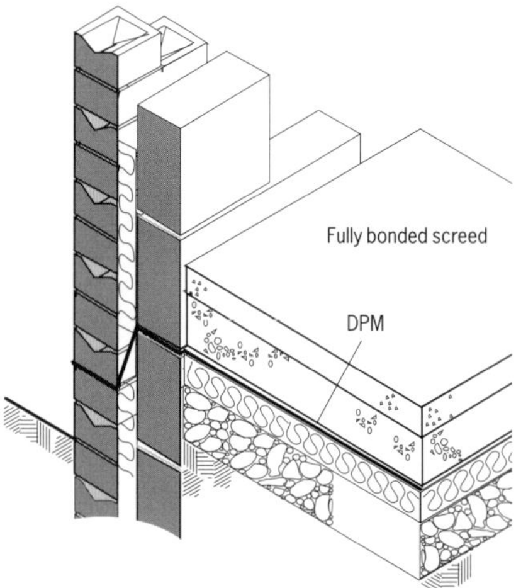

Figure 4.3
A fully bonded screed on a concrete base

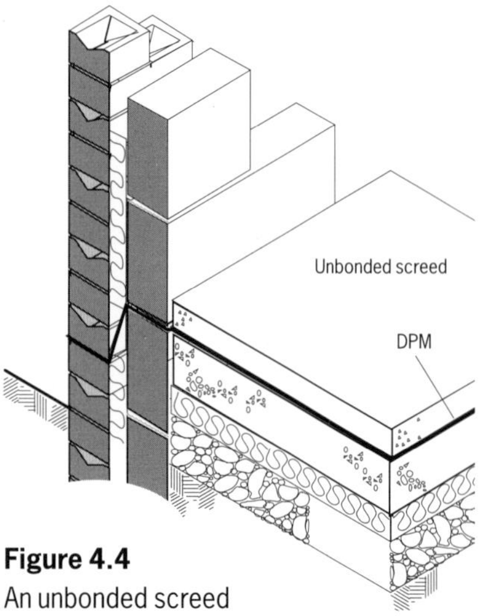

Figure 4.4
An unbonded screed on a concrete base with the DPM under the screed

Detailing

The concrete bases described in Chapters 2.2, 2.3 and 3.1–3.3 all provide a suitable base for cementitious screeds.

Cementitious screeds do not usually perform well on timber joisted and decked floors, especially older ones, and for that reason are rarely seen. Such screeds can only be unbonded, and their mass usually discourages their use in these situations.

Services

Services are best routed away from screeds if at all possible; for example it may be possible to use hollow skirtings to accommodate service runs, thus avoiding the floor altogether (Figure 4.5).

Where building services are to be installed under screeds they should preferably be laid in ducts within the base structure. If there is no alternative but to lay them in the screed itself, the overall depth of the screed must be increased by the depth of the particular service accommodated. Otherwise the screed will fail. BS 8203[35] suggests that the screed depth above inserted pipes should be a minimum of 25 mm and reinforced, but BRE has found that this can prove insufficient.

In any event, under local Water Byelaws, water services must be laid in ducts to enable them to be repaired or replaced as appropriate (Figure 4.6).

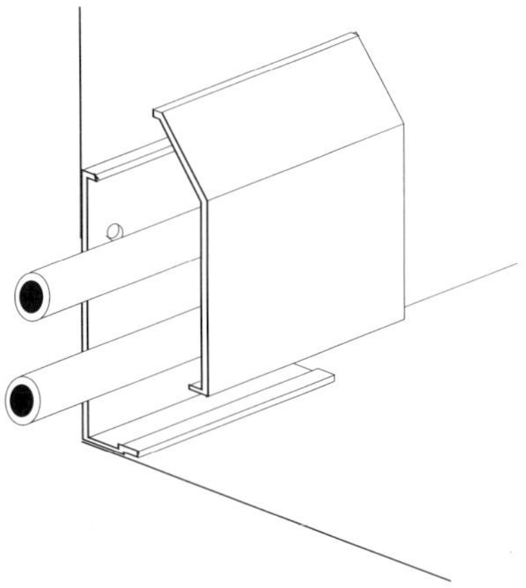

Figure 4.5
Services are best routed away from screeds if at all possible

Main performance requirements and defects

Choice of materials

Most screeds for relatively light loadings, such as are to be found in houses, will be made only from cement and sand, without additives or reinforcement. BRE recommends mix proportions of 1:4 by weight (approximately 1:3.5 by volume) for normal use, although mix proportions between 1:3 and 1:4.5 by weight are also suitable. The 1:3 mix should be specified for use where heavy traffic is anticipated.

To control drying shrinkage for screeds over 50 mm thick, a coarse aggregate of small size may have been incorporated. These are sometimes referred to as fine concrete screeds. Traditionally they were mixes of 1:1.5:3 cement:sand:10 mm single size aggregate and could be only laid with a workability similar to that of concrete. They require extra water compared with the semi-dry sand and cement screeds. However the extra water required to make the mixes more workable took away all the benefits of using the larger aggregate to control shrinkage. Consequently it has become much more common to replace only sufficient sand by 10 mm aggregate that will enable the screed to be laid semi-dry and finished to a good standard. The overall cement:aggregate ratio aimed at has been about 1:4.5 to 1:5 by weight. With most aggregates, mixes in the range of 1:3.5:1 to 1:3:1.5 cement:sand:10 mm aggregate have been found to work well. Aggregate gradings and particle shape determine how much 10 mm aggregate can be used and the screed still be laid semi-dry.

A fine concrete screed of 1:3:1 with 6 mm single size aggregate has also been used, though it should be noted that the sand itself can contain particles up to 5 mm size.

Prior to 1965, sands for screeds were commonly specified to BS 1199 and 2000[195]. At that time it was realised that these sands were too fine for screeds and more use was made of fine aggregates complying with BS 882[196]. This latter standard

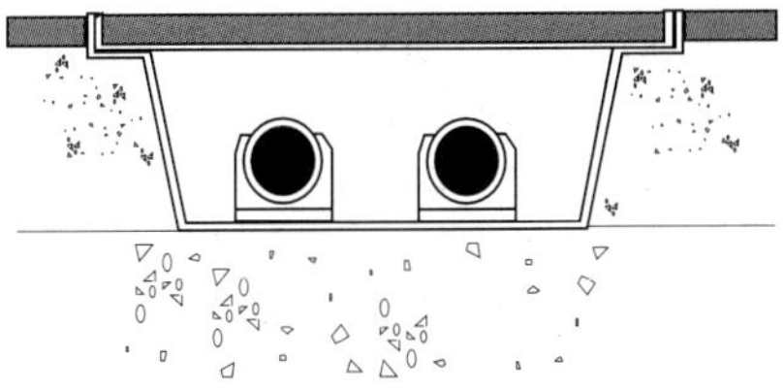

Figure 4.6
New and replacement water services must be laid in ducts

progressively became the dominant one for specifying sands for screeds.

Sands to grading limits zones 1 and 2 were capable of producing high quality screeds. However, when BS 882 was revised in 1983 these categories were modified and the new grading limits – commonly specified as C or M – became less appropriate for high quality screeds: grade C is too coarse at its top end and M too fine at the lower end.

For the highest quality screeds, sand grading limits should be specified according to sieve sizes to BS 410[197] †.

Most sand and cement screeds are laid at a consistency commonly referred to as semi-dry. This term defies precise definition. The required consistency is usually judged by pressing a ball of the mixed screed material in the hand: if it is too wet, water exudes through the fingers; if too dry, the screed material falls apart when the hand is opened. If the water content is right, then the screed material should hold together as a ball when the hand is opened. The water:cement ratio will thus depend on sand grading and particle shape, but will generally lie between 0.35 and 0.45.

Aggregates for fine concrete screeds should also comply with BS 882; the 10 mm single size should also comply with the grading limit of Table 3 in BS 882.

Steel fabric for reinforcement of screeds should comply with BS 4483[147]; the lighter D designated meshes are those normally used. Small 50 × 50 mm meshes can also be used. Any reinforcement should be placed towards the centre of the screed depth. The purpose of reinforcing is primarily to control shrinkage cracking. Wire netting is sometimes specified. Furthermore, though it is not suitable as reinforcement, it is commonly used under floating screeds to protect the insulation during the laying process (see Chapter 4.2).

† Further advice should be obtained from BRE Advisory Service or by reference to the latest version of BS 8204:Part 1. This standard is undergoing revision. A draft for comment is recommending different grading limits from those in BS 882. Also, there is a CEN draft which gives different limits again. What the situation will be in the future is far from clear.

Strength and stability

Where the screed is to be laid over precast suspended concrete floors, adequate surface preparation to receive a bonded screed may be difficult to achieve; the use of an unbonded screed in these circumstances may be preferred. Modern shotblasting machines have improved greatly the preparation of these bases.

The use of bonding agents and laying of bonded screeds are covered in the workmanship section later in this chapter.

Non-structural screeds, whether bonded, unbonded or floating, add to the overall stiffness of a floor, and can be taken into account in structural calculations.

The necessary thickness of screeds has been dealt with already under the characteristic details heading earlier in this chapter.

Dimensional stability, deflections etc

Coefficient of linear thermal expansion per °C: 10 to 13 × 10^{-6}.

Reversible moisture movement: 0.02–0.06%.

Drying shrinkage: usually 0.03–0.04%.

In the 1970s there was a noticeable increase in the number of cases of rippling of thin floorings over discontinuities and irregularities in screeds and, to some extent, over screedless floors (Building Research Station Current Paper CP 94/74[198]). Most of the failures occurred in new hospitals. Rippling in most cases was observed within a period of one to four weeks after the flooring had been laid. In some cases the rippling could be detected only when viewed from a low level against the light. In others large ripples or rucks were distinct, up to 25 mm high and 50 mm wide. At all sites examined, the ripples were found to be immediately above daywork joints or cracks in the screed.

Also, irregularities in the surface of screeds tend to show through certain thin floor finishes.

All concrete slabs and screeds made with Portland cement and aggregates shrink as they dry. Since the upper surface loses moisture more rapidly than the lower levels, it is the upper surface that usually shrinks first. The result is a tendency for the screed to curl upwards at the edges of bays, resisted by the degree of bonding to the subsurface and by the total thickness of the screed. Thus, bonded screeds have less tendency to curl than unbonded screeds. Also, thick unbonded screeds tend to curl less than thinner ones as screeds with greater thickness have more strength to resist the curling forces (Figure 4.7).

Drying shrinkage can also lead to the development of shrinkage cracks. If shrinkage is excessive, this

Figure 4.7
A core has been taken to investigate the reasons for cracking in, and lifting of, a proprietary cementitious screed

in turn can lead to curling at these cracks. The risk and extent of curling and shrinkage cracking is influenced by mix proportions, water content, thickness, curing and drying times, and bond strength.

The aim of most good screed specifications is to keep the drying shrinkage of the material and its effects to a minimum. This is achieved by keeping the water content to a minimum by selecting a well graded coarse screeding sand and adding a water reducing agent. Curing should be carried out over at least four days, but preferably seven, by covering with polyethylene sheet after the screed is laid, and then allowing the screed to dry slowly.

Traditionally, to limit shrinkage problems, screeds were laid in bays. Bay sizes recommended were generally about 5 × 4 m. However, difficulties were encountered at bay joints where curling and lipping had occurred. In the late 1960s, the practice was changed and it was recommended that screeds should be laid in the largest areas possible without joints. This method leads to much reduced curling, as the screed cannot curl until it has cracked, by which time most of the drying shrinkage has taken place. Aggregate interlock tends to prevent lipping at cracks. Of course, laying screeds in large areas does lead to random shrinkage cracks; where these occur they can easily be repaired before covering with flooring.

There is still some merit in laying heated screeds, and some screeds for brittle finishes, in bays.

Thermal properties

Thermal conductivity of dense sand and cement screeds: 1.83 W/mK.

See also the same section in Chapter 3.1.

Control of dampness and condensation

The DPM in a groundbearing floor slab and screed system can be laid in a variety of positions. In the case of a fully bonded screed the DPM must be laid under the slab, although for an existing screed a surface DPM can be used. The exception to this is the use of an epoxy DPM, which can act as a bonding agent and be laid as a sandwich.

Fire

Non-combustible.

Sound insulation

See the same sections in Chapters 2.2 and 2.3.

Durability

The semi-dry method of laying floor screeds was introduced in the late 1960s and quickly found widespread acceptance. It enabled the screed layer to finish the screed with a flat well-closed-in surface to receive the thinner sheet and tile floorings then being introduced. Poor workability, particularly if mixes were a little too dry, led to poor compaction. This could mean that a dense well compacted surface as thin as 5 mm would overlay a weak friable layer which extended through the remainder of the thickness of the screed. In service, these screeds, when covered by thin sheet and tile floorings, can break down to form depressions in the floorings. This became a very common type of failure and the depressions became known as elephants' footprints. The BRE screed tester was developed explicitly to test screeds before they were covered with floorings to prevent this type of failure. See the feature panel on the BRE screed tester at the end of this chapter.

Where screeds broke down in service, it was often found that not only were the screeds poorly compacted but the components had not been thoroughly mixed together. Balls of cement were often found, the sand between the balls containing little cement. Subsequently it was found that free-fall concrete mixers which were used to mix semi-dry screed material produced inconsistent material. The cement was not distributed throughout the mix but was concentrated as round pellets. Forced action mixers produced well blended material and these are now recommended for mixing semi-dry screeds.

Case study

Leaking pipe buried in floor screed

BRE Advisory Service investigated the case of dampness in screeds and the lower parts of walls. Moisture content values obtained from drilled samples showed the lower parts of the internal walls to be very wet. A hole cut in the floor screed filled with water. The atmosphere in the building was very humid. Operation of the ceiling heating produced condensation on some of the walls.

The source of the water was found to be a leaking pipe buried in the floor screed. No structural repair work was required – only repair of the leaking pipe and a prolonged period of drying out followed by redecoration. Dealing with the water leak and improvements to the heating system should have helped eliminate the condensation problem.

Case study

Moisture trapped in a screed

The perimeter strip of a screed had proved unsatisfactory and had been renewed. It showed dampness some time after completion of the work, in particular affecting the textile floor covering. High readings were obtained using a conventional moisture meter. Further investigation by surface hygrometers, by the Bretek high frequency moisture meter and by drilled samples led to the conclusion that the original slab was at a low and acceptable moisture content. The new perimeter strip was slightly damp, with more moisture present near the surface than deeper down. The likelihood of surface condensation being in part responsible for this effect could not be discounted, but the client was advised that a screed can take a long time to dry down to a residual moisture level; therefore he should wait at least six months before covering the floor with a carpet.

Case study

Poor bond of tiles to floor substrates in a hospital

BRE Advisory Service investigated the failure of floor screeds and tiling in two operating theatres of a hospital. In the first operating theatre, part of the screed was shown to be of a poor quality. The movement of the operating table had resulted in the disintegration of the area of screed beneath it. Some of the antistatic PVC tiles had ridged and blistered due to a poor bond between the tiles and the screed. It was suggested that the screed should be replaced with a sand and cement screed laid to the recommendations of British Standard code of practice BS CP 204:Part 2[199]. (This was the relevant standard at the time. It has now been withdrawn.) When the screed had dried sufficiently the antistatic tiles were able to be replaced. In the second theatre the screed had to be removed and replaced with antistatic precast terrazzo tiles bedded on a sand and cement mortar 12–25 mm thick. This allowed the operating theatre to be brought back into use more quickly than the first theatre because a long drying time was not necessary.

Case study

Cement content deficient in screeds in an office block

Floor screeds in offices which had been occupied for only three years started breaking up over large areas. The screeds were very weak and friable and had the appearance of being deficient in cement. This was confirmed by chemical analysis which showed that three samples had mix proportions between 1:18 and 1:22 by weight. The BRE screed tester showed that the screeds would continue to break down in service. The screeds had to be taken up and relaid.

Case study

Minor cracking in unbonded screed in an airport building

BRE Advisory Service was asked to report on the quality and likely performance of the floor screeds at an airport building. Although there was some cracking it was not excessive for unbonded construction. The hollowness which was present in parts of the floor was not associated with any visible curling or lipping at cracks and joints. Measurements taken using the BRE screed tester showed the soundness of the screed to be good. It was considered that the screeds would perform well in service. No remedial action was recommended.

A screed often cracks where services are laid within the thickness of the screed, particularly where pipes have not been laid at sufficient depth below the screed surface (Figure 4.8).

In the BRE site inspections on quality in traditional housing (*Quality in traditional housing*, Volume 2: An aid to design[9]), some 38 mm thick unbonded screeds were found to be breaking up after just three years, in spite of being reinforced. They were replaced with 60 mm deep screeds.

One problem which has been seen relatively widely in the past is where cementitious screeds were used in swimming pools. These have been subjected to sulfate attack induced by poor pH control of the pool water.

Maintenance

Generally, maintenance of screeds should not be necessary, though some replacement may be required where patches of inferior screed have broken down under heavy impacts.

Work on site

Storage and handling of materials

See the same section in Chapter 2.2.

Restrictions due to weather conditions

Too rapid drying out in hot weather is a potential cause of failure. Precautions must be taken.

See also the same section in Chapter 2.2.

Workmanship

It has been estimated that the labour content in laying fine concrete screeds is about 25% greater than in laying sand and cement screeds, if the correct proportions of water are to be maintained. Using a rounded aggregate will give improved workability over crushed rock aggregate, and allow water proportions to be maintained nearer to their correct level.

For a bonded screed, whether it is laid on the top of an old or a new concrete slab, it will be necessary to prepare the surface of the slab to receive the screed.

In the case of a new slab, all surface laitance must be removed. Surface retarders and subsequent washing is sometimes specified for a new slab, but care should be taken to remove all cement from exposed coarse aggregate. A much more reliable method is to scabble or shotblast the surface of the slab to expose the coarse aggregate and to provide a key.

For an older slab, the surface must be abraded, either with a scabbler or by shotblasting.

The surfaces of all slabs should then be soaked with water to make sure that no suction remains and bonding coats applied. These may be either:

- cement and water slurries of the consistency of pouring cream or
- proprietary resin based bonding agents (some bonding agents do not require the slabs to be soaked with water)

In the case of the cement and water slurries, it is most important to work in very small areas, and to lay the screeds on the still wet slurries. It is helpful to add a bonding polymer to the mixes, though shorter wet times may result.

In the case of the proprietary bonding agents, it is important that the manufacturers' instructions on minimum and maximum times

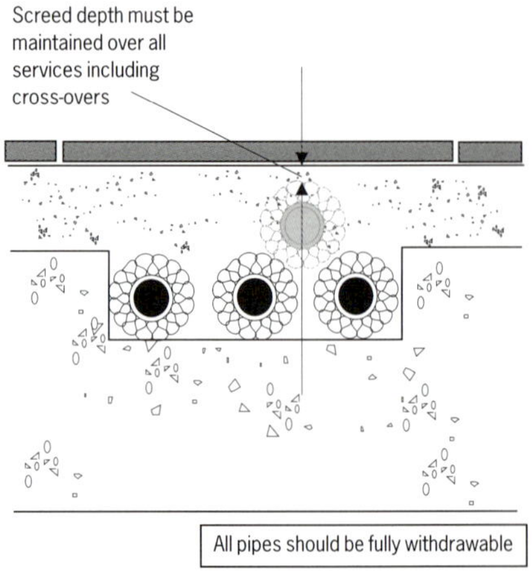

Figure 4.8
Screed depth must be maintained over services

before laying the screeds are observed.

Polyvinyl acetate bonding agents are not recommended for floors since they are unsatisfactory in wet situations. Polymer dispersions (eg styrene-butadiene rubber and styrene acrylics), and epoxy dispersion polyamines or solventless epoxies are satisfactory for floors which may become wet, though the last category is the more expensive. Solventless epoxies, however, will also act as DPMs.

Practices that are most likely to increase drying shrinkage, and curling of sand and cement screeds, are the use of:

- sand which is too fine
- too high a water:cement ratio
- insufficient curing
- too rapid drying

Both daywork and batten formed joints between bays of screeds should be straight butt joints only. There may be some merit in coating the vertical face of a set butt joint with a cement slurry before laying the adjacent area of screed up to it.

There is no longterm difference between the performance of batten formed joints and 'wet joints' in screeds. Wet joints are formed by laying relatively narrow (say 300 mm) widths of screed to level, spaced bay widths apart; then, after the initial set has taken place but before the battens have dried out (that is to say, within one hour), filling in and tamping to level between them. The screeds must be thoroughly compacted.

Tamping with a screeding rule is unsatisfactory and may lead to insufficient compaction of the screed.

Mixing on site

Batching should preferably be by weight rather than by volume since it is not easy to estimate the moisture content of damp sands; also to allow for bulking.

Water:cement ratio is crucial to performance, though it is unusual to find screeds being laid too wet since these mixes are difficult to finish to level. On the other hand, sufficient water must be used to enable full compaction to be achieved. The empirical test in common use (as already described in this chapter in the section on choice of materials for structure) is to squeeze a handful of the mix – if it holds together and no excess water is squeezed out, then the mix is about right.

Although it has already been said earlier in the chapter, it is worth repeating that, to obtain good durability, site mixing should be by pan or paddle mixer – ordinary concrete mixers are not capable of mixing adequately the drier mixes of screeds, and cement rich pellets are not broken up, leading to weak mixes (Figure 4.9). It is recommended, particularly, that free-fall mixers are not used for this purpose since they do not distribute cement efficiently in screed mixes. Screed pumps, which have forced action mixing, will be satisfactory.

Laying technique

Wooden battens are sometimes screwed down to the concrete base. This technique is inadvisable where there is a DPM immediately below the unbonded screed, and either loose-laid or 'wet' battens should be used.

Curing and drying

Screeds should be covered with a suitable curing membrane (for example polyethylene sheet) for a period of not less than four days, but preferably seven, in order to allow it to develop its full strength. All foot and vehicle traffic must be kept clear.

If the screed has been laid in frosty weather in a building under construction, further protection may be required.

Once curing is completed, the impervious membrane should be removed. Drying should be allowed to proceed slowly as forced drying can lead to excessive drying shrinkage, causing cracking and curling. About one third of the mixing water used in a screed is used to hydrate the cement. Most of the remaining water must be allowed to escape before laying moisture sensitive floorings. The time required to do this is often underestimated (see Table 1.3 in Chapter 1.4).

With heated screeds, the heat should not be turned on until the screed has been allowed to dry naturally for at least four weeks. When the heating is first turned on, it should be at a low level, only increasing gradually.

Accuracy

Criteria for flatness of screeds are given in BS 8203 and BS 8204: Part 1[159]. The normal criterion used for flatness is a maximum deviation of 5 mm in a 3 m long straightedge

Figure 4.9
A section through a badly mixed sand and cement screed. The large item is unmixed cement. Other balls of cement are visible

laid in contact with the finish (Figure 4.10), though this can be reduced to 3 mm where greater accuracy is required. There is also a utility standard, where surface flatness is not critical, of ±10 mm.

For levelness, an overall deviation of ±15 mm for normal purposes is suggested for large floors.

Further aspects of workmanship are covered in BS 8000:Part 9[(194)].

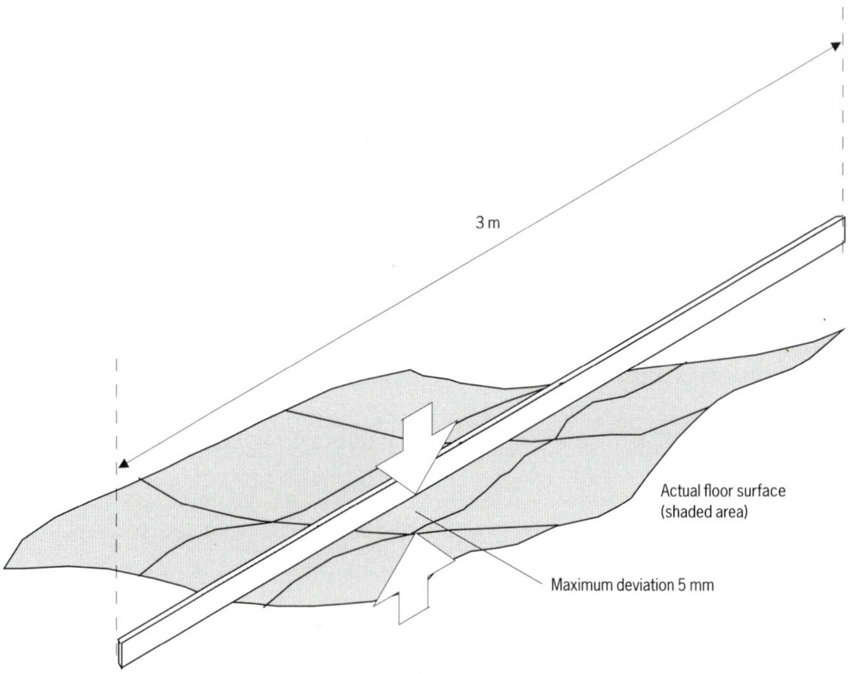

Figure 4.10
The normal standard of accuracy for flatness of a screed is a maximum deviation of 5 mm in a 3 m long straightedge laid in contact with the finished surface

The BRE screed tester

To identify screed failures, suspect areas can be tested by using the BRE screed tester before laying the flooring (Figure 4.11). Since 1978, the Building Research Station (as BRE then was) developed the screed tester † to check the soundness of screeds and issued Current Paper CP 72/78[(200)], many tests and observations have been made using the equipment by a number of major authorities, including the former Property Services Agency (PSA), the former Cement and Concrete Association (C&CA) and BRE Advisory Service. The test is now commonplace and has been included in a number of British Standards – even the smallest building and plastering contractors, anxious to prove the quality of their screeds, use it before the floor is accepted by a client. Other non-destructive methods including the Schmidt rebound hammer, ultrasonic pulse velocity (UPV) and gamma-ray attenuation density determinations have been proposed, but have not been widely accepted.

Figure 4.11
The BRE screed tester in use

† The BRE-designed screed tester is manufactured by Wexham Developments Ltd. This equipment and the procedures for its use are described in BS 8204:Part 1.

The tester works by a 4 kg annular weight falling a distance of 1 m down a rod and striking the collar of an anvil, the face of which is in contact with the screed surface. The contact area is 500 mm^2. Normally four blows are necessary to get an effective result. The depth of the resulting indentation is a measure of the screed's ability to carry traffic. Measurements are made to the nearest 0.1 mm using a depth gauge.

Observations and measurements taken with the screed tester normally show quite conclusively which screeds would, if left, give trouble and ultimately need to be replaced.

Indicative values (as shown by depths of indentations) for the performance of solid bedded cementitious screeds under test are:

- very heavy traffic or impacts where subsequent disruption cannot be tolerated: 3 mm (Category A)
- heavy wheeled traffic or impacts in public areas: 4 mm (Category B)
- foot traffic and light wheeled traffic: 5 mm (Category C)

See also BS DD 230[(201)] which is based on BRE work.

Inspection

Screed mixes which contain excessive amounts of cement or water are more liable to crack and curl than those correctly specified, and those with less than the correct amount of cement may develop insufficient strength. If supervision of mix proportions or laying operations is known to be or suspected of being inadequate, provided the screed is of a bonded or unbonded type it can be tested in situ with the BRE screed tester (see the feature panel opposite).

The hygrometer method should be used to assess when a screed is sufficiently dry for laying moisture sensitive floorings. Meters which measure the moisture content of screeds are of dubious value in assessing whether or not a screed is sufficiently dry.

Existing screeds failing in service and new screeds which do not meet the assessment limits specified by the BRE screed tester can both be treated in the same way. The tester, which is normally applied at 1 m intervals, can be used to find the extent of defective areas by testing at more closely spaced intervals. The extent of remedial measures depends mainly on the relative size of the defective area and the cost of the work; if the defective area is large, it may be better to replace the whole of the screed. Rapid drying proprietary screeds are available for use where extended drying times are not acceptable. Alternatively, low viscosity epoxy resins are available for strengthening poor screeds. They are poured onto screed surfaces and impregnate the screeds to depths of around 30 mm before setting. They are useful where floors need to be brought back into use quickly or where dust from the removal of defective screeds would be unacceptable.

The problems to look for are:

All cementitious screeds

◊ friable screeds which have been poorly compacted and not mixed thoroughly
◊ screeds showing elephants' footprints being of insufficient strength (check with the BRE screed tester, Figure 4.12)
◊ marine deposits in sands
◊ services not laid in ducts
◊ mix proportions not between 1:3 and 1:4.5 by weight (to be established by laboratory examination)
◊ upwards curling at perimeters
◊ cracking
◊ irregularities in thin floorings
◊ detachment of moisture susceptible floorings

Fully bonded screeds, as for all cementitious screeds plus:

◊ laitance not removed
◊ scabbling of old slabs not adequate
◊ screeds not between 25 and 40 mm thick
◊ thickness of screeds over buried services not adequate

Unbonded screeds, as for all cementitious and fully bonded screeds plus:

◊ screeds not greater than 50 mm thickness

Figure 4.12
Indentations after testing with the BRE screed tester

Chapter 4.2 Floating screeds: sand and cement

Floating screeds are used mainly to enable installation of one of two types of insulation:

- thermal insulation
- impact sound insulation

The required properties of these materials forming the layer immediately below the screed are very different.

There is some indication that, whatever the quality of the specification for floating screeds, their installation on site does not always achieve a satisfactory standard.

Characteristic details

Basic structure

The screed

Sand and cement screeds laid immediately over compressible layers like expanded polystyrene or glass fibre are classed as floating screeds, and should have thicknesses of not less than 65 mm for lightly loaded floors and 75 mm for more heavily loaded floors (BS 8204:Part 1[(159)]). There is probably a stronger case for floating screeds to be composed of fine aggregate concrete rather than plain sand and cement in order to have greater control over curling and shrinkage.

There was a revision to the recommended thicknesses of floating screeds around 1987 following recognition of the greater risk of damage to floating screeds of less than 65 mm thickness in areas of high floor loadings.

The resilient layer

With floating screeds used for thermal insulation purposes, it is important that the insulation board or mat should be sufficiently dense to support the likely loads when the floor is in use.

For floating screeds used for impact sound insulation purposes, the board or mat should be sufficiently resilient to prevent impacts being transmitted to the structural floor beneath, checking with the manufacturer that any expanded polystyrene is not too stiff for this application.

Above the mass-spring-mass resonance frequency, the resilient layer isolates the structural base from vibrations in the floating screed. For effective operation, this resonance frequency must be low, and certainly below 100 Hz.

Detailing

When detailing floating screeds, it is important that contact between the screed or flooring surface and any other part of the fabric by which impact sounds can be transmitted is avoided. This particularly applies to all edges of the screed; for example small gaps should be left under skirtings (Figure 4.13). In addition, all penetrations of the surface (eg those caused by central heating pipes) should be suitably sleeved.

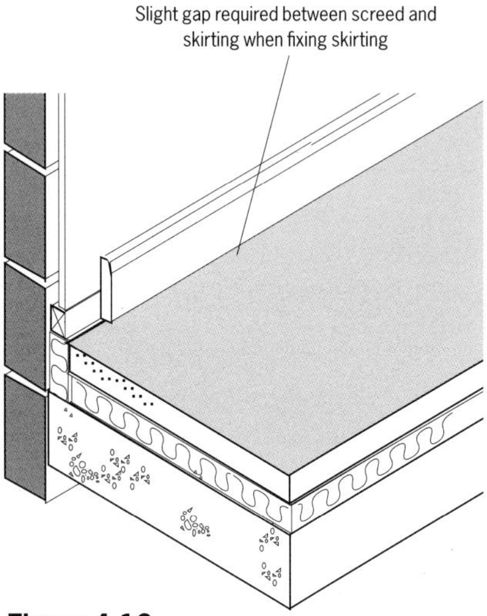

Figure 4.13
Floating screeds should have no solid contact with any other part of the structure (eg under skirtings)

Main performance requirements and defects

Choice of materials

Mixes described in Chapter 4.1 are suitable also for floating screeds.

Strength and stability

Before 1996 there was no readily available method of assessing the strength of floating screeds as the standard BRE screed tester was not suitable. When used it would often fail a floating screed which was fit for its purpose by punching out a truncated cone of the screed beneath the anvil. When asked to assess the acceptability of floating screeds, BRE has used a number of variants of the tester. Following an extensive research project, a modified procedure has been introduced which has been published by BSI as a Draft for Development (BS DD 230[201]). However the procedure and interpretation of results are subject to limitations.

See also the same section in Chapter 4.1.

Dimensional stability, deflections etc

Coefficient of linear thermal expansion per °C: 10 to 13 × 10^{-6}.

Reversible moisture movement: 0.02–0.06%.

Drying shrinkage: usually 0.03–0.04%.

Reinforcement of floating and unbonded screeds may sometimes be necessary. The reinforcement is normally placed in the centres of screed depths and is there to limit the size of shrinkage gaps and to prevent lipping. Two kinds of errors in placing reinforcement may become apparent in failed screeds:

- lack of adequate lapping of individual sheets of mesh
- voids where dry mixes fail to penetrate under the mesh, particularly at overlaps

Where the reinforcement is at the bottom of a screed which is laid over thermal insulation, it may have been placed in that position for the express purpose of providing some protection to the insulation from barrowing in of the wet screed. It does not matter if the reinforcement is not fully enclosed.

Housing Association Property Mutual data shows that around 1 in 3 of reinforced screeds has not fully complied with requirements.

Chicken wire will often be found installed within floating screeds for sound insulation purposes. It should have been placed immediately above the resilient quilt, its purpose being to protect the quilt during the screed laying operation and not to reinforce the screed.

Thermal properties

Thermal conductivity of sand and cement screeds: 1.83 W/mK.

Thermal conductivity of glass wool quilt: 0.04 W/mK.

Control of dampness and condensation

See the same section in Chapter 2.2.

Fire

See the same section in Chapter 2.2.

Sound insulation

Generally, floating screeds will improve both the impact and airborne sound insulation of floors, though in practice will add little improvement in impact sound to some of the more resilient surface finishes such as cork, textile or rubber.

In a survey in which field measurements were taken of the sound insulation of floating floors with a solid concrete structural base, airborne sound insulation performance was found to be better than impact performance relative to the standard performance curves. While there appeared to be little doubt that all types measured were capable of easily attaining the performance standard for impact insulation when properly constructed, a considerable proportion of floors with a screed had adverse deviations (AADs) for impact insulation exceeding 55 dB. Large AADs were thought to indicate that unsuitable materials had been used for the resilient layers or that the layers had been bridged (Figure 4.14).

BRE has also taken field measurements of the sound insulation of floors with floating screeds on hollow precast bases. Many variations are possible in the

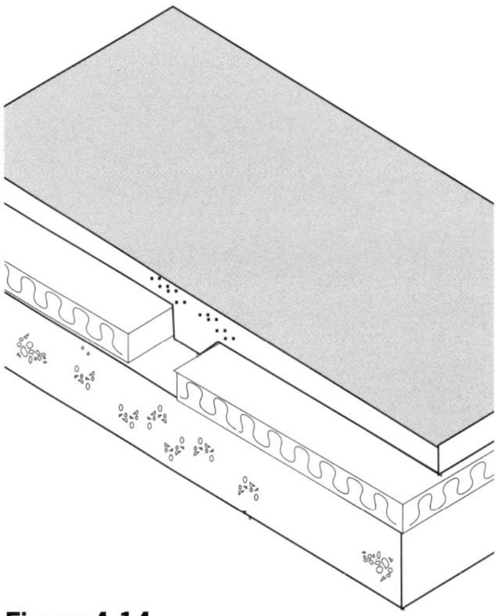

Figure 4.14
Gaps between sections of quilting allow the screed to touch the slab below impairing the effectiveness of the sound insulation

design of floating floors with hollow precast bases; the amount of data available from the measurements can be placed into three groups:

- floors with fibre resilient layers and bases consisting of hollow beams
- floors with fibre resilient layers and bases consisting of beams and hollow concrete or clay blocks
- floors with plastics resilient layers and any type of hollow base

In spite of these differences in floor construction, the ranges of performance for the three groups were found to be very similar.

It has already been emphasised that there must be no contact between the floating screed and the concrete substrate (Figure 4.14 on page 167). This means in practice that the wet screed must be prevented from flowing into any gaps in the insulating layer; this can be achieved by installing a separating layer of building paper or polyethylene. Around 1 in 10 installations do not comply with this criterion, and therefore the performance of the floating screed in this case will be deficient.

In site inspections, BRE staff have many times recorded poor performance from floating floors where they are interrupted by services which have not been properly sleeved or isolated (Figure 4.15).

There are two schools of thought as to whether it is permissible or even desirable to build a partition off a floating floor. Both BS 8233[202] (Clause 14.5.5.1.3) and *Principles of modern building*[6] say it should not be done. However, building a partition off the structural floor will simply allow the transmission of impact noise on the partition to travel through the structural floor to the storey beneath. The reason put forward for not building it off the floating screed is that it might overcompress the quilt. That argument is feasible if the partition was heavy, such as plastered brick or block, but most partitions, in modern housing at least, are lightweight, consisting of cored plasterboard or stud and plasterboard. Of course if the partition is built on top of the floating screed, there is a risk of cracking at the junction of the partition with the ceiling above and at the junction of the partition with the perimeter loadbearing wall. The BRE view now is that building lightweight partitions off floating screeds can be the lesser of two evils.

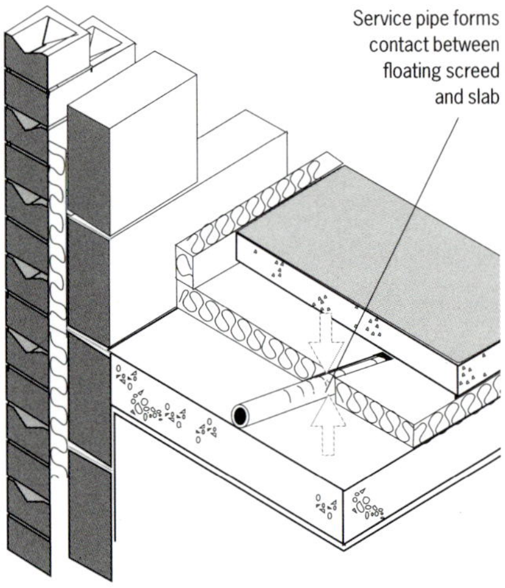

Figure 4.15
Service pipes should not interrupt the resilient layer

Case study

Lack of adequate compression of a quilt from too dry a screed

When the floors of a north of England hospital were examined five years after entering service, the surface layers, some of cellular backed PVC and others of linoleum, were found to have distorted. The construction of the floors was of a structural base covered with a resilient insulating quilt and topped with a cement screed. The quilt should have compressed to a third of its expanded thickness (ie to 12 mm) when under the specified thickness (60 mm) of screed.

Examination of the screed and underlying quilt was undertaken on site and in the laboratory. It was found that the quilt had not been adequately compressed because the screed was not heavy enough and it (the screed) had been mixed with insufficient water, leaving dry components at the bottom. Although the top of the screed was good, the lack of compaction below had caused movement which had resulted in cracking. The screed was also too thin.

There was evidence, further, that the composition of the screed was not appropriate for its purpose. Concrete containing a smaller aggregate would have performed better in the circumstances. Design and construction faults were identified, and total replacement of both the screed and the flooring seemed the only viable solution.

Durability

The isolation afforded by the quilt is adversely affected by the continuous static load of furniture and the dynamic load of people walking on the floor. Ultimately, and possibly after long service, the efficiency of the insulation will be reduced, and the floor may need to be relaid.

Maintenance

Maintenance should be unnecessary, though some replacement may be necessary where patches of inferior screed have broken down under heavy impacts.

When the flooring is to be renewed, and the resilient layer and screed are still satisfactory, the surface will require adequate preparation to receive the new flooring; in particular, taking care to ensure that new adhesives are compatible with any remains of the old.

Work on site

Storage and handling of materials

See the same section in Chapter 2.2.

Restrictions due to weather conditions

See the same section in Chapter 2.2.

Workmanship

Floating screeds laid on resilient quilts are very difficult to consolidate adequately. For this reason it can help if they are laid in two layers, with the second layer being laid within 30 minutes of placing a raked lower layer.

Further aspects of workmanship are covered in BS 8000:Part 9[194].

Inspection

In addition to those listed under all cementitious screeds in the inspection section at the end of Chapter 4.1, the problems to look for are:

- ◊ thicknesses of screeds less than 65 mm for lightly loaded floors and 75 mm for the more heavily loaded floors
- ◊ underlays lacking adequate compressibility and recoverability
- ◊ screeds making inadvertent contact with substrates through open joints

Chapter 4.3 Levelling and smoothing underlayments

Levelling and smoothing underlayments are needed where a base is of insufficient quality to accept an intended thin flooring without the defects showing through (Figure 4.16). Underlayments are invariably proprietary and a relatively recent addition to the floor layer's choice of materials and techniques.

Characteristic details

Basic structure

Self-levelling underlayments are based on mixtures of powder and water which need a minimum of trowelling to achieve a smooth surface without bubbles. Other underlayments, based on mixtures of powder and natural rubber latex or synthetic emulsion, require more trowelling to achieve a smooth surface. All types are laid in thicknesses of just a few millimetres and therefore cannot be expected to remove major imperfections in the substrate.

The compounds generally produce a hard smooth surface which covers minor imperfections likely to show through thin coverings. They can also be used to provide slightly absorbent surfaces to allow better adhesion for fixing materials laid on other relatively hard non-absorbent surfaces.

Some latex underlayments can be used over old adhesives to provide a smooth surface for new floorings.

Figure 4.16
Pouring an underlayment

Main performance requirements and defects

Choice of materials

Both self-levelling and trowelled types of underlayments can be laid on concretes or cementitious screeds. They require adequate preparation of the base to achieve good adhesion, with the latex based mixtures possessing the better adhesive qualities. Neither type of mixture is suitable for use on timber boarded substrates.

Strength and stability

Proprietary mixes do differ in their hardness characteristics and manufacturers should be consulted where high impact resistant surfaces are required. The most common defect to arise with the water/powder mixes is that of too high a suction in the substrate which removes water needed for hydration of cement. This results in a relatively weak surface. Some products contain water-retaining additives which inhibit the loss of water required for hydration.

Dimensional stability, deflections etc

Underlayments are stable when laid according to manufacturers' instructions.

Thermal properties

Underlayments are usually too thin to be of much significance.

Control of dampness and condensation

Some preparations retain good wet strength, but others do not. Information should be sought from manufacturers.

Fire

Non-combustible.

Sound insulation

The use of underlayments will make little difference to the sound insulation characteristics of floors.

Durability

There are two main potential problems.

- Weakness in underlayments, including by adding too much water to mixes, can contribute to rippling.
- Overlatexing leading to soft underlayments which may be susceptible to plasticiser migration from floorings.

Maintenance

Maintenance of underlayments should not be required. When floorings are removed for renewal, underlayments can become damaged in which case they will need also to be replaced by the same or a compatible material.

Work on site

Storage and handling of materials

Some underlayment emulsions have limited shelf lives. They should be stored in a frost-free environment.

Workmanship

Surfaces to be covered should be free from grease and dust. If necessary, major hollows or imperfections should be patched before the smoothing compound is laid.

A substrate which is too porous can jeopardise the self-levelling properties of a mixture and some additional trowelling may be needed. When recommended by suppliers, primers should be applied to substrates to increase adhesion, prevent loss of water, and prevent bubbling by displaced air in porous materials. Bubbling must be removed by careful trowelling.

Adding more than the recommended amount of latex to a latex based mix to improve the flow will compromise durability.

Inspection

The problems to look for are:

◊ underlayments detaching
◊ underlayments having weak surfaces
◊ rippling of finishes
◊ lack of durability

Case study

Rippling over a levelling compound

The flooring of wards in a hospital had shown slight rippling. The flooring consisted of flexible PVC sheet to BS 3261:Part 1 [203] Type A, 2.5 mm thick, stuck with a resin rubber emulsion; also sheet linoleum to BS 810 [204] (now replaced by BS 6826 [205]) of 2.5 and 3.2 mm stuck with a lignin paste over a latex levelling compound over a 60 mm bonded screed.

The ripples, generally not exceeding 0.3 mm in height, had appeared both in straight lines, and in a random manner across the floors. Removal of the flooring at the random ripples revealed similar random cracks in the substrate beneath. The straight cracks occurred over daywork joints.

Rippling in a screed is produced by three conditions acting in combination:

- the screed requiring a long drying time, usually nine months or more
- shrinkage cracking, curling and loss of bond occurring during the drying period
- the surface of the screed being treated with a water based levelling compound and the flooring then fixed with a water based adhesive

Where rippling from the described mechanism occurs it does so within 28 days of laying the flooring. Often small ripples can occur, but remain undetected until the floor has been polished; they then become noticeable against back lighting. This has misled some observers to report that the ripples have appeared some months after the flooring was laid, but it is doubtful whether this is the case. Ripples appear usually within 7 to 14 days and rarely after 28 days as a result of this mechanism. Ripples existing before the 28 days do not get worse after this time.

The remedial work at the hospital involved removing the sections of affected flooring, filling with an incompressible material (epoxy resin is usually recommended) and re-laying the floor.

Case study

Sulfate attack in a levelling compound

PVC sheet flooring laid over a thin synthetic anhydrite screed in a hospital unit had become defective, probably because moisture had entered the screed from leaking water supply pipes, and also because water from the shower room had penetrated to the adjacent bathroom via the joints in the ceramic wall and floor tiles. The moisture in the gypsum screed had allowed sulfate attack to occur on the levelling compound resulting in it losing strength. The adhesive between the screed and the PVC sheet had also failed.

The solution involved removing all material affected, providing adequate protection to the floor from water leaking from the shower rooms, and completely replacing the screed and flooring.

Chapter 4.4 Lightweight screeds

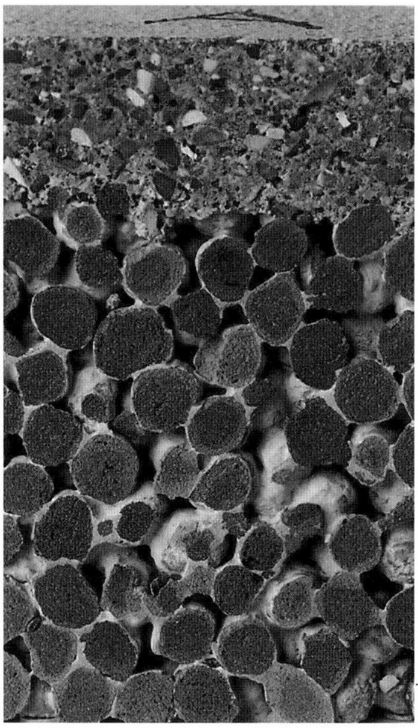

Figure 4.17
A section through a good quality no-fines lightweight aggregate screed with a dense sand and cement topping

Lightweight screeds are used for many reasons including achieving better thermal insulation or simply saving weight. Since the body of a lightweight screed is less robust than a solid sand and cement screed, providing an extra and more durable topping to resist the in-service loads is necessary (Figure 4.17).

Characteristic details

Basic structure

There are two basic types of lightweight cement based screed:

- lightweight aggregate
- aerated concrete

Lightweight aggregates absorb moisture in the mixing process and very high water:cement ratios are required to obtain workable mixes. As with dense screeds, therefore, the drying times tend to be prolonged. On the other hand, aerated screeds, containing much less water per unit volume, tend to dry more quickly.

Because of the reduced impact resistance when compared with dense sand and cement screeds, it is necessary to provide a thin topping of sand and cement. This may confuse later inspection.

Neither type can be used as the wearing surface of a floor.

Main performance requirements and defects

Choice of materials

A variety of lightweight aggregates can be used, though the most popular include sintered pulverised fuel ash (PFA) and expanded clay proprietary aggregates similar to those used for making lightweight concrete blocks.

Strength and stability

Both types of screed described in this chapter are capable of being used as light or medium duty substrates.

With respect to the ability of lightweight screeds to carry traffic, one of the main factors on which performance depends is the quality of the upper surface or topping; the BRE screed tester can be used to assess the resistance of lightweight screeds to indentation. Indicative values of depth of indentation (as given in the feature panel on the BRE screed tester in the inspection section of Chapter 4.1, but repeated here for convenience) of lightweight screeds under test are:

- very heavy traffic where subsequent disruption cannot be tolerated – 3 mm (Category A)
- heavy wheeled traffic in public areas – 4 mm (Category B)
- foot traffic and light wheeled traffic – 5 mm (Category C)

Categories B and C listed in the screed tester feature panel should be readily achieved; Category A can be achieved if particular attention is given to the mixing and cement content of the lightweight layer and to the thickness of the sand and cement topping.

Dimensional stability, deflections etc

Both types of screed also tend to shrink more than dense screeds, and bonding to the substrate is therefore much more critical to the avoidance of cracking and lifting. Actual amounts of shrinkage depend on the materials used.

Thermal properties

Lightweight cementitious screeds are not appropriate for use with underfloor heating because of their insulation value, although they may prove useful in other circumstances.

Control of dampness and condensation

These screeds must be left until excess water has dried out before laying flooring. Table 1.3 in Chapter 1.4 will give general guidance on drying times for lightweight aggregate screeds; the times shown can be reduced for aerated screeds.

Fire

Non-combustible.

Sound insulation

The cementitious screeds in this category are not suitable for floating floors.

Durability

Provided the screeds (and particularly the toppings) are of adequate thickness, mix designs are appropriate, and workmanship is of good quality, there is no reason to expect any less inherent durability than for dense sand and cement screeds. However, heavy impacts which punch through floorings and then through the toppings can cause premature failure.

Maintenance

Lightweight screeds are more prone to impact damage than dense sand and cement screeds. If the damage is widespread, there has been a failure of specification, and a stronger replacement may be called for instead of patching.

Work on site

Storage and handling of materials

Lightweight screeds are supplied and laid by specialist contractors.

Restrictions due to weather conditions

Precautions will be necessary in hot weather to prevent a screed drying out too rapidly.

Workmanship

If good quality is to be achieved, it is essential that screeds in this category are laid by specialist contractors.

A common problem with lightweight aggregate screeds has been caused by having too much water in the cement slurry used to bind the pellets together. This has led to cement slurry draining from the upper part of the screed leaving the pellets only weakly bonded together.

Inspection

The problems to look for are:

- ◊ screeds of insufficient strength showing elephants' footprints (check with the BRE screed tester)
- ◊ thin dense toppings of sand and cement not adequate
- ◊ top layers of screeds weak; sand and cement topping becoming hollow and detached
- ◊ irregularities in thin floorings
- ◊ moisture susceptible floorings detaching from bases

Case study

Disintegration of a lightweight screed in a university building

BRE Advisory Service was called in to examine the condition of a floor in a university building. In general the floor finish of PVC tiles on a lightweight cellular concrete screed was in an acceptable condition, but occasional detachment and breaking of the corners of individual tiles were observed along with localised disintegration of the lightweight screed beneath them. Most of the defects could be attributed to the manhandling of heavy objects over the floors and not to any inherent problem with the tiles or the adhesive. The screed also was behaving as well as could be expected in spite of the fact that it was incorrectly specified. Remedial measures included patching and levelling the screed, and replacing damaged or loose tiles together with a recommendation for good housekeeping. Further damage of the tiles would indicate a need for a more robust screed.

Chapter 4.5 Screeds based on calcium sulfate binders

Where an existing screeded floor needs to be replaced, screeds based on calcium sulfate binders offer the opportunity to install thermal insulation since they can be laid much thinner than conventional screeds. The thinner screed and a thin layer of thermal insulation may be laid within the same overall thickness as the thicker cementitious screed. This can only be done, however, if there is absolute certainty that the floor will not become damp.

These screeds may also be useful for replacement work when time scales are very short since strength develops quickly. A floor can normally be used after 24 hours which may be crucially important where the building needs to be occupied quickly.

Figure 4.18
A pumpable anhydrite screed being laid (Photograph by permission of Isocrete)

Characteristic details

Basic structure

There are two kinds of screeds based on calcium sulfate binders:

- anhydrite
- hemi-hydrate

Anhydrite flooring consists of crushed rock anhydrite with ground synthetic anhydrite as a binder. It was first introduced on a large scale in Germany, but later was manufactured in the UK. When first introduced into the UK it was mixed with water and an accelerator (hydrated lime and potassium sulfate) and laid by trowel, effectively, as a thin screed, or what might be called a plastered floor. It has mainly been laid in thicknesses of around 15 mm – that is to say significantly thinner than cementitious screeds – but sanded mixes have been thicker. Although the initial set is rapid, ultimate strength builds a little more slowly than that of the hemi-hydrates. By the early 1960s the mix usually used was synthetic anhydrite with sand. It was laid similarly to a sand and cement screed, commonly unbonded using a separating layer of waxed paper, to a thickness usually between 20 and 40 mm. In about 1990, self-levelling pumpable anhydrite screeds became available (Figure 4.18).

Anhydrites have generally been used in domestic floorings, and in other situations only where traffic is very light.

Hemi-hydrate flooring (the alpha hemi-hydrate) is more closely related to the wall plasters (the beta hemi-hydrates) than to the anhydrites. It is quicker to set, but will not set so hard as anhydrite. Mixtures which flow have been produced, making laying somewhat easier.

Main performance requirements and defects

Choice of materials

Anhydrite screeds may be laid instead of sand and cement screeds except in areas which are wet or where the screeds may become wet.

Strength and stability

All the types of screed described in this chapter are capable of providing light and medium duty substrates.

Anhydrite has been laid successfully on sand over a troughed steel floor deck.

Indicative values of depth of indentation (as given in the screed tester feature panel in Chapter 4.1, but repeated here for convenience) for the performance of bonded or unbonded anhydrite and hemi-hydrate screeds under test are:

- very heavy traffic where subsequent disruption cannot be tolerated: 3 mm (Category A)
- heavy wheeled traffic in public areas: 4 mm (Category B)
- foot traffic and light wheeled traffic: 5 mm (Category C)

Anhydrite used as a flooring, bound with oils, is suitable for domestic and certain other areas subjected only to very light usage. It is not suitable for wheeled traffic.

Dimensional stability, deflections etc

Coefficient of linear thermal expansion per °C: 12 to 20×10^{-6}.

Drying shrinkage: 0.001% – in other words negligible.

Thermal properties

Thermal conductivity of gypsum hemi-hydrate screeds: 0.46 W/mK, and of anhydrite screeds: around 1.6 W/mK. None are warm to the touch.

Anhydrite screeds are very appropriate for use with underfloor heating. However, heating appliances, such as boilers, should not be placed over anhydrite screeds if they are likely to heat the screeds to more than 70 °C when strength begins to decline.

Control of dampness and condensation

Since practically all the water used in the mix for a semi-dry anhydrite screed is used in the setting process, there is very little excess water remaining to dry out. Nevertheless, it is usual to allow a minimum drying time of three weeks to be on the safe side. A calcium sulfate screed must be laid on a base protected by an effective DPM since it rapidly loses strength in the presence of water.

Case study

Failure through sulfate attack of a composition block floor over an anhydrite screed

BRE Advisory Service was called in to examine a composition block floor laid over an anhydrite screed on a suspended structural floor in a newly built school in the north of England. Underfloor heating coils were present and had been used. Three months after the school opened the blocks had bulged and arched, and the undersides of the tiles and the cementitious bedding mortar were seen to be very wet – globules of water were present.

Earlier checks by other investigators had established that there were no leaks in the heating system and, by the time the BRE investigator had arrived on site, the whole of the block flooring had been removed.

When the samples of blocks and bedding mortar were examined, most were shown to be adhering well together. However, the underside of the bedding was covered with a 0.5 mm deposit of white powder. This powder was examined in the laboratory and found to contain significant amounts of the mineral ettringite, showing clearly that sulfate attack had taken place.

Arching and bulging are usually caused by expansion of one of the layers at or near the surface. It was possible that there had been a slight contribution from moisture movement of the blocks themselves once the bond had been broken, though it was unlikely to have initiated the failure. In this case, though, it was sulfate attack on the mortar bed which was the prime cause.

For sulfate attack to take place, three constituents are needed:

- soluble sulfate salts
- ordinary or rapid hardening cement
- persistent wetness

Soluble sulfates were present in the anhydrite screed, and Portland cement in the bedding mortar. But where had the water come from? Four possible sources were investigated:

- from the exterior – groundwater or rain penetration
- spillages or leaks from plumbing
- condensation
- excess construction water

Water from the exterior was eliminated. Not only was the floor suspended, but there was an effective DPM installed. This source could be discounted.

It was reported that there had been no leaks from the plumbing system, and there had been no spillages (eg of cleaning water).

If condensation had occurred in this type of floor construction (composition block), the condensation would have occurred on the tops of the blocks and the sealing coat would have prevented it reaching the undersides of the blocks.

That left construction water. For anhydrite screeds laid semi-dry, the rate of drying would require more than 20 days. Less time than that had been allowed before laying the flooring, and it was therefore very unlikely that the screed was fully dry at the laying stage. There were also indications that the bedding mortar had been mixed on top of the screed, thereby allowing even more water into the screed. Since anhydrite screeds have a higher suction than cementitious screeds, it is possible that the bedding mortar was made a little wetter than normal to overcome the suction.

The BRE investigator concluded that the construction water could easily account for all the water found.

Since this investigation had been going on for most of the summer holidays, little time was left (18 days) to install replacement flooring. To undertake the remedial work within the short timescale, the recommendations were:

- keeping the screed as dry as possible
- reinstating the patches damaged by the investigation of possible plumbing leaks
- coating the surface of the screed with three coats of polymer emulsion, acrylic polymers or SBR (styrene-butadiene rubber), using the last coat as a bonding agent
- bedding the replacement composition blocks in sulfate resisting Portland cement mortar after mixing the mortar away from the screed
- sealing the blocks only after the bedding mortar had dried

If client confidence had been lost in the anhydrite screed or composition block flooring, consideration could have been given to replacing the screed with an early drying cementitious one and an alternative flooring. This would have tended, however, to impinge on the tight replacement timetable.

These screeds are therefore unsuitable for use in wet areas such as kitchens and washrooms.

The drying time of proprietary pumped self-levelling anhydrite and hemi-hydrate screeds is somewhat longer than those laid semi-dry.

Ceramic tiles can be laid on calcium sulfate screeds provided they are dry and will remain so. Tiles are best fixed with thin bed proprietary adhesives, but manufacturers' advice on priming the surfaces of the screeds must be strictly followed.

The use of cementitious bedding mortars for floorings directly over gypsum screeds must only be contemplated if there is **absolute certainty that the floors are dry and will remain dry**, as the case study on page 175 shows.

Fire

Non-combustible.

Sound insulation

Both anhydrite and hemi-hydrate screeds are suitable for floating floors.

Durability

Anhydrite screeds should not be covered with any flooring involving cementitious mixes or beddings unless suitable precautions are taken to prevent the interaction of sulfates on the cements. Primers are available for applying to screeds before laying underlayments.

Maintenance

Gypsum based screeds should be kept dry at all times.

Work on site

Restrictions due to weather conditions

Precautions will be necessary in hot weather to prevent an anhydrite screed drying out too quickly, as with a cementitious screed.

Workmanship

Screeds in this category must be laid by licensed contractors. Even so, as the telephone exchange case study aside shows, the margin between success and failure can sometimes be slender. The gypsum screeds should be laid on a separating layer of polyethylene.

Inspection

The problems to look for are:

- ◊ screeds generally breaking up
- ◊ screeds of insufficient depth over buried services
- ◊ wearing surfaces in moderate or densely trafficked areas breaking up
- ◊ screeds laid directly onto cementitious bases not being permanently dry
- ◊ cementitious material in contact with gypsum floors breaking up
- ◊ sanded semi-dry mixes not adequately compacted

Case study

Breakdown of synthetic anhydrite screeds in a hospital

The floor of a hospital had been laid with PVC tiles or textiles on synthetic anhydrite screeds. In several heavily trafficked areas the floorings had lifted and the screed appeared to have broken down (Figure 4.19).

Figure 4.19
Breakdown of a synthetic anhydrite screed leading to lifting of PVC tiles. The upper surface of the screed has become detached along with the adhesive and tiles

Examination of samples of the screed indicated that it was dry, eliminating rising damp as a cause of breakdown. In any case the breakdown was not consistent with that caused by moisture. Hard nodules of anhydrite binder and aggregate indicated that the binder was in a satisfactory condition. The only logical conclusion was that the screed was not properly compacted at the time of laying, and was therefore inherently weak – just as a conventional sand and cement screed would have been in similar circumstances.

Case study

Breakdown of synthetic anhydrite screeds in a telephone exchange

BRE Advisory Service was invited to investigate floorings which had given problems in a telephone exchange. The trowelled anhydrite semi-dry screeds had been sanded and were unusually thick (75 mm), but in spite of this had broken down in some areas.

Two tests were carried out:

- impact tests by the BRE screed tester
- chemical analysis

The tests using the BRE screed tester showed that most results after the fourth blow of the impactor were under 3 mm indentation, indicating that most of the screeded areas had achieved satisfactory strength. However, a few cases exceeded 3 mm and were beyond the margin of acceptability.

The chemical analysis showed that the mix proportions were approximately of the right order – 2.5:1 sand:synthetic anhydrite. The analysis also showed, though, that an insufficient amount of activator had been added to the mix making it weak. The lesson to be learnt was that full quality control on site was critical to performance.

Chapter 4.6 Matwells, duct covers and structural movement joints

Matwells are set into floorings in order to accommodate the thickness of mats or other forms of flooring used in entrances to prevent wetness and detritus from the exterior being brought into the interior of a building (Figure 4.20).

Ducts and duct covers enable building services to be installed within floors to provide access for repairs and replacement of pipes and conduit. Indeed, Water Byelaw 58 requires that all water services should be readily accessible for repair, removal and replacement. No joints are allowed which are not readily accessible by removable duct covers.

To prevent disruption and unintended gaps in floor coverings, structural movement joints in coverings are made to coincide with movement joints in the floor structure below.

Characteristic details

Basic structure

Matwells

Matwell frames are normally made of metal angle or tee section, welded or otherwise jointed into rectangles to suit appropriate sizes and thicknesses of matting. They should be fixed firmly to bases through the horizontal flanges of the frames. Most examples will be found fixed by cramps or holdfasts into the base, but some may have been fixed only by adhesion of the overlapping shoulder of the screed to the base. They serve to armour and protect both the matting and the adjacent flooring and screeds from scuffing and impacts from traffic.

Duct and inspection covers

Duct and inspection covers are normally fabricated from steel or aluminium or, perhaps, brass. They are made in short lengths and, in many cases, recessed to accommodate floorings to match the surroundings. Internal drainage manhole covers must incorporate seals.

Structural movement joints

These are positioned to coincide with structural movement joints in the base and are normally anchored within the screed. Each side of the joint will be armoured with a metal angle fixed to the base in similar manner to matwells described above. Alternatively the armouring may consist of a strip of metal anchored back into the screed with wire ties. Metals may be stainless or galvanised steel or brass. Occasionally, the armouring may consist of plastics strips. The sides of structural movement joints too are vulnerable to scuffing if inadequately fixed. The level of finish of the joint and any possible cover strip will depend on the flooring.

Detailing

Matwells

Barrier matting at the entrances of buildings needs to be sufficiently large to remove rainwater and accompanying detritus from the soles of footwear (Figure 4.21 on page 178). Otherwise floorings will quickly deteriorate (Figure 4.22 on page 178).

The upstand of the angle or tee forming the edge of the matwell should be capable of being set to finished floor level, ±1 mm (Figure 4.23 on page 178).

Figure 4.20
A domestic matwell (with the mat partly removed) set into a floor of ceramic tiles

Figure 4.21
Extensive barrier matting, which has been provided for this building entrance, reduces wear and soiling on adjacent floorings

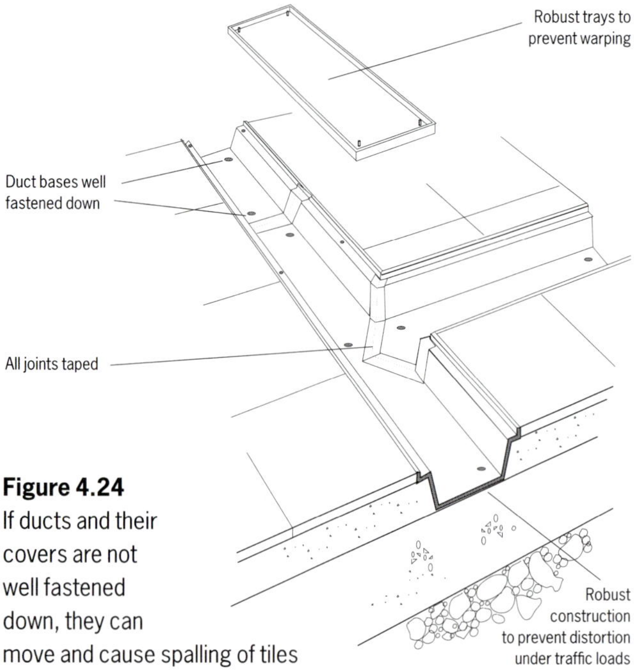

Figure 4.24
If ducts and their covers are not well fastened down, they can move and cause spalling of tiles

Figure 4.22
This barrier matting is totally inadequate and should extend well into the building. The linoleum is heavily stained by water and dirt brought in by foot traffic

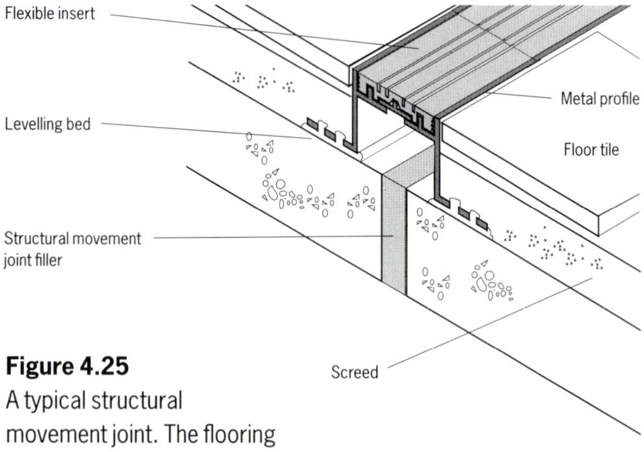

Figure 4.25
A typical structural movement joint. The flooring must be interrupted to accommodate a joint with a flexible insert. The flexible insert will usually wear and split, and need periodic replacement

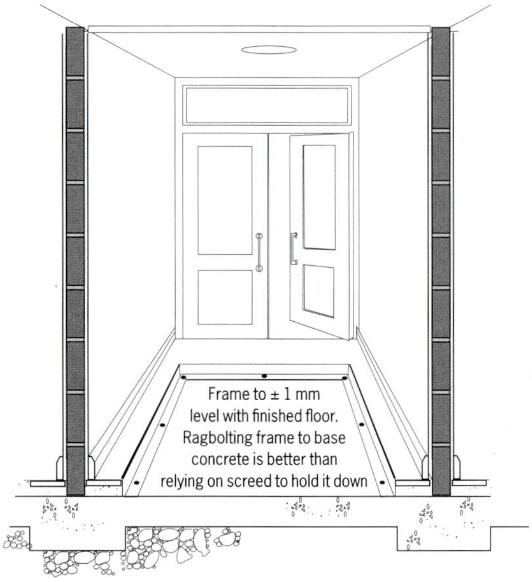

Figure 4.23
Matwell frames should be firmly fastened to the substrate

Figure 4.26
These upper and lower covers have been raised to give access to telecommunications wiring. The edges of the flooring tiles are beginning to spall in spite of the substantial brass section frame

Ducts and covers
Both ducts and their covers set within floors need to be sufficiently robust to resist distortion under loads, particularly where the floorings are thin or brittle (Figure 4.24).

Structural movement joints
BRE investigators have encountered a number of examples where attempts have been made to carry floorings over structural movement joints without making sufficient provision for the inevitable disruption when the building moves. The flooring must be interrupted to provide a joint with sufficient movement accommodation factor – movement accommodation factor, or MAF, is a measure of the ability of a movement joint filler or sealant to accommodate tensile strain (*Cracking in buildings*[148]). A number of standard or proprietary solutions are available (Figure 4.25).

Main performance requirements and defects

Choice of materials
Angle section, channelling and strip for matwells, ducts, structural movement joints etc are normally made in metal, either galvanised mild steel or non-ferrous metal (eg aluminium or brass).

Strength and stability
Many types of cover are not designed to develop their full strength until filled with screed and, consequently, they could be easily damaged when being installed. There have been problems with trays being too shallow; even though they are correctly filled with screed and flooring, they deflect in service causing tiles to spall at their edges.

Dimensional stability, deflections etc
Coefficient of linear thermal expansion per °C:

- cast iron 10×10^{-6}
- mild steel 12×10^{-6}
- aluminium 24×10^{-6}
- brass 21×10^{-6}
- ebonite 65 to 80×10^{-6}
- GRP (glass-fibre reinforced polyester) 20 to 35×10^{-6}

Reversible moisture movement of metals is 0%. However, there could be moisture movement of the infilling in trays. See the chapter appropriate to each flooring material.

It is important that structural movement joints remain free to allow the adjacent parts of buildings to move towards or away from each other, and their components to expand and contract. Joints should not be filled with incompressible material or detritus subsequent to installation, their integrity being maintained under maximum shrinkage.

Thermal properties
Metal ducts and internal inspection covers for drains and for access to services will tend to form cold bridges through the upper layers of floors if they are accommodated as interruptions to any thermal insulation layer. This is only likely to be a nuisance if they occur at the building's perimeter where the wall is also inadequately insulated.

Control of dampness and condensation
It is good practice to seal all joints in ductwork laid within the floor depth with adhesive tape. Pipes carrying cold water from outside the heated enclosure are liable to condensation when they reach the heated enclosure of the building. Corrosion of metals may ensue, therefore, if they are not insulated and covered with a vapour control wrapping.

Fire
Where a duct penetrates any fire resisting enclosure (eg a protected escape route), dampers or some other form of protection will be required within the duct to prevent fire spread.

Sound insulation
Ducts penetrating separating walls will be a source of noise transmission. Their installation should be avoided if at all possible, but it may be possible to obtain some improvement by dense packing within the duct.

Durability
Durability will depend, to a large degree, on the density and character of traffic; in particular whether there is any wheeled or castored traffic (Figure 4.26). Plastics armouring strips will be less durable than metal, and consideration will need to be given to the disruption caused when their replacement becomes necessary.

Maintenance
Matwells
Although the mats at entrances will need frequent inspection and maintenance, the armouring of the wells themselves should need little attention if correctly fixed.

Ducts and covers
Rocking duct and inspection covers, and their frames, need attention. Otherwise, apart from resealing after access has been completed, little attention should be required (Figure 4.27 on page 180). Metal ducts have been seen to work loose if the bounding screed breaks down, and brittle tiles set within covers may need replacing where the covers are insufficiently robust (Figure 4.28 on page 180).

Structural movement joints
These joints probably only need attention if the original specification was inadequate to accommodate the movements experienced in service, but it does depend on design. Flexible strips may need periodic renewal.

Figure 4.27
A well-sealed duct cover in PVC safety flooring in a wet area

Figure 4.28
Lack of robustness in a duct access tray has caused the tray to deflect. The brittle flooring tile edges have been compressed leading to spalling

Work on site

Inspection

The problems to look for are:

Matwells

◊ breakup of floorings alongside matwells
◊ compaction of screeds adjacent to frame edges not adequate
◊ adjacent floorings not level

Ducts and covers

◊ joints not sealed with adhesive tape
◊ condensation and corrosion
◊ dampers and fire stopping absent from ducts passing under compartment walls
◊ brittle floorings not adequate to resist deflections
◊ adjacent floorings not level

Structural movement joints

◊ movement joints filled with grout or detritus or other incompressible material
◊ joint filling material not retaining movement capability and, if plastic or elastic, not hardened
◊ sides of joints (armouring) worn or displaced
◊ adjacent floorings not level

If cracks appear in floorings parallel to and short distances away from structural movement joints, investigations should include establishing the positions of daywork joints in substrates.

Chapter 5 Jointless floor finishes

This chapter deals with floorings which are formed in situ from materials laid in molten, liquid or plastic form. The thicker types are normally trowelled and the thinner ones painted or poured to cover comparatively large areas without joints.

Since the finished surfaces of the materials are formed in situ, the quality of workmanship is crucial to the acceptability and durability of the flooring.

This book is not the place to deal with ancient floor finishes such as grip mentioned in the introductory chapter. Local 'recipes' for mixes show considerable differences, and those wishing to replicate such finishes in reconstructions will need to undertake research.

Figure 5.1
A jointless floor finish in a warehouse. These situations provide some of the most onerous service conditions for floorings

Chapter 5.1 Concrete wearing surfaces

It was pointed out as early as 1959 (Building Research Station Factory Building Studies No 3[206]) that, despite the ease with which in situ concrete finishes could be laid, many of them fell far below acceptable standards of quality. To judge from the number of instances of poor performance which came to the notice of the BRE until around the early 1980s, matters had not improved perceptibly within those 20-odd years. The case study aside epitomises the kinds of problems which used to occur.

However, as the charts in Figure 0.15 show, there has been a significant decrease in problems between the 1970s and the early 1990s (the segments for concrete in the charts also include granolithic topping problems). This improvement has been brought about by a number of factors which include:

- improved methods of laying concrete floors (eg by the long strip method, including the use of vibrating beams, laser screed machines and vacuum dewatering)
- better knowledge of concrete mix design
- appropriate use of additives
- provision by ready-mix suppliers of concrete conforming to specification
- better methods of compaction and finishing, including power floating and power trowelling.

All these methods provide increased abrasion resistance compared with hand trowelling.

Figure 5.2
Concrete which has been finished to provide a wearing surface in a warehouse

Case study

Disintegration of a concrete floor in a warehouse after a few months of use

The concrete floors in the warehouse of a transport depot in a London suburb, which had been in use for only a few months, were showing signs of disintegration. The concrete was 200 mm thick and the mix was 1:2.6:4.5. There was a large amount of dust present from the breakup of the surface. Power trucks with pneumatic tyres were generally used, but there was also much traffic from small trolleys with hard wheels. The concrete appeared to be as required by the design, but it was totally inadequate for its task. The cause of the deterioration was mainly inadequate compaction of the slab which made it vulnerable to the heavy traffic.

There have also been improvements brought about by a better understanding of the need for curing, and the entry into the market of specialist and competent contractors.

Characteristic details

Basic structure

The essential requirements of a concrete floor are covered in Chapter 3.1. This Chapter (5.1) deals only with floors where the concrete is finished to form the wearing surface (ie to take traffic directly).

Traditionally, plain concrete was used to provide floors in a wide range of buildings (eg workshops, stores and warehouses). The concrete was brought to a smooth finish by floating and hand trowelling shortly after placing and compacting the concrete. These floors were commonly made from mix proportions of 1:2:4 Portland cement:fine aggregate:coarse aggregate. Constituents were mixed on site. Close control was not considered to be of crucial importance with the result that many mixes were over-watered, leading to excessive bleeding and laitance on the surface. However, most floors of plain concrete had sufficient abrasion resistance to withstand foot traffic and trolleys with rubber wheels, but they were inadequate for those circumstances where steel wheeled trucks were used and for which granolithic concrete was the preferred option.

Modern mix design and finishing can produce concretes with excellent abrasion resistance suitable for the heaviest traffic. Table 5.1, adapted from BS 8204:Part 2[192], relates the mix design and finishing technique for a concrete floor to the use to which the floor is intended to be put.

An alternative sometimes used for those floor areas receiving heavy wear is to pave them with precast concrete paving slabs, similar to those used on external pavements. These have considerable advantages in industrial areas, where overnight repair means that traffic disruption is minimal, if at all. The cost of course is higher than for in situ concrete, but the premium may be worth paying.

BS 8204:Part 2, published in 1987, is the standard appropriate for concrete wearing surfaces. Floors constructed prior to 1987 should have complied with BS CP 204: Part 2[199].

Finishing

The surfaces of slabs which receive no further applied finish are normally finished by one of three methods:

- power floating (Figure 5.3)
- power trowelling
- hand trowelling

All three methods, applied by skilled operatives, are capable of providing satisfactory wearing surfaces. Special laying and finishing techniques are available for producing very flat floors for special cases; for example in high rack warehouses (The Concrete Society Technical Report No 34[193]).

Rather than having a high strength concrete throughout the thickness of the slab, the concrete can be topped with a sprinkle finish. For this a lower grade of concrete is used and a special hard aggregate, sometimes with cement, is sprinkled onto the still-plastic material and trowelled in. Thicknesses of sprinkle finishes commonly vary from 1 to 5 mm.

Main performance requirements and defects

Appearance and reflectivity

New Portland cement concretes give reflectances of around 0.45. They can become, however, very dirty and are vulnerable to staining.

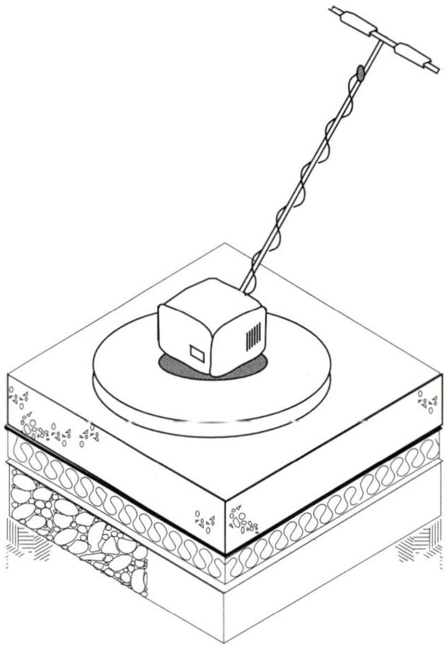

Figure 5.3
Power floating a concrete surface usually provides adequate wearing qualities, depending on the suitability of the mix

Table 5.1
Selection of specification for a concrete floor according to use

Performance required	Use	Type of concrete and grade	Finishing process
Severe abrasion and impact	Very heavy duty engineering workshops etc	Special mixes, proprietary sprinkle finishes	Trowelling, depending on degree of compaction needed
Very high abrasion; steel wheeled traffic and impact	Heavy duty industrial workshops, special commercial etc	High strength toppings and other special mixes C60+ (granolithic concrete)	Trowelling two or more times followed by curing
High abrasion; steel or hard plastics wheeled traffic	Medium duty industrial and commercial	Direct finished concrete C50	Trowelling two times followed by curing
Moderate abrasion; rubber tyred traffic	Light duty industrial and commercial	Direct finished concrete C40	Trowelling two times followed by curing

Choice of materials
In this situation the concrete slab also, of course, is a base. Therefore the selection of hardcore, blinding and DPM must be undertaken precisely to ensure uniformity of support and protection against rising damp.

See also the section on choice of materials in Chapter 4.1.

Strength and stability
Good. See also Chapters 1.1 and 4.1.

Dimensional stability, deflections etc
Coefficient of linear thermal expansion per °C: 10 to 13 × 10^{-6}.

Reversible moisture movement: 0.02–0.06%.

Wear resistance
Abrasion resistance varies between poor and very good depending on the type of aggregate, mix, curing, finishing etc. Impact resistance is normally good, and longterm resistance to indentation very good.

Wear resistance is greatly influenced by the quality of finishing and the strength of the concrete to produce a dense, tight surface layer. Good power floating, followed by power trowelling, can greatly enhance the wearing qualities of a concrete floor.

Lower grade coarse aggregates in concrete do not necessarily have a detrimental effect on abrasion resistance of concrete floors. Lower grade fine aggregates in concrete, however, are generally detrimental to the abrasion resistance of concrete floors.

Even the best finish to in situ concrete, with surface hardeners, will produce dust sooner or later; and where the environment demands a dust-free finish, as in certain manufacturing processes, it is better to seek alternative finishes. Surface hardeners, though, cannot restore wearing qualities to a poor mix (Figure 5.4).

Slip resistance
The slip resistance of smooth concrete surfaces is fair to good, depending on surface characteristics. Surfaces are slippery when wet unless a textured finish is applied or slip resistant aggregate used.

Coefficients of friction are, for smooth finish:
- dry > 0.75
- wet 0.2 to 0.3

for carborundum finish:
- dry > 0.6
- wet 0.3 to 0.5

Figure 5.4
Poor quality concrete will not provide a suitable wearing surface. In this example a floor laid to falls for drainage purposes has worn badly and will need to be refinished

Case study

Curling of concrete slab edges in large bays

BRE investigated curling at the concrete bay edges in the floor of a goods receiving area at a hospital. Examination of the floor revealed a combination of three factors: the misplacement of steel reinforcement, concrete being laid in bays slightly larger than recommended in the relevant standard (BS CP 204), and inadequate curing. The curling at the bay edges had happened during drying, but it was found to be worse in floor areas where the concrete was thinner than specified. Further drying shrinkage movement taking place subsequent to the time of examination was likely to be negligible. Remedial measures that were suggested included drilling holes at regular intervals along the line of the joints at hollow and curled edges, and injecting a grout or epoxy resin. When dry, the curled edges should then have been ground before laying a new floor finish.

Control of dampness and condensation
Concrete is unaffected by water and can be used, therefore, in wet areas, provided no chemicals are transported by the wetness.

Thermal properties
Thermal conductivity of concrete: 0.1–1.8 W/mK, depending on density.

Warmth to touch
Very poor.

Fire
Non-combustible and no flame spread.

Suitability for underfloor heating
Suitable for all likely temperatures, but concrete will be adversely affected by accelerated drying shrinkage which must be taken account of at the design stage.

Sound insulation
Sound absorption is very poor. Airborne sound insulation depends largely on the mass of the structural floor. Unmodified cementitious screeds can be used in a floating floor but performance depends on overall thickness of the screed.

Durability

Portland cement concrete surfaces have the following chemical resistances to various liquids:

- water — very good
- organic acids — poor
- hydrochloric acid — very poor
- sulfuric acid — very poor
- nitric acid — very poor
- alkalis — good
- sulfates — poor
- mineral oil — good, but will stain
- animal oil — poor
- organic solvents — good

Maintenance

It is difficult to maintain concrete surfaces in anything resembling their original appearance since even the most dense concrete will absorb liquids and become stained. Scrubbing with hot water and detergents will remove dirt. The surface may be sealed with a hardening agent or a resin from time to time to prevent or reduce dusting.

Repairs to surfaces depend to a large extent on the bond achieved between old and new work. The old concrete should be cut down to a sound surface, and undercut to provide a key. The area should be well wetted to reduce suction and a cement slurry mopped over the surfaces to be filled. Proprietary bonding agents are beneficial.

Work on site

Storage and handling of materials

See the same section in Chapter 2.2.

Restrictions due to weather conditions

See the same section in Chapter 2.2.

Workmanship

Failure of replacement floorings on old concrete bases can often be attributed to inadequate preparation of the surface of the concrete before laying the new flooring. The feature panel in the same section of Chapter 2.2 describes suitable techniques for reducing the risk of problems occurring.

Deterioration in concrete surfaced floors usually results from:

- badly graded aggregate
- inadequate control of mix proportions
- too much water in the mix
- inexpert power or hand finishing
- inadequate curing

Sometimes the aggregate settles or the mix is trowelled too early and a smooth finish with surface laitance is produced. Too wet a mix produces high drying shrinkage with consequent cracking and curling at bay perimeters, as well as weak dusty surfaces.

Further aspects of workmanship are covered in BS 8000:Part 2:Section 2.1[152] and Section 2.2[153]. BS 8000:Part 9[194] also applies.

Inspection

In addition to the appropriate items listed in the inspection section at the end of Chapter 4.1, the problems to look for are:

- ◊ marine deposits in sands
- ◊ services not laid in ducts
- ◊ mix proportions not between 1:3 and 1:4.5 by weight (exact proportions to be established by laboratory examination)
- ◊ cracking
- ◊ poor wear resistance
- ◊ slipperiness
- ◊ breakdown of finish
- ◊ upwards curling at perimeters
- ◊ dusting
- ◊ surface 'pop-outs' (Figure 5.5)
- ◊ spots of low abrasion resistance

Figure 5.5
A 'pop-out' caused by pyrites in aggregate

Chapter 5.2 Polymer modified cementitious screeds

These screeds are commonly used where thin toppings of around 8 to 10 mm are required to be laid onto concrete bases, although they can range in thickness up to 40 mm. They normally have good bond strength to the concrete base and good abrasion resistance. Chemical resistance is enhanced by the incorporation of polymers.

Characteristic details

Basic structure

In the past cementitious screeds have been mixed with a variety of modifying agents to produce finished floors. The modifiers have included natural latex and synthetic rubbers of various kinds, PVA (polyvinyl acetate), SBR (styrene-butadiene rubber) and bitumen emulsions.

PVA has been rarely used since the mid-1970s because of its sensitivity to water; SBR and acrylic polymers are now much more common.

Mixes incorporating 10–25% rubber latex have been used where finishes having low permeability to water were desired. Aggregates and fillers may comprise a wide variety of materials – cork, rubber, hardwood, sand, slate and wood flour. Some may contain ground asbestos. The most common aggregates found will be sand and fine granite or other stone chippings.

Bitumen modified mixes have frequently been used for patching floors in industrial premises where appearance was not a priority, but, occasionally whole floors may be found where the material has been waxed and polished, and used as the final surface. The mix is essentially a sand and cement mix gauged with bitumen emulsion.

The standard appropriate to the materials described in this chapter is BS 8204:Part 3[54]. In previous years they were partly covered by BS CP 204:Part 2[199].

Detailing

Although these floorings are most often laid as wearing surfaces, they can be used as screeds to receive other floorings. They are very useful in making up thicknesses of between 10 and 25 mm. Coves and skirtings can be formed in the materials.

Main performance requirements and defects

Appearance and reflectivity

Appearance depends largely on the aggregates and fillers used. Sometimes the surface is ground to expose decorative aggregates. They can become drab in the absence of suitable maintenance.

Bitumen modified cements are usually black or very dark grey. Modified screeds containing other polymers are usually dark grey and can easily be mistaken for other cementitious materials not containing polymers.

Reflectivity is usually fair at around 0.45, but depends on cleanliness.

Choice of materials for substrate

Most mixes adhere well to any firm substrate, though concrete is the preferred one. Although the materials have been used on timber boards, they are likely to be disrupted over tongued-and-grooved or plain edge board joints, so a suitable underlay is required. Expanded metal was formerly recommended for reinforcement, though even with this the materials could not always cope with movements in boarded floors.

Strength and stability

Resistance to indentation is usually good, but the bitumen mixes may only give fair performance. Overlatexing can lead to softer underlayments. See also Chapter 1.1.

Since many of these floorings are laid relatively thinly, the bond strength to the base can be critical. Bond strength should not be less than 1.0 N/mm^2 and preferably reach 1.5 N/mm^2 where subjected to wheeled traffic.

Dimensional stability, deflections etc

Coefficient of linear thermal expansion per °C: 20 to 40×10^{-6}.

Cement rubber latex floorings have a relatively high shrinkage rate, but disruption of the surface is rare because of the material's resiliency.

Wear resistance

Resistance depends largely on the aggregate or filler, though most mixes will give good performance. Higher proportions of latex in the latex modified mixes will give improved wear resistance up to the point where the material softens.

Slip resistance

The slip resistance of surfaces is normally good or very good, depending on additives. Bitumen modified mixes are slippery when wet unless a textured finish is applied or slip resistant aggregate used.

Coefficients of friction are, for smooth finish:

- dry > 0.75
- wet 0.2 to 0.3

for carborundum finish:

- dry > 0.6
- wet 0.3 to 0.5

Control of dampness and condensation

Although the finishes are all relatively impervious to water, they cannot be regarded as equivalent to DPMs. Just how impervious depends on the type and amount of polymer incorporated. In the years following the 1939–45 war it used to be common practice not to lay a DPM below these floorings. More recently the DPM has usually been placed under the main slab. These floorings are bonded to the slab and it is therefore not possible to lay DPMs immediately below them unless they, the DPMs, are of the epoxy bonded type.

PVA modified mixes are less resistant to water than natural latex modified mixes, but are more resistant to oil.

Bitumen modified mixes are susceptible to softening by oils and greases.

Thermal properties

Thermal conductivity of polymer modified cementitious screeds: 0.1 to around 1.8 W/mK, depending on density.

Warmth to touch

Fair or good to fair.

Fire

Non-combustible.

Suitability for underfloor heating

Suitable.

Sound insulation

Sound absorption is very poor. Airborne sound insulation depends largely on the mass of the structural floor. Modified screeds can be used in a floating floor; performance depends on overall thickness of the screed.

Durability

Plasticiser migration may occur if these products are used below flexible PVC floorings and too much latex has been added to the mixture.

Bitumen emulsion cements should not be used as screeds immediately below sheet and tile flooring. Instead they should be separated by laying 3 mm of a suitable underlayment on top of the screed.

Maintenance

Washing using a mild detergent is the preferred method of cleaning. Scrubbing should be avoided. Ground surfaces of the latex modified mixes can be polished with a wax emulsion. Solvents should be avoided.

Work on site

Storage and handling of materials

See the same section in Chapter 2.2.

Restrictions due to weather conditions

See the same section in Chapter 2.2.

Workmanship

The strength of the bond to the base concrete is critical. It is also essential that the base concrete has sufficient strength and that preparation is thorough. In case of doubt, the pull-off bond strength should be measured and not be less than 1.0 N/mm^2.

BS 8000:Part 9[194] and BS 8204: Part 3 are relevant.

Inspection

The problems to look for are:

- poor wear resistance because lower grade fine aggregates used in mix
- slipperiness
- finishes breaking down
- finishes of latex modified screeds breaking down
- screeds breaking up
- reduced or missing polymer content
- bitumen emulsions softened by solvents

Chapter 5.3 Granolithic and cementitious wearing screeds

Traditionally, granolithic finishes have been used where a high degree of abrasion resistance is required. Indeed, the derivation of the term is from the granite aggregates used. Typical applications would be communal areas in blocks of flats, particularly entrance halls and lift lobbies, and in industrial areas where a hard wearing but relatively inexpensive surface was sought.

Granolithic paving was formerly laid either as a topping on fresh concrete or as a screed, either bonded or unbonded. It is no longer laid as unbonded.

Sand and cement screeds are not usually considered to be suitable as wearing surfaces. However, if they are laid slightly wetter than normal levelling screeds (ie not semi-dry), they can provide moderate abrasion resistance. The addition of small amounts of 10 mm single sized aggregate to the mix will enhance the wear properties and reduce drying shrinkage problems. They will often be found where a high degree of wear resistance is not required and, in these situations, they can perform well.

Since 1985, granolithic flooring has lost market share to power floated or power trowelled directly-finished concrete slabs which provide comparable durability to granolithic finishes but without so many defects.

Characteristic details

Basic structure

Granolithic concrete wearing surfaces are essentially concrete which contains suitable granite or other crushed hard rock aggregate, such as quartzite, to provide enhanced wear resistance. The maximum size of aggregate was commonly 10 mm, but up to 14 mm has been used. It can be found in two forms:

- as an integral topping to an ordinary concrete mix, laid before the base has achieved initial set – known as monolithic construction
- as a separate granolithic screed – known as separate bonded construction

When laid as a topping on fresh concrete, thicknesses have typically been of the order of 15 to 20 mm.

Prior to 1987, the thickness of separately bonded granolithic screeds exceeded 40 mm as this was the minimum recommended. Since then the optimum recommended thickness has been between 25 and 40 mm. This change has arisen because it has been found that granolithic screeds exceeding 40 mm are much more likely to debond and become hollow because of drying shrinkage.

Up to 1987, granolithic concrete was sometimes laid as an unbonded topping. It was recommended that the minimum thickness should be 75 mm and that bays should be kept small. However, there were many failures due to shrinkage cracking and curling so that after 1987 this method was no longer recommended. Since then, where bonding has not been possible (eg over DPMs or old bases which are too weak or contaminated with oil), it has been recommended that a new concrete slab of at least 100 mm thickness is provided and the granolithic topping laid monolithically with it.

Many variations of sand and cement screeds and fine concrete screeds will be found. They normally consist of Portland cement and sand, sometimes with a coarse aggregate up to 10 mm which will often be local gravel aggregate or even pea shingle.

The relevant Standard is BS 8204:Part 2[192] but many existing screeds will have been laid to the requirements of BS CP 204: Part 2[199].

Main performance requirements and defects

Appearance and reflectivity

Some granolithic surfaces may have been coloured by the use of coloured cements or pigments, and some may have a slight reddish colour imparted by the colour of the fine aggregate used.

Reflectivity is usually poor at around 0.25.

Choice of materials for substrate

A concrete base should have a minimum cement content of 300 kg/m^3.

See also the same section in Chapter 4.1.

Strength and stability

Good. The base should have a characteristic strength of 35 N/mm^2 (BS 8204:Part 1[159]).

See also the same section in Chapters 1.1 and 4.1.

Figure 5.6
A granolithic surface on a staircase

Dimensional stability, deflections etc

Coefficient of linear thermal expansion per °C: 10 to 13 × 10^{-6}.

Reversible moisture movement: 0.02–0.06%.

Wear resistance

Wear resistance is the prime requirement for granolithic toppings. Abrasion resistance is good to very good depending on the type of aggregate, quality of finishing and curing. Impact resistance and longterm indentation are normally very good (Figure 5.6).

Slip resistance

The slip resistance of surfaces is good when dry but it can be poor when wet, particularly when worn smooth.

Where enhanced slip resistance is required, an abrasive grit can be incorporated into the finish. Surfaces which have become slippery may be mechanical roughened or chemically etched.

Coefficients of friction are:

- dry > 0.5
- wet 0.2 to 0.3

Control of dampness and condensation

These screeds are unaffected by dampness though they may be subject to condensation just as any other solid floor, particularly in unheated buildings or at the perimeters of heated buildings.

Thermal properties

Thermal conductivity of granolithic screeds: 0.8 to around 1.8 W/mK, depending on density.

Warmth to touch
Poor.

Fire
Non-combustible.

Suitability for underfloor heating
Suitable, although it can show excessive drying shrinkage caused by the underfloor heating drying out the material too quickly.

Sound insulation

Sound absorption is very poor. Airborne sound insulation depends largely on the mass of the structural floor. Granolithic finishes have been used in floating floors but they are not recommended by BRE.

Durability

The two factors which most often lead to the early failure of granolithic concrete are excessive drying shrinkage leading to cracking and curling, and poor preparation of the base in bonded construction.

Life expectancy will be around 50 years for good quality granolithic flooring, though rather less if specification and mixing is of inferior quality. Round particles exposed by wear create a slippery surface.

One possible cause of failure due to increased shrinkage is that too rich a mix had been used. Although the shrinkage rate of the finish will always be higher than that of the base, if the difference is too great, poor adhesion will lead to debonding, cracking (Figure 5.7) and curling (Figure 5.8).

Other factors affecting durability include:

- water:cement ratio
- proportion of dust in the mix
- adequacy of thickness
- curing
- adequacy of bonding

Granolithic concrete mixes have poor workability – they are said to be 'harsh' – because of the angular shape of the crushed coarse aggregate. Also they are very rich in cement content; mixes are commonly 1:1:2 cement:fine aggregate:coarse aggregate.

Traditionally granolithic concrete

Figure 5.7
Granolithic flooring cracking in a loading bay

Figure 5.8
Curling of a granolithic screed at the edges of bays alongside a duct

was made from crushed fine and coarse hard aggregates. However, excessive dust (with particle size < 100 μm) in the fine aggregate required extra water for workability which subsequently led to excessive drying shrinkage; cracking and curling was a common defect. Rapid drying of granolithic concrete leads to a moisture gradient being set up across the thickness of the slab. Consequently the top section shrinks and leads to curling. To overcome this deficiency, crushed fine aggregate is replaced by sand from natural disintegration (ie screeding sand) which does not contain excessive fines.

For other aspects of curing, and for thickness and bonding, see the appropriate sections in Chapter 4.1.

Maintenance

Maintenance is normally limited to removal of ingrained dirt by scrubbing with hot water and a neutral detergent. The surface should not be polished.

Repairs to worn patches by epoxy resin bonding agents and skim coats of granolithic is feasible, but such repairs will be visually obtrusive.

Case study

Cracking in a granolithic floor due to drying shrinkage

Although a granolithic floor finish in an industrial building was showing cracks, generally a good bond had been achieved between the topping and base slab. The specified mix proportions for the topping were stronger than recommended in BS CP 204:Part 2. During drying, shrinkage of the cement rich mix had caused cracking in the floor finish, the extent of which had been increased due to laying in bays larger than recommended in the code. Interruptions in the finish such as manholes, gullies etc had further aggravated the situation. It was recommended that at hollow and curled edges the cracks were filled with an epoxy resin and, to improve the overall appearance of the floor, completed with coloured grout to match the existing finish. The curled edges should have been ground down to a level surface.

Case study

Cracking and hollowness in a granolithic floor due to a weak bond

The cause of cracking and hollowness in the proprietary granolithic floor finish to the basement of a public building was investigated and found to be attributable, mainly, to poor adhesion between the topping and concrete base. The weak bond, caused by inadequate preparation, had been unable to resist the drying shrinkage stresses of the cement rich granolithic topping. Remedial measures suggested were removing the topping and thoroughly roughening the concrete base; then laying a new granolithic topping on a properly prepared surface. Cracks in the topping in other less badly affected areas should be filled with styrene-butadiene rubber bonding agent.

Work on site

Storage and handling of materials

See the same section in Chapter 2.2.

Restrictions due to weather conditions

See the same section in Chapter 2.2.

Workmanship

A granolithic finish depends very heavily on high quality workmanship to achieve a dense wearing surface. It cannot be emphasised too strongly that in the successful laying of granolithic screeds the preparation of the base, the careful selection of aggregates and the need for adequate curing are of paramount importance.

Other points which need critical supervision include:

- strength of base
- mix proportions
- water content
- compaction
- trowelling skill and timings
- size of bays
- thickness of screeds

BS 8000:Part 9[194] and BS 8204: Part 2 are relevant.

Case study

Granolithic flooring cracking for a variety of reasons

The floor of a factory in the north east of England had given continuous trouble from the time of occupation up to the date of the investigation three years later. The problem consisted of cracking of the granolithic concrete floor and, in places, corruption of the surface by pop-outs. The National Coal Board had given assurances that the problem was not caused by mine subsidence.

The reasons for the problems were considered to be fourfold.

- The design of the floor, quality of aggregate and standards of workmanship at the time of construction were all found to be poor.
- An integral waterproofer had been used in the construction of the concrete base, causing an unsatisfactory bond with the granolithic screed surface and lack of restraint on horizontal movement of the surface layer.
- The bays were planned to be around 17 m^2 each but generally this had been greatly exceeded.
- There were found to be small pieces of coal mixed in with the aggregate which had given rise to the disintegration of the surface. (The aggregate was probably contaminated before delivery.)

Although the cracks were unsightly, they did not inhibit the manufacturing activity. (Indeed, all the bays were being used by the occupiers.) It was decided, therefore, that they should be left as they were.

The areas where the surface had broken up needed to be removed and relaid.

Inspection

The problems to look for are:

◊ friable screeds
◊ poor wear resistance
◊ slipperiness
◊ finishes breaking down
◊ cleaning practices inappropriate
◊ cracking and curling
◊ dusting
◊ pop-outs of coal, lignite and pyrites

Chapter 5.4 In situ terrazzo

In situ terrazzo has been used successfully in cases where the quality of workmanship could be guaranteed, and where movements in the structure of the building could be sufficiently guarded against so as not to disrupt the terrazzo. It must be said, however, that many installations, even those in prestigious buildings, do suffer from cracking. This is one reason why terrazzo tiles, which are manufactured under controlled conditions, have taken the lion's share of the market (see Chapter 7.4).

Characteristic details

Basic structure

Terrazzo is composed of a mixture of fine aggregates of marble or other decorative rock bound with one of a number of coloured cements. The finish is ground down after curing to expose the decorative aggregates.

The terrazzo is normally laid on a 'green' cementitious screed to a thickness of not less than 15 mm and in bays of about 1 m^2 separated by strips of ebonite or plastics, but older floors may have metals such as brass. Where aggregates in the mix exceed 10 mm, the thickness of the material is increased.

Occasionally it has been laid monolithically with the base which allows larger areas without joints.

The standard appropriate to the materials described in this chapter is BS 8204:Part 4[(55)]. The material was covered formerly by Section 3 of BS CP 204:Part 2[(199)].

Main performance requirements and defects

Appearance and reflectivity

Terrazzo exhibits a very wide variety of colours and patterns depending on the aggregates used.

Reflectivity can be very high depending on the aggregates. Cream shades give reflectances of around 0.45, while darker shades around 0.25.

Choice of materials for substrate

Terrazzo should only be laid on concrete bases, and certainly not on timber bases.

Strength and stability

The indentation resistance of terrazzo is very good though it may be cracked by impacts. It is also susceptible to cracks in the substrate.

See also Chapter 1.1.

Dimensional stability, deflections etc

Coefficient of linear thermal expansion per °C: 10 to 13 × 10^{-6}. Where marble is the aggregate, the lower values will be relevant.

Reversible moisture movement: 0.02–0.06%.

Since most terrazzo is bound with cements, the flooring is prone to shrinkage cracking (Figure 5.9). Cracking is very often seen in older buildings with large unbroken areas of flooring, although the cracks will most likely be of longstanding duration and usually reflect cracks in the base. In modern installations the material is laid in small bays interrupted by metal joints which mitigate the cracking problem.

In order to provide particular appearances, large aggregate is sometimes used which produces harsh mixes with high water contents – this inevitably leads to high shrinkages.

Figure 5.9
Shrinkage cracking in an in situ terrazzo half landing

Case study

Cracks in in-situ terrazzo flooring in a bank

The cracking consisted of slight random shrinkage cracks or straight cracks across the width of the narrow sections of flooring. Although the screed on which the terrazzo was laid was found to be rather weak and friable, the terrazzo topping appeared to be of good quality and well laid. The cause of the failure was that the drying shrinkage was not restrained by the screed, being of poor quality. However, in this case it was concluded that replacement of whole areas of the flooring on the grounds of further deterioration could not be justified, and a cosmetic filling of the cracks was recommended.

Wear resistance

Terrazzo is normally very hard wearing, though to some extent this depends on the aggregate used. Soft marbles wear more quickly than the cements, so dished aggregate particles are common.

Slip resistance

Slip resistance depends on the type of surface finish (see the maintenance section later in the chapter). Some cleaning agents make terrazzo slippery. The fine grit finish obtained by grinding is generally not found to be slippery; however, in use, the surface tends to wear to a very smooth surface which is very slippery when wet. A non-slip finish can be maintained by using a fine abrasive during maintenance.

It is possible to obtain a highly polished appearance by using special polishing compounds, but these tend to give slippery surfaces in the wet. Terrazzo can be exceptionally slippery when wax polished.

Polish should be avoided on surfaces adjacent to terrazzo as it can be transferred by treading. On stairs, slip resistant nosing is necessary.

Coefficients of friction are:

- dry > 0.5
- wet 0.2 to 0.3

The dry value may be improved to 0.6 and the wet to 0.5 by incorporating carborundum in the finish.

Control of dampness and condensation

Depending on the construction of substrates, terrazzo finishes will tend to 'sweat' in unheated buildings when warm fronts follow cold. Resistance to surface water is very good.

Thermal properties

Thermal conductivity of terrazzo: there is no precise information but probably around 1.0 to 1.8 W/mK.

Warmth to touch

Very poor.

Fire

Non-combustible.

Suitability for underfloor heating

Suitable for all likely temperatures.

Sound insulation

Sound absorption is very poor. Airborne sound insulation depends largely on the mass of the structural floor. Terrazzo can be used in a floating floor – the design is critical and performance depends on overall thickness of the screed – but it is not recommended by BRE for this application.

Case study

Cracks in in-situ terrazzo flooring in a hospital

At first sight, the flooring appeared to be a good quality, well compacted, in situ terrazzo. The surface was smooth and, over most of the flooring area, appeared to be giving good service. However, a closer examination revealed the following defects:

- pitting 1 mm in diameter in the matrix caused by air bubbles in the mix
- irregular shaped fretting and pitting in the aggregate and in the matrix, particularly at panel edges
- cracks at interfaces with the plastics dividing strips caused by shrinkage of the cement rich mix
- rough irregular patches caused by deficiency of the aggregate
- shrinkage cracks in skirtings formed of in situ terrazzo – the skirtings were very long and thin
- patch repairs of resin containing marble aggregate were failing
- some panels had curled and had risen 2 or 3 mm above adjacent panels, causing fracturing across corners
- discoloration by cleaning agents containing phenols or cresols
- variability in the colouring of the aggregates used

None of the defects seen were unusually severe for in situ terrazzo, though they were clearly unacceptable for the situations where the terrazzo had been installed. It is to avoid such deterioration that terrazzo is normally specified in tile or precast panel form.

In order to keep the hospital functioning, it was recommended that a temporary flooring of PVC was laid over the least affected areas. For the remainder there was no alternative but to remove the finish and re-lay the floor using preformed terrazzo tiles.

Durability

Life expectancy will be between 50 and 65 years for a good quality flooring, though it will be less in heavily trafficked areas. In one example where in situ terrazzo had been used on a staircase in a railway station, it had worn to an unacceptable degree in less than ten years of use.

Acids and alkalis will attack both the matrix and the marble aggregate, leaving rough pitted surfaces. Some disinfectants containing phenols tend to stain.

Maintenance

Terrazzo may be cleaned with water and a fine abrasive powder. Although there is very little evidence to support the view that sulfates in cleaners have caused problems in in-situ terrazzo flooring, it is recommended that cleaners should be sulfate free. Neutral detergents may be used, but disinfectants and soaps avoided. Wax polishes should also be avoided. Discoloration due to chemical reactions of iron in the mix with phenols and cresols may be impossible to remove, even by regrinding.

Work on site

Storage and handling of materials

See the same section in Chapter 2.2.

Restrictions due to weather conditions

Terrazzo should not be laid in high temperatures, strong sunlight or strong draughts since premature drying out may result in surface crazing.

Workmanship

The terrazzo mix should be laid on a high quality screed while the screed is still 'green'; if the screed is up to two days old, a neat cement slurry should be used to provide a bonding agent. Terrazzo should not be laid on older screeds without further advice.

If shrinkage cracking is to be minimised, the terrazzo should be laid in bays not exceeding 1 m^2, subdivided by strips of jointing material preferably having similar wear characteristics to the finished terrazzo. This probably implies a metal such as brass rather than a softer material such as plastics.

The surface should be trowelled and rolled so that it makes good contact with the substrate, though not so vigorously as to bring cement laitance to the surface.

Coarse grinding to expose the aggregate is undertaken after about four days, then filling and fine grinding after a further three days. This will produce a fine grit finish. Terrazzo should not normally be sealed. Where sealing is essential, seals based on acrylic resins have been found suitable. Surface hardening agents based on magnesium or zinc silico-fluorides have sometimes been used.

Adequate curing and slow drying of the terrazzo is essential if crazing is to be minimised. In this respect, the building should not be heated for a period of six to eight weeks after grinding. The longer the curing period and the slower the drying period, other things being equal, the better the final result.

Inspection

The problems to look for are:

- staining
- poor wear resistance
- slipperiness when dry
- slipperiness when wet
- finishes breaking down (eg pitting)
- cracks running along the plastics dividing strips
- sweating (condensation) in unheated buildings or buildings open to the atmosphere (for the most part unavoidable)
- pitting caused by air bubbles in the mix
- rough irregular patches
- shrinkage cracks in skirtings
- panels curling above adjacent panels and fracturing across the corners of the panels
- crazing

Chapter 5.5 Synthetic resins

Resin floorings are very much a development since the early 1960s. They provide the specifier with a variety of materials which can be modified to achieve a very wide range of performance characteristics, albeit that such versatility and performance can be expensive.

These floorings do provide an enhanced resistance to a wide range of chemicals and, being jointless, also offer a hygienic solution for buildings in the pharmaceutical and food industries.

Characteristic details

Basic structure

There are two main types of in situ flooring based on synthetic resins and polymers used as binders.

- Polymer modified cementitious floorings. These comprise emulsified polymers and hydraulic cements which, when combined in certain proportions, produce the best characteristics of the resin and cement components. The main binding agent is the cement, the properties being modified by the polymer. These are dealt with in Chapter 5.2 and are not repeated here.
- Resin bound aggregate flooring. These are commonly syrups which cure (crosslink) by reaction of the chemical components. Hydraulic cements are rarely present. Curing happens after the addition and mixing of a catalyst or a cross-linking agent. This is commonly called the curing agent and its type depends on the resin system.

The resins used are epoxy resin, polyurethane, polyester and acrylic (methyl methacrylate). They are often referred to as thermosetting resins. Except where they are used as coatings, they are always mixed with substantial quantities of aggregates which include sand, quartz, carborundum and bauxite.

There are three basic types of resin bound aggregate floorings.

- Resin screeds. These are relatively stiff mixes with the material being applied by trowel or sledge followed by power trowelling (Figure 5.10). The mix proportions are often in the range of 1:5 to 1:9 resin:aggregate. The aggregate size can range up to 5 mm and some decorative screeds have even larger aggregates. Because the resin content is rarely sufficient to fill all the pore space between the aggregate particles, these floorings are often finished with a seal coat of the same resin to provide a hygienic surface. With careful selection of resin, hardening system and aggregates, resin floors can provide excellent chemical and abrasion resistance, and be used for very heavy duty areas. They are most often laid between 4 and 6 mm thick.
- Self-levelling resin floorings. These are very fluid materials, as their name implies (Figure 5.11). They contain much finer aggregate and provide very smooth non-porous hygienic surfaces. Often laid in the range 1–3 mm thick, they provide excellent floors in clean rooms, pharmaceutical factories, food processing areas etc. While these very smooth floorings can be

Figure 5.10
A sledge-applied epoxy resin screed flooring being laid (left) and the same flooring being power trowelled (right) (Photographs by permission of Flowcrete Systems Ltd)

Figure 5.11
An epoxy self-levelling flooring (Photograph by permission of Flowcrete Systems Ltd)

slippery when wet, a degree of slip resistance can be engineered into the material by sprinkling the surface with fine aggregate before it sets.

- Coatings and seals. These are dealt with in Chapter 5.6.

Main performance requirements and defects

Appearance and reflectivity

Base resin and hardener systems tend to vary in colour from clear through to white or straw yellow. Most systems are pigmented, a wide range of colours being available. Cream and grey shades give reflectances of around 0.45, darker shades around 0.25.

Choice of materials for substrate

Synthetic resin floorings can be laid successfully on a wide variety of primed substrates, but concrete is by far the most common. The alkali content of concretes can inhibit hardening of some resins, and high moisture content can inhibit adhesion, so the longer the concrete is left before applying the flooring the better.

Some thermosetting resins require the concrete substrate to be below a certain moisture content before they can be successfully laid.

Strength and stability

BS 6319:Part 2[207], Part 3[208], Part 4[209] and Part 7[210] provide methods of test for compressive strength, flexural strength, bond strength and tensile strength. Although results from these tests may give the specifier some indication of relative performance, care has to be taken that the resin product is appropriate to a particular application.

The minimum compressive strength of a concrete base should be 25 N/mm^2 but higher strength will be required for some formulations.

Dimensional stability, deflections etc

Improved impact resistance may be gained in some cases by specifying a thicker than normal layer, for example up to 10 mm instead of 2 to 5 mm, where the forces can be dissipated over a greater mass of material. Also, it will be necessary to increase the depth of the finish adjacent to a structural movement joint (Figure 5.12).

The BRE screed tester is insufficiently rigorous to enable differentiation between resin surfaces.

BS 6319:Part 11[211] deals with creep and Part 12[212] with shrinkage and coefficient of thermal expansion.

Slip resistance

Coefficients of friction are:

- dry > 0.7
- wet 0.2 to 0.3

The value for wet resins may be improved to 0.3–0.5 by the incorporation of carborundum.

Like most floorings, thermosetting resins are usually safe when dry. Some of the self-levelling types produce very smooth surfaces and can be dangerous when wet, but special grades are available which either contain hard slip resistant aggregates as fillers or have them sprinkled onto the surface before they set to produce fine or rough textures. Many of these products have excellent slip resistance when wet.

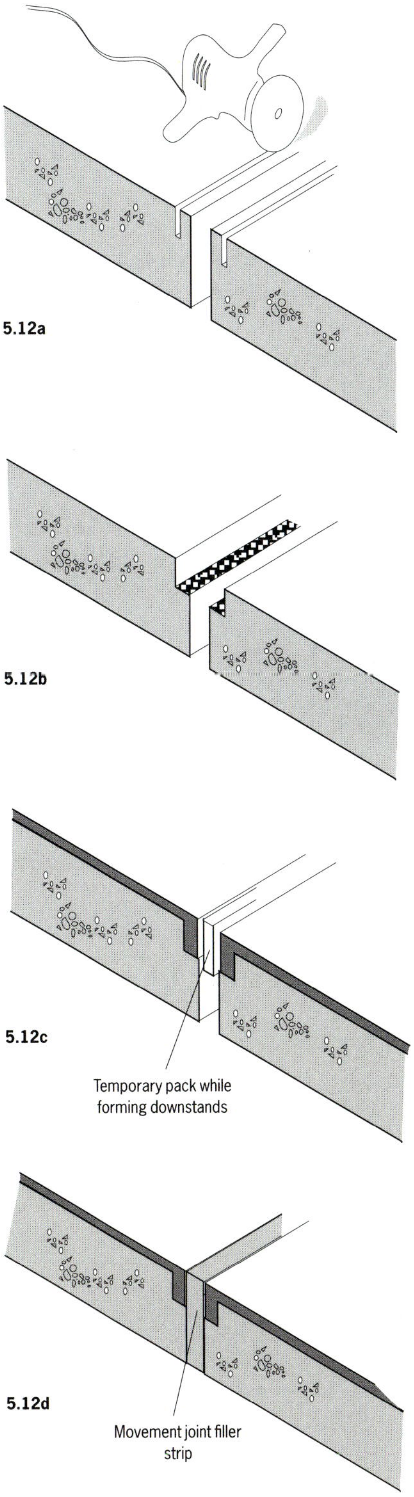

Figure 5.12
Providing a movement joint in a synthetic resin flooring (a – d)

Figure 5.13
An epoxy coating wearing and peeling

Figure 5.14
Epoxy resin rippling over a crack in the base. This type of problem can be caused by thermal or moisture movement

Control of dampness and condensation

Special formulations of epoxy resins have found widespread use as surface DPMs. They adhere well to damp-but-surface-dry concrete at the time of application. They can be used to provide a DPM where the moisture condition of the concrete is up to 92–93% relative humidity – a measurement which must be taken by the hygrometer method. They are rarely specified for new construction but are used when a building without a DPM undergoes a change of use (eg factory-to-office and farmhouse conversions) or to control excess constructional water in thick concrete bases when insufficient time is available for them to dry.

Thermal properties

Care should be used when specifying resin surfaces in kitchens where temperatures in the vicinity of ovens and heating appliances may cause softening of the surface.

Thermal conductivity of synthetic resins: 0.23 W/mK.

Warmth to touch

Resin floors provide no thermal insulation and behave similarly to other hard floorings such as concrete and ceramic tiles.

Fire

Resin floorings are unsuitable for use on floors within firefighting shafts.

Suitability for underfloor heating

Resins are rarely used over underfloor heating although there is no reason why they should not, provided the bases are sufficiently strong, crack free and dry. The surface temperatures should not exceed 28 °C.

Sound insulation

Sound absorption is very poor. Airborne sound insulation depends largely on the mass of the structural floor. Synthetic resins can be used over a screeded floating floor but performance depends on overall thickness of the screed.

Durability

The most common problem with a resin floor is detachment from the base due to poor preparation of the surface on which it has been laid (Figure 5.13). Since the resin fully bonds with the substrate, any cracking in the substrate will inevitably affect the flooring (Figure 5.14).

Epoxy floorings are, in general, not suitable for use in floors where there is a risk of thermal shock, especially where the flooring consists of more than one coat and the coefficient of thermal expansion of the topping is greater than that of the base coat.

Case study

Blisters in a resin floor covering

BRE Advisory Service was asked to monitor the treatment of a very large resin floor covering which started blistering some days after laying.

When drilled the blisters were found to contain an aqueous solution. After lengthy investigation, osmosis was found to be the most likely cause of the problem. It was resolved by laying a specially formulated permeable epoxy screed which would allow dissipation of any subsequent osmotic pressure, followed by a surface sealing coat. See the feature panel on osmosis in Chapter 1.9.

Case study

Hot water spillages and chemical attack on flooring in a brewery

BRE was called in to investigate the failure of a polyester flooring in a new bottling hall at a brewery. Boiling water spilling onto the flooring had caused thermal shock in the resin coating which did not have sufficient adhesion to the substrate to resist thermal movements. The adhesion of the resin was tested and the low pull-off strengths achieved caused concern about its overall quality. No immediate action was called for, but it was suggested that the floor should be kept under close observation for a sensible period of time.

Because of the repeated failure of the flooring in areas exposed to chemical attack, it was necessary also to replace relatively small areas under or around some of the plant with a ceramic tile flooring system which would be more corrosion resistant.

However, they have excellent resistance to a wide range of chemicals including alkalis and moderately concentrated inorganic acids. Most formulations break down rapidly in the presence of some organic solvents like acetone and chlorinated solvents. Early products disintegrated rapidly when in contact with weak organic acids like lactic acid and acetic acid. Modern formulations can resist these acids. Generally, epoxies soften at temperatures between 60 and 80 °C and should not be used where liquids above these temperatures can be spilt on the floor.

Polyester floorings possess good chemical resistance but they are susceptible to attack by alkalis, especially hot solutions.

Polyurethanes generally have much better resistance to temperature and some can be used where liquids exceed 100 °C.

Most self-levelling thermosetting resins can suffer blistering caused by osmosis. See the feature panel on osmosis in Chapter 1.9.

Maintenance

Epoxy resin surfaces should perform well between 50 and 70 °C and some polyurethane resins between 100 and 120 °C. Care should be exercised when using steam cleaning apparatus on floors in industrial premises to avoid distortion of the surfaces.

The use of solvent based cleaning pastes could lead to progressive deterioration of certain types of resin flooring.

Work on site

Where existing concrete is being covered by resin flooring, it is essential to carry out a full site survey to note any previous contaminants. These must be removed by chemical or mechanical means to ensure a good bond. Some organic contaminants can be burnt off the surface of the concrete.

Restrictions due to weather conditions

In order for epoxy flooring to harden satisfactorily within 24 hours, it should not be laid in atmospheric

Figure 5.15
Applying an epoxy floor coating by squeegee and roller (Photograph by permission of Flowcrete Systems Ltd)

Case study

Difficulties in cleaning resin flooring in a textile factory

Three problems arose concerning epoxy resin flooring at a textile factory where fabric manufacturing machinery was moved around within the workshop. The first problem was related to the effect of dirt and machine oil on the performance of the flooring; the second about the evenness of the surface of the floor; and the third concerned discontinuity in the appearance of the flooring. There was no obvious reason to suppose that the floor had not been well laid or that it exceeded the tolerances to which this type is normally laid: that is to say between 3 and 7 mm thick.

The first problem was examined by applying detergent to a small section of the floor which had comparatively little contamination and cleaning off the dirt with water. The surface matched the original sample of flooring well, and there was no evidence of dirt and oil penetration. It was not surprising that there had been a considerable buildup of dirt and oil in the vicinity of the machines as no effort had been made to prevent or catch oil drips, nor to prevent employees treading the oil around generally. A carbon tetrachloride substitute was used to remove the heavy oil and dirt buildup near to the machines. Subsequent cleaning with detergent as before proved that there was no penetration of, nor damage to, the floor surface. General rehabilitation of the floor would be a job for specialist cleaning contractors using powerful detergents or solvents.

The second problem, of unevenness of surface, was found to be no more than would result from applying the tolerances normal to the laying of the concrete underfloor slab – under a 3 m straight edge there were two small areas of depression of up to 6 mm. Small score marks resulting from the original trowelling should have been coated with resin on first observation, but to have done it at that stage, some time after the floor had been laid, would have compromised the slip resistant nature of the floor.

The third problem, of the discontinuity of the flooring, was again found to be within normal descriptions of 'jointless' work. The only evidence of discontinuity in its appearance was at daywork joints, and was acceptable within the criteria normally applied.

temperatures below 12 °C. Laying some epoxy floorings in conditions of high humidity can lead to surface blooming.

Workmanship

For in situ resin floorings to be applied successfully, being relatively thin, they must be bonded firmly to a sound base (Figure 5.15 on page 197). Priming for epoxy finishes with a mixture of resin and hardener, without fillers, is required; a topping coat is added before the primer has hardened.

New concrete bases must be left for an adequate length of time before laying synthetic resin flooring (see Table 1.3 in Chapter 1.4). Surface laitance should be removed by shotblasting or scabbling.

Where construction joints run through the base, the flooring should also be jointed and pointed with a more flexible mixture of the same type as the rest of the flooring (ie with less filler content).

Inspection

The problems to look for are:

◊ finishes too soft
◊ impact resistance poor
◊ slipperiness
◊ finishes breaking down
◊ finishes showing bubbling
◊ finishes showing lack of adhesion
◊ contamination (eg by oils and fats)

Case study

Flaking of an epoxy finish in a warehouse due to inadequate surface preparation

Failure of a proprietary epoxy floor finish in a new warehouse was investigated by BRE Advisory Service. The epoxy coating had flaked from the concrete slab in areas of heavy fork lift truck traffic. The investigators found that the loss of adhesion of the finish was due to inadequate preparation of the concrete surface and the concrete being too wet when the floor coating was applied. They suggested that loose and poorly bonded areas of the coating should be removed and the concrete surface be either mechanically hacked or, preferably, shotblasted to form a good key. The concrete slab should then be tested to ensure it had dried sufficiently before the same proprietary finish was reapplied.

Chapter 5.6 Paints and seals

Floor paints are normally used over very large surfaces as a relatively inexpensive means of improving appearance in the short term. Seals are applied to concrete floorings as a means of surface hardening and to prevent dusting, again as a short term measure. In many instances the terms floor paint and floor seal are interchangeable.

Characteristic details

Basic structure

Paints are normally one of the following main types:

- polymer based emulsions (eg acrylics)
- bitumen based emulsions (very rarely encountered)
- resin based solvent mixes
- oleoresinous paints
- one and two-pack polyurethanes
- two-pack epoxies
- chlorinated rubber

With floor paints the polyurethanes are the type most frequently used. Some preparations are available with fine aggregates to enhance slip resistance or, where traffic is light and stability is adequate, to improve resistance to bacteria. Chlorinated rubber paints also have been widely used on floors.

Jointless in situ PVA (polyvinyl acetate) floors will also be found. They were introduced during the 1950s. The material contains a mixture of PVA emulsion with fillers and pigments, and was laid cold in several thin coats. The setting process involved only the evaporation of the water content. This type of flooring was widely used during the 1950s and early 1960s, but since then has been specified only rarely.

The first seals used on timber floors were solutions of shellac in methylated spirits, the basis of French polishing, or varnishes in various forms, some of which contained stains. These seals were, for the most part, superseded by the oleoresinous seals which, although still used, have, in turn, been largely replaced by the polyurethanes.

For concrete, seals or surface hardening agents are normally one of the following principal types:

- sodium silicate solutions
- silico-fluoride solutions
- tung oil with resins
- one and two-pack polyurethanes (usually pigmented)
- epoxies (usually pigmented)

Solutions of sodium silicate, magnesium or zinc silico-fluoride applied to the surface of a good quality concrete floor will often increase its resistance to abrasion and dust formation. This treatment is of little use, though, if the surface is of poor quality.

Seals based on tung oil blended with phenolic resins have been extensively used in the past, with some success in reducing dusting.

Main performance requirements and defects

Appearance and reflectivity

Paints are available in a wide range of colours. They can discolour under certain conditions of use. Cream shades give reflectances of around 0.45, darker shades around 0.25.

Seals can significantly affect the appearance of a floor, especially the pigmented types. Polyurethanes on timber give a very high gloss which becomes matt where trafficked.

Choice of materials for substrate

As is normal with paint systems, the condition of the substrate has a significant effect on performance. Thin coatings cannot be expected to restore the integrity of pitted, worn or friable surfaces; indeed, they will readily detach from the high spots. Seals, too, are often expected to work wonders with dusting surfaces, whereas their capabilities are limited by the condition of the substrate. When timber is resealed, it is normally first sanded. Other floorings, such as linoleum, may be also sealed (eg with acrylics).

Strength and stability

Two-pack paint mixes usually produce tougher finishes than one-pack. They adhere by penetration of the pore structure. Adhesion is affected by the presence of contaminants such as oils or greases on the surface, leading to selective detachment.

The sodium silicate and silico-fluoride seals act by filling the pores of the concrete or screed, coupled with some chemical reaction with the free lime in the cement to harden the surface. They are not used on timber.

Dimensional stability

Provided paints and seals adhere strongly, their movements are controlled by the substrate.

Case study

A seal on a warehouse floor that could not overcome defects in the substrate

Inspection of a proprietary concrete sealer at an industrial warehouse, which included a sales area, showed that the seal had been worn away in areas that had been subjected to foot and light trolley traffic, mainly between the storage racks and at the cash outlets. From the site observations and information received from the original concrete supplier, the concrete laid for the warehouse floor had insufficient strength and abrasion resistance. It was necessary, therefore, to apply an improved wearing surface to the whole of concrete floor of the warehouse. There were a number of possibilities but they would all involve, to a varying degree, some disruption to the operation of the warehouse.

The surface failed for one of two reasons: either the substrate was friable or the coating was too thin to withstand the traffic. It was important for the client to realise that the life expectancy of any seal was very limited.

Wear resistance

The polymer emulsions are not as good as one-pack polyurethanes which, in turn, are not as good as two-pack epoxies and two-pack polyurethanes. The last two give the highest abrasion resistance. PVA emulsions absorb water, and their resistance to wear is considerably reduced when they are wet. Because of their small thickness, all paints and seals on floors will have a limited life before wearing through.

Slip resistance

Special textured floor coatings are available which provide varying degrees of slip resistance in wet conditions. Information should be obtained from manufacturers.

Control of dampness and condensation

Because PVA absorbs water, it cannot be used in situations where the floor is likely to become wet; nor can it normally be used where there is no DPM in the substrate. Some of the polyurethanes and epoxies can give reasonable performance if laid on dry substrates even if there is no DPM. None can be laid in wet conditions.

Thermal properties

Thermal conductivity of paints and seals: 0.23 W/mK, although in this context the property is somewhat academic.

Warmth to touch

These surfaces are so thin that they add nothing to the thermal performance of the floor.

Fire

Paints and seals do not materially affect the performance of a screed in fire.

Suitability for underfloor heating

All paints and seals (except bitumens) can be used.

Sound insulation

Paints and seals make negligible difference to the floorings to which they are applied.

Durability

Paints not specifically designed for use on cementitious surfaces usually meet with disaster sooner rather than later. They tend to saponify (become soap-like) and break down because of the alkalinity of the substrate.

The durability of floor paint systems depends as much on preparation of the surface and degree of traffic as on their formulation. (One domestic concrete floor which had been inspected by BRE, painted with chlorinated rubber to prevent dusting and to improve appearance, had given good service for at least five years with no wear showing, even in trafficked areas.) Lack of adequate preparation can lead to patchy wear in a comparatively short

Figure 5.16
Almost certainly dampness has caused this painted finish to detach from the base

Case study

Waterproofers unable to accommodate shrinkage of the substrate in a hotel garage

The floor of a hotel garage had a history of dampness resulting from wet and snowladen cars entering the building. The dampness was treated first by applying silicone to the floor, then with a proprietary paint. Both attempts at waterproofing treatment proved unsuccessful; the paint was also peeling and had turned black. The continued dampness was attributed to the waterproofers being unable to accommodate the drying shrinkage movement of the concrete garage floor. Coatings over existing cracks were themselves liable to crack. The loss of adhesion and colour change were attributed to earlier silicone treatment and mould growth respectively.

Remedial measures recommended included thorough cleaning followed by methods to seal the cracks and applying a suitable finish such as traditional mastic asphalt or a proprietary paint. Some newly available coatings, specially formulated for car park conditions, have been shown to have improved extensibility over cracks.

time (Figure 5.16).

Since these coatings are all very thin, the life expectancy in heavily trafficked areas is always limited. Therefore, in choosing a coating for trafficked areas, the ability to recoat is vital. Some polyurethanes, for example, do not tolerate a recoat.

Sodium silicate solutions are effective for comparatively short periods and retreatment is normally needed at yearly intervals. They do very little for high quality concrete floors and do not improve very dusty surfaces, but they can enhance the dusting and abrasion qualities of medium quality concretes.

Some coatings (eg polyurethanes and epoxies) have good resistance to a wide variety of chemicals.

Maintenance

It is important that paint systems on existing floors are correctly identified when the time comes for recoating. It may be necessary to examine samples of old coatings in the laboratory to determine the compatibility of proposed paint formulations.

If paint systems are used in industrial locations, hot water or steam cleaning should be limited to polyurethanes.

Work on site

Storage and handling of materials

There is a very wide variety of paints and seals, and general guidance is not possible. However, manufacturers' instructions should be observed, especially with regard to shelf life or pot life after mixing.

Restrictions due to weather conditions

Different restrictions apply to different products, and the manufacturers' instructions should be ascertained. For example, PVA emulsions should not be applied in temperatures less than 10 °C.

Polyurethanes are sensitive to damp conditions during the application process.

Some epoxies can take on a whitish bloom if laid in conditions of high humidity. This can also happen if the surface is wetted before it is fully cured.

Workmanship

The application of dustproofing solutions and sealers can be useful, but they cannot improve friable concrete. Two or more applications of sodium silicate solution are required, allowing each coat to dry for 24 hours and washing with clear water before applying the next coat.

Three applications are normally required for magnesium or zinc silico-fluoride solutions at 24 hour intervals.

Where acid etching of the floor surface is considered as a measure to improve the adhesion of paints and seals, thorough rinsing of the surface is imperative. A better method is to lightly shotblast the floor; this operation ensures that the floor remains dry.

Where any two-pack materials are used, they need to be thoroughly mixed, otherwise treacle-like patches will result.

Inspection

The problems to look for are:

- ◊ wear resistance poor
- ◊ slipperiness
- ◊ finishes breaking down
- ◊ finishes showing bubbling
- ◊ finishes showing lack of adhesion
- ◊ contamination (eg by oils and fats)

Chapter 5.7 Mastic asphalt and pitchmastic

Bitumen – the basis of mastic asphalt – has been used in building construction for around five thousand years, though perhaps more for its cementitious and waterproofing qualities than for its qualities as a flooring.

Mastic asphalt has been successfully used as underlays for a variety of floorings such as rubber, linoleum, wood block, and cork and ceramic tiles. The grades are exactly the same as for finished floorings. Where used as an underlay for very thin floorings, a levelling layer of latex may be required. Adhesives used for floorings must be compatible with the asphalt.

Pitchmastic, although much used in the past – some floors have survived – is not used now. However, confusion with mastic asphalt may arise, since pitchmastic was laid in similar thicknesses to mastic asphalt and its appearance can be similar.

Characteristic details

Basic structure

Mastic asphalt

Mastic asphalt is prepared from mixtures of bituminous binders and inert minerals, usually in the form of aggregates. Jointless mastic asphalt flooring is normally laid in thicknesses from 15 to 50 mm.

The appropriate standards for asphalt floorings are:

- BS 8204:Part 5[56]
- BS 6925[39], Type F1076 for limestone aggregate or Type F1451 for coloured aggregate
- BS 1447[213]. Material meeting this standard should be used for waterproofing balconies and exposed rooftop access ways classified as floors

For waterproofing duty, the appropriate grades are Type R988 or T1097 specified in BS 6925.

Four grades of mastic asphalt are available for materials covered by Types F1076 and F1451 of BS 6925:

- Grade I: special hard flooring 15–20 mm thick – used for domestic-type floors where indentation under prolonged loading is possible
- Grade II: light duty flooring 15–20 mm thick – used for domestic and industrial floors with no risk of indentation or point loads
- Grade III: medium duty flooring 20–30 mm thick – used for industrial floors which do not carry wheeled traffic.
- Grade IV: heavy duty flooring 30–50 mm thick – used for industrial floors carrying wheeled traffic, or floors which are subjected to severe wear or heavy point loads. These floorings can be armoured with metal grilles buried in the asphalt.

There are special grades of mastic asphalt for applications such as resistance to chemicals, and for anti-static or spark-free floors. Most acids will attack the limestone aggregate in mastic asphalt, though some enhancement of acid resistance is possible in particular grades; this is mainly obtained by using sands and slate dusts to replace the limestone. Anti-static applications can incorporate metal grille conductors, and sparks are avoided by removing grit from the mix.

Mastic asphalt is normally laid on a separating layer, except for some applications of Grade IV, particularly where the asphalt is to be laid on a porous or cracked surface.

Pitchmastic

Pitchmastic is similar in composition to mastic asphalt, but the significant difference is that low or high temperature coal tar pitch with flux has been used as the binder instead of bitumen. Some grades have been modified by the addition of a proportion of lake asphalt. Sometimes the pitchmastic was gauged with a small proportion of sawdust which was claimed to give a warmer floor.

Pitchmastic should not have been laid on a timber base because of its brittle nature; it is more brittle than mastic asphalt.

Detailing

Up to about 1980 it was common to apply the adhesive directly to the surface of mastic asphalt where the asphalt was used as an underlay for sheet and tile or for textile floorings. Whether water or solvent based, time had to be allowed for the solvent or carrier in the adhesive to evaporate before floorings could be laid. If this was not done, either poor adhesion resulted or the mastic asphalt could be softened by the solvents.

Since the early 1980s it has become recommended practice to overlay the mastic asphalt with a

Figure 5.17
A heavily indented asphalt floor. The damage had been caused by overnight parking of wheeled electric mechanical handling equipment during recharging

3 mm thickness of a suitable underlayment to act as a 'scavenger' for the solvent. An underlayment also prevents plasticiser from PVC floorings and foamed PVC backed carpets migrating downwards to soften the asphalt.

Skirtings can be formed in mastic asphalt, though it may be necessary to use two coats and to reinforce the upstand with expanded metal lathing (EML).

Two coat work may also be required when laying paving grade asphalts on balconies where additional grit may be added to the top layer to improve wear resistance.

Main performance requirements and defects

Appearance and reflectivity

Mastic asphalt floorings are normally either black or red, though examples in other dark colours may be found in old floors. Surfaces are sand rubbed, crimped or floated without sand while they are still warm.

A black or dark red surface colour gives relatively poor reflectivity at around 0.10.

Choice of materials for substrate

Mastic asphalt can be laid on many kinds of substrate though special treatment may be required for floors which have been contaminated with oils, fats or greases. On metal substrates, a bitumen primer may be needed. The most common base is concrete (or a screed), and a separating membrane (nowadays based on glass fibre) should be used.

Strength and stability

Mastic asphalt and pitchmastic are susceptible to longterm point loads. Indentation can be a problem, particular in well heated buildings. It is the one thing that discourages the wider use of these materials.

See also Chapter 1.1.

Dimensional stability, deflections etc

Coefficient of linear thermal expansion per °C: 30 to 80 × 10^{-6}.

Reversible moisture movement: 0%.

Mastic asphalt and pitchmastic, depending on grade, are relatively accommodating of small imperfections in the substrate. However, mastic asphalt and pitchmastic floorings should not be carried over movement joints in the structure of the building.

Wear resistance

Mastic asphalt

Abrasion resistance is fair to very good, impact resistance fair to very good, and longterm indentation poor to fair, depending on grade (Figure 5.17).

Pitchmastic

Abrasion resistance is fair to good, impact resistance fair to good, and longterm indentation poor to fair.

Slip resistance

The slip resistance of surfaces of mastic asphalt and pitchmastic varies from fair to good, but frequent polishing tends to make it slippery. Where either is used on stair treads, suitable non-slip nosings are required. These act as armouring as well as providing slip resistance.

BS 8204:Part 5 calls for a slip resistance value greater than 39 in both wet and dry states. The test regime is specified in Annex B to the standard. Values need to be corrected for surface temperature.

A crimped surface, or one incorporating special chippings, may offer improved slip resistance.

Coefficients of friction are:

- dry > 0.7
- wet 0.2 to 0.3

A sand rubbed finish will give higher values in the wet but may become polished in service.

Control of dampness and condensation

Mastic asphalt forms a very good barrier to rising damp. Vapour resistivity is also good (not less than 100 000 MNs/g). Because of this vapour resistance property, it is commonly used under moisture sensitive floorings where no other DPMs exist.

Thermal properties

Mastic asphalt softens in a heated environment, affecting its resistance to indentation. Some flooring grades are also susceptible to cracking at low temperatures, and paving grades should be used therefore in unheated buildings.

Thermal conductivity of mastic asphalt and pitchmastic: approximately 0.50 W/mK, depending on specification.

Warmth to touch

Mastic asphalt is moderately warm to the touch – about midway between concrete and timber.

Case study

Foul smells from an organic underlay in a sports hall

A mastic asphalt sports hall floor, 20 years old, had blistered and cracked causing localised breakdown of the surface. The hall was out of use at the time of inspection pending resolution of the problem and repairs.

The essential components of the problem were:

- a vegetable fibre insulating layer (in this case jute felt within bitumen)
- moisture, either remaining from the time of construction or arriving subsequently as rising damp
- pressure on the surface through use for its designed purpose.

Gases, methane and others – and foul smelling – were generated by the jute felt rotting under anaerobic conditions. This had caused expansion between the base and the surface layers resulting in irregular lifting of the surface layer. Pressure from above (ie from sporting activities) then caused the floor surface to crack and blister.

The problem could be handled by correcting each outbreak of cracking and blistering as it occurred, or by completely removing the organic components and replacing with non-biodegradable materials. These latter materials have been recommended for use since the British Standard code of practice giving preference for glass fibre isolating membranes was first published in 1977 (BS CP 204: Part 2[199], amendment slip no 1 – the code of practice has been withdrawn).

Fire

The limestone aggregate content of mastic asphalt confers a degree of protection against combustion, although mastic asphalt is itself, of course, combustible. While there is no test for surface spread of flame for floorings (except for textile floorings described in Chapter 6.1), mastic asphalt is not normally considered to constitute a risk. However, it cannot be used in firefighting shafts, and may be advised against for escape routes since it could ignite or melt.

Suitability for underfloor heating

In general, mastic asphalt and pitchmastic floorings are not suitable for underfloor heating, though examples have been seen by BRE investigators. Even the hardest grades soften at ambient temperatures above around 30 °C. The effects of underfloor heating are nullified by these floorings which also act as insulants.

Mastic asphalt can be used, however, as a DPM below floors which are heated, provided the asphalt is protected by a layer of thermal insulation.

Sound insulation

Mastic asphalt provides a slightly resilient surface which provides some help in reducing impact noise. Airborne sound insulation depends on the overall mass of the floor.

Durability

Jute based sheathing felt was used extensively in the past as a separating layer in mastic asphalt ground floors, but it has given problems. Moisture in the slab causes jute based felt to degrade accompanied by the production of gases (eg methane) which disrupt the mastic asphalt. Glass fibre tissue of 50 to 70 g/m^2 is inert and obviates this particular problem; therefore glass fibre should always be used.

Mastic asphalt surfaces (depending on grade) have the following resistances to various liquids:

Case study

Blistering and brown liquid in a mastic asphalt floor of a school

The surface of flooring over a concrete slab in a West Country school had disrupted. An organic separating felt and an 18 mm mastic asphalt DPM above the slab formed a bed for a variety of floorings.

There were three symptoms of the problem:

- blistering and cracking of the asphalt layer which caused the disruption of the surface finishes – wood blocks and needle punched carpet in particular, but also linoleum
- seepage through the cracks of a brown liquid which left a dark stain
- foul smelling gases

Surfaces other than the wood blocks, carpets and linoleum showed no signs of problems; PVC flooring, noticeably, remained unaffected.

The problem arose through the presence of water, either left from the construction of the concrete slab or from subsequent rising damp from the ground. This led to the decomposition of the natural fibres, probably jute, in the layer above the concrete. In the case of some surfaces, water did not penetrate the top layer and gases escaped horizontally from the edges of the flooring.

Advice contained in BS CP 102[40], which recommended the use of non-organic glass fibre insulating layers, was not observed when the floor was constructed. Piecemeal repair as necessary with an appropriate non-organic underfelt was the solution recommended.

• water	very good
• dilute acids	fair (good for acid resistant grades)
• alkalis	good to fair
• sulfates	very good
• mineral oil	poor
• animal oil	poor
• organic solvents	poor

It was possible in the past to obtain oil resistant mastic asphalts, but good performance of an existing asphalt floor cannot now be replicated in replacement construction since oil resistant grades are no longer available.

Alternate hot and cold water discharges over mastic asphalt can cause cracking.

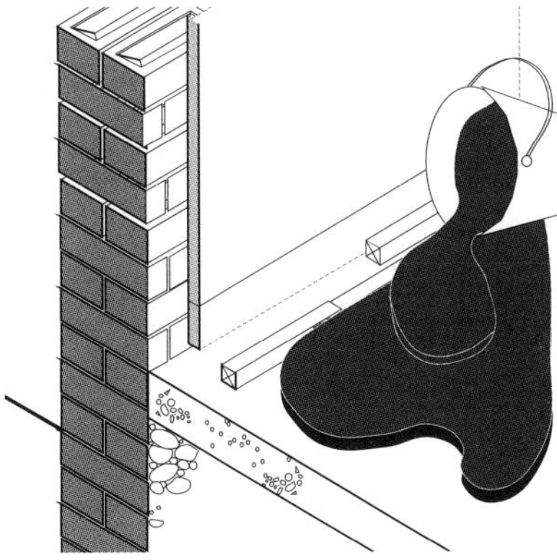

Figure 5.18
Laying mastic asphalt flooring (special hard flooring grade) against a batten at an abutment

Pitchmastic surfaces have the following resistances to various liquids:

- water — very good
- dilute acids — fair
- alkalis — good to fair
- sulfates — very good
- mineral oil — fair
- animal oil — poor
- organic solvents — poor

Maintenance

Light scrubbing with warm water and a detergent will remove most superficial dirt. Cleaning agents and polishes containing solvents should be rigorously avoided. Oils, fats and greases must be removed immediately.

The use of solvent based seals and cleaning pastes have led to progressive deterioration of mastic asphalt and pitchmastic floorings. Only emulsion waxes should be used.

Work on site

Storage and handling of materials

Great care is required to prevent asphalt heating apparatus from igniting flammable parts of the structure.

In any repair work, blow torches should not be used on asphalt since the material will be carbonised. If asphalt has been overheated during the laying process, a more brittle material results. It is possible to modify the asphalt by the addition of a suitable polymer to make it less vulnerable to overheating.

Restrictions due to weather conditions

Special hard floorings of mastic asphalt should not be laid in ambient temperatures below 10 °C. Also, precautions have to be taken to avoid these grades being placed tight to abutments or service perforations. It is preferable to place battens at the positions shown in Figure 5.18, and lay the remainder of the surface up to the abutments in a second operation after removal of the battens. Such precautions are unnecessary with other flooring and paving grades of mastic asphalt.

Workmanship

Mastic asphalt floors are normally laid in a single coat, except where armouring is incorporated or where they are specified for use in certain wet areas. Daywork joints in new floorings, or the edges of cut out patches in defective old floorings, should be softened with hot asphalt before completion of new work.

Mastic asphalt floorings can be expected to be laid to the following standards of accuracy (permissible departure from a 3 m straightedge laid in contact with the floor):

- special or high standard floors of Grades I and II: 3 mm
- normal standard floors of Grades I to III: 5 mm
- floors of Grades III and IV, and paving grade where surface regularity is not critical: 10 mm

Mastic asphalt, where used as a finish, should be laid late in the construction process (at the same time as other floorings, eg PVC) so that the surface does not become contaminated. Mastic asphalt floorings do not need to be cured but can be put into use as soon as they have cooled to room temperature.

Special attention must be paid to the following:

- correct laying of underlays
- maintenance of adequate thicknesses
- accuracy of levels
- avoidance of surface irregularities
- avoidance of overheating (maximum temperature 230 °C)

Inspection

The problems to look for are:

- ◊ indentations
- ◊ slipperiness
- ◊ finishes too soft
- ◊ asphalt finishes lacking adhesion
- ◊ floorings lacking adhesion
- ◊ finishes cracking
- ◊ foul smells

Chapter 5.8 Magnesium oxychloride (magnesite)

This chapter deals primarily with magnesite as the final finish to a floor. Magnesite has also been used as a screed and there is no essential difference in the applications.

The magnesite type of floor finish has been much used in the past, and examples may still be found in both the domestic field and in certain industrial applications, particularly in the food processing industries where resistance to organic solvents has been required. Magnesite floorings are not currently in common use, going into decline in the 1960s. They were very popular in the interwar period, and also extensively used as floor finishes in local authority housing between 1945 and 1960 (Figure 5.19). However they were not much used in building types other than housing because of potential breakdown with moisture or following impacts. Some examples were coloured using brown or green pigments.

Where the magnesite has been protected by an effective DPM, it can be overlaid with sheet or tile floorings. However, where there is any doubt about the effectiveness of the DPM, this should not be done. If magnesite is wetted and remains wet, it disintegrates.

Figure 5.19
Magnesite flooring laid on a staircase. The tread and nosing inserts were cast in as the magnesite was laid

Characteristic details

Basic structure

Magnesite flooring is made from a mixture of calcined magnesite and magnesium chloride solution, with various fillers such as wood flour and sawdust, and pigments. Sand was sometimes added and, on rare occasions, asbestos fibre. The relevant standard for the material was BS 776:Part 2[214], now withdrawn. Suggestions for the composition and laying of these floorings were for many years incorporated into various editions of BS CP 204:Part 2[199], but this also has been withdrawn. Single coat work was laid between 10 and 25 mm thick, but two coat work could have thicknesses up to 50 mm. Concrete was the most common base to which the mixture was bonded.

Light industrial premises with timber boarded floors were improved sometimes by fixing down galvanised wire netting and screeding with magnesite, and then often overlaying with linoleum.

Detailing

Coves and skirtings can be formed in magnesite laid over a cementitious render coat. Contact with calcium sulfate plasters should be avoided (Figure 5.20).

Main performance requirements and defects

Appearance and reflectivity

Magnesite is one of the more difficult floorings to recognise with certainty on site without the benefit of laboratory examination of samples.

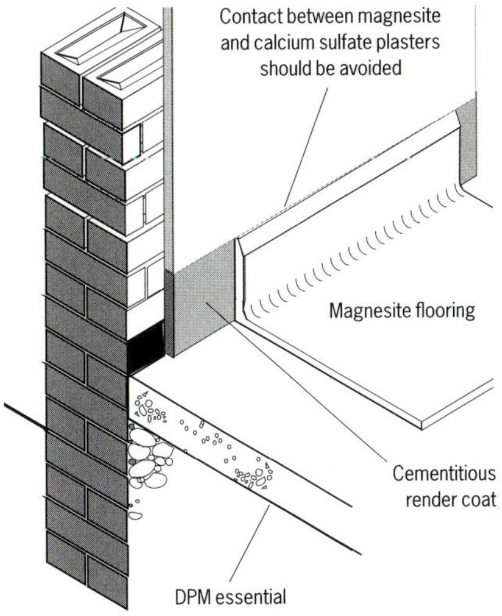

Figure 5.20
Typical edge detail of a magnesite floor. The magnesite may be in one or two coats

Colour is no certain guide. While many examples of magnesite floorings were coloured dull red or pink, they can be confused with pigmented cementitious screeds (Figure 5.21), though the presence of sawdust fillers may make them recognisable. It was commonly pigmented to either a brick red or a straw yellow, but sometimes magnesite mortars of two different colours were partially mixed together to produce mottled effects.

Some mixes may be marbled, with reds, greens and whites used to produce the marbling effects.

Reflectivity is fair for most mixes at around 0.25.

Choice of materials for substrate

Magnesite floorings have been laid successfully on a very wide variety of substrates including new and existing concrete and timber. Timber needs a mechanical key, often formed in the past with galvanised wire netting fixed with clouts. Naturally the greater the degree of movement in the substrate, the less durable will be the magnesite flooring.

Strength and stability

Resistance to impacts is fair.

Dimensional stability, deflections etc

Common defects include:

- loss of bond to the substrate
- curling
- cracking

Because the set and hardened product has low drying shrinkage, these defects tended to be less significant than with other cementitious toppings. On rare occasions a magnesite floor, which had given good service for many years, would expand and bulge upwards. The cause of this has never been fully understood.

Wear resistance

Abrasion resistance is fair to good, impact resistance fair to good, and longterm indentation good.

Slip resistance

The slip resistance of untreated surfaces is fair, though wax finishes will increase slipperiness. No values can be given.

Control of dampness and condensation

Magnesite can only be used in dry situations. Magnesite floorings are very vulnerable to dampness (especially rising damp); those without effective DPMs and suffering rising damp will probably have failed and been removed long ago if covered by an impervious flooring. If a magnesite floor becomes damp and remains so, the oxychloride reaction is reversed: the

Case study

Slipperiness of a magnesite floor in a factory

The frictional characteristics of a repaired and extended magnesite floor in a factory were tested by BRE Advisory Service. Coefficient of friction measurements were taken with the Transport Research Laboratory skid tester. These measurements showed that there was a big difference between the slip characteristics of the new floor, where it had been trafficked, and the old floor. This was thought to be caused by the type and size of fillers in the different areas of the floors. Comparison of the coefficients of friction obtained with the value recommended for a safe floor indicated that, for soft rubber soles on the trafficked areas in the extension, the floor was on the borderline of safety; for leather soles, though, the values were below those recommended.

The building owner was advised that slightly roughening the floor by sanding might be sufficient to overcome the problem. If this was not successful, a further possibility was to apply to the sanded surface an epoxy coating containing an abrasive grit, although this would alter the appearance of the floor.

Figure 5.21
This flooring may look like magnesite but it is, in fact, a pigmented sand and cement screed

material rapidly loses strength (Figure 5.22) and, in the worst cases, may form a mush. The reversed reaction releases magnesium chloride which rapidly corrodes metals; if it penetrates concrete bases it can affect reinforcement.

Water can reach magnesite in a number of ways such as spillage, plumbing leaks, and construction water during major refurbishment of a building. It was not uncommon, though, for it to be laid in ground floor situations without a DPM, any rising moisture vapour diffusing through the flooring and evaporating away without harm. But if such a floor is covered by an impervious flooring, such as PVC or rubber backed carpet, moisture can build up in the magnesite with adverse results. Magnesite screeds should not be covered with a new DPM either, since in these circumstances they will also suffer accelerated deterioration. All magnesite screeds should therefore be completely removed before installing new surface DPMs.

Magnesite can undergo a phenomenon known as 'sweating' in moderately humid conditions. This is characterised by beads of magnesium chloride forming on the surface and is not the result of condensation. Sweating is to some extent a property of the material since magnesium chloride, an essential ingredient, is very hygroscopic; because it is held in the material in slight excess, it readily takes up moisture from humid air.

Magnesium chloride may also migrate into adjacent walls and will cause dampness there because of its hygroscopicity.

The slight excess of magnesium chloride ensures that magnesite flooring is electrically conducting. As a result, moisture meters of the resistance type cannot be used to assess the moisture condition of this flooring. Even when bone dry, most meters give nearly a full scale deflection (Figure 5.23).

Thermal properties

Thermal conductivity of magnesite: no information is available but probably around 0.5 W/mK.

Warmth to touch

Magnesite made with wood fillers is reasonably warm to the touch but mineral fillers produce a finish which is no warmer than cementitious finishes.

Fire

Incombustible.

Suitability for underfloor heating

Magnesite is totally unsuitable because of the need to protect metals from the risk of corrosion from the highly corrosive chlorides produced in the presence of dampness.

Figure 5.22
Magnesite flooring breaking up following prolonged wetting

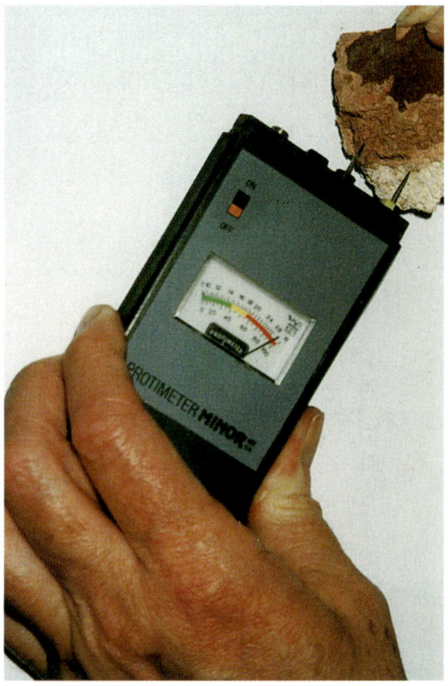

Figure 5.23
Full scale deflection of an electrical resistance moisture meter on a bone dry sample from a two coat magnesite floor. Wet or dry, the meter would give the same reading

Sound insulation

The airborne sound insulation qualities of a floor screeded with magnesite will probably be very similar to one screeded with a cementitious screed, though no actual measurements are available.

Durability

Magnesite tends to expand under the influence of free lime as an impurity causing cracking. Since it has only moderate strength, and is laid bonded and relatively thin, it will reflect movements in the base on the surface. It tends to break up in sympathy with movements in the substrate (Figure 5.24).

In the absence of these influences, a good quality magnesite floor will give good service for 50 years. Resistance to wear in domestic situations is good, but it is unsuitable for very heavy wheeled traffic.

Chlorides may migrate from magnesite screeds into structural concrete below, and accelerate corrosion in any reinforcement. Magnesite will corrode all mild steel products with which it comes into contact, including gas and water pipes.

Figure 5.24
A shrinkage crack in magnesite flooring

Case study

Magnesite too weak for office conditions

A magnesite screed in a mill which had been converted to offices failed. This led to loosening, curling and fracturing of a flexible vinyl tile flooring. The magnesite had been laid on a timber floor and was found to be too weak for its purpose as a subfloor for the floor tiles. Recommendations were made for remedial action by replacement of the screed with one more suited to an office environment.

Maintenance

Magnesite floorings should preferably be cleaned with warm water only and dried thoroughly. Alkalis should be avoided. The surface can be waxed or treated with a drying oil, but waxing may increase slipperiness.

Work on site

Restrictions due to weather conditions

Magnesite requires curing. It should not be allowed to dry rapidly, hardening for at least three days before light use becomes possible. Curing will take longer in cold weather.

Workmanship

Replacement with magnesite

As with many other floorings, skilled workmanship is a crucial factor in the performance of a magnesite floor. Complete mixing of the ingredients is vital, paying particular attention to the density of the magnesium chloride solution and its proportion in the mix, and to cleaning the substrate before laying the magnesite. Too high a proportion of magnesium chloride will increase the risk of sweating of the flooring. Special precautions are needed to avoid it coming into contact with the structure of the building. All metal components must be protected (eg with bitumen paint).

Replacement with alternative floorings

Most magnesite floorings were installed before the 1960s and by the end of the 1990s would probably be nearing the end of their useful lives. If they are unaffected by moisture, and can be guaranteed to remain so – that is to say they are on a suspended floor or protected by an effective DPM – they can be used as a screed to receive other floorings. However, **if there is any doubt about moisture protection, magnesite floorings should be replaced.**

Minor cracks can be repaired with cementitious latex mixtures but it is rarely possible to match the colour or the texture.

Because the common thickness of magnesite is between 10 and 20 mm, there have been difficulties in replacement because most other screed materials cannot be used at these thicknesses. Furthermore, a DPM has to be installed. See also the section on surface DPMs in Chapter 1.5.

Powdered asbestos may have been used as a filler in some examples of magnesite flooring, and appropriate care should be exercised in removing these finishes.

Inspection

The problems to look for are:

- ◊ dampness. (Magnesite cannot be tested with electrical moisture meters as the material is electrically conducting)
- ◊ screeds generally breaking up

Identification of magnesite flooring might need to be confirmed by laboratory examination.

Chapter 6 Jointed resilient finishes

This chapter deals with thin sheet and tile floorings as described in BS 8203[35] which also gives full installation instructions for all resilient floorings except textiles.

The most common type of failure with all these jointed resilient floorings is insufficient time being allowed for construction moisture in screeds and concrete bases to dry out. None of these floorings should be laid on new construction until moisture condition has been established by the hygrometer method, and shown to be 75 to 80% RH (relative humidity) or less; and none should be laid in a ground floor situation without the protection from moisture provided by an effective DPM.

Since the late 1980s BRE has investigated a number of sheet and tile failures which have resulted from materials being laid at temperatures well below those recommended by BS 8203.

It is not good practice to stick any of the floorings described in this chapter onto any others of the same or similar materials; various chemical and physical interactions can produce unwelcome effects such as plasticiser migration which lead, in turn, to shrinkage or swelling.

Maintenance of sheet and tile floorings is dealt with in BS 6263: Part 2[114].

BS 5442:Part 1[112] describes adhesives for all sheet and tile floorings in this category.

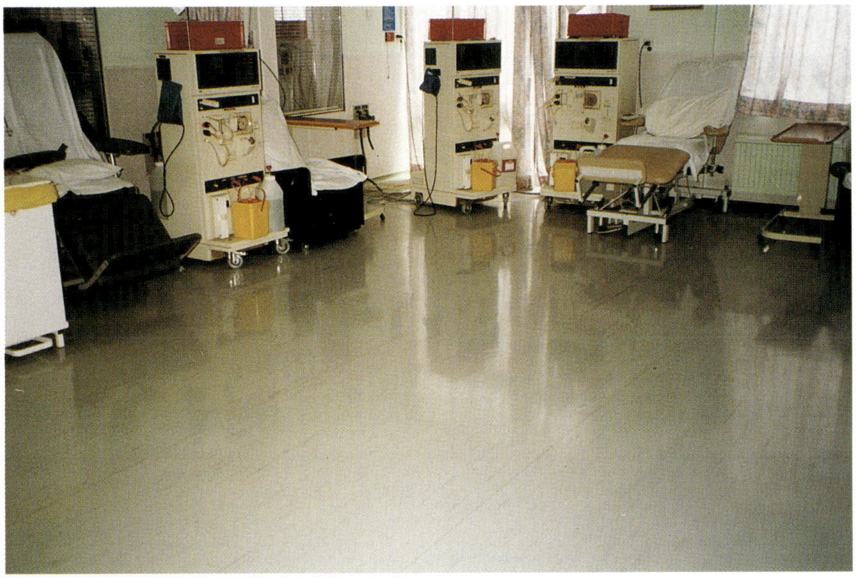

Figure 6.1
Flexible PVC flooring with welded joints in a dialysis unit of a hospital

Chapter 6.1 Textile

Where textiles are stuck to the substrate, and they are effectively therefore part of the building contract, they are included in this chapter. Loose textiles (eg loose-laid carpets) fall outside the scope of this book.

Even with this limitation, however, the range of materials is enormous; all that can be done here is to cite a few of the more common varieties together with some of their main characteristics.

Characteristic details

Basic structure

Textile floorings usually take the form of carpets of which there are many varieties and qualities. Materials used include wool and cellulose, and various combinations of synthetic fibres such as acrylics, polypropylene and nylon for piles, with polypropylene, nylon and polyesters used for backings. Some carpets are reinforced with glass fibre.

Most carpets are laid with an underlay or incorporate an underlay material in their construction.

Pile carpets made in the UK are either Wilton or Axminster weave. Wilton weave produces the higher density of pile: the yarn is woven as loops into the backing, and the loops are then cut to become the pile. Axminster, which is generally less dense, is cut on the loom during weaving with the pile held in place only by the weft of the carpet.

In tufted carpets the tufts are stitched to a jute or synthetic backing and locked into place with an adhesive. A wide variety of grades is available.

Fibre bonded carpets, often based on polypropylene or nylon, have become popular. They can offer considerable advantages from the point of view of ease of cleaning, and give excellent abrasion resistance.

Carpet tiles also have become more common, with loose-laid tiles giving significant advantages over stuck tiles when it comes to replacement. Nevertheless, sticking may be necessary at points where traffic routes change direction.

Carpets laid as part of the building contract are normally fixed by gripper battens round the perimeter of the area to be covered; or they can be stuck which may give problems when they need to be replaced.

Figure 6.2
A carpeted bar in a pub: one of the most onerous of service conditions

Main performance requirements and defects

Appearance and reflectivity

Such a wide variety of designs is available that comment on appearance is not practicable except to note that some dyes used for textile floorings will fade.

Light coloured textiles and carpets give reflectances of around 0.45, medium shades around 0.25, and dark shades around 0.10.

Choice of materials for substrate

Where textile floorings are to be laid on groundbearing bases, the bases must incorporate an effective DPM. For suspended floors, many substrates are suitable.

Dimensional stability, deflections etc

Some textiles will tend to stretch under continuous traffic (Figure 6.3).

Figure 6.3
This carpet has stretched and wrinkled because of an unsatisfactory underlayment. The underlayment was breaking up under the effects of foot traffic

Usually the fault is caused by inadequate care taken in laying the adhesive or with other methods of fixing to the floor. It is not possible to provide data on stretching in carpets.

Other textiles will shrink as they dry out, both synthetics and natural fibres. The problem relates mainly to the instability of the backings used.

Wear resistance

With woven carpets, denser piles give greater durability. Flocked carpets can also give good performance in areas of heavy traffic.

Resilient underlays of appropriate thickness are important in promoting longevity.

Slip resistance

Textile coverings will usually present no problems. Where a carpet adjoins a different floor finish at a doorway, a dividing strip or threshold should be provided.

Coefficients of friction are:

- dry > 0.7
- wet 0.4 to > 0.6

Static electricity

Certain kinds of textile flooring, particularly those containing man-made fibres, in common with certain kinds of vinyl and other plastics, have been prone to problems caused by the buildup of static electricity from friction. The phenomenon occurs in conditions of low humidity, and can be uncomfortable to occupants and disastrous to microelectronics or to atmospheres where flammable gases can accumulate. This is mainly a problem with new carpets and is particularly evident during the first heating season. As the carpet becomes more soiled, conductivity improves and static is therefore less of a problem.

The problem can be ameliorated by incorporating conducting materials within the flooring.

Control of dampness and condensation

An effective DPM is essential for most kinds of textile flooring, particularly those made with natural fibres. Textile floorings made with natural fibres will settle down to moisture contents of around 15 to 18%, while synthetics can be as low as a few per cent.

It is also important that bases and screeds are given sufficient time to dry before the floorings are applied (see Table 1.3 in Chapter 1.4).

Treatment of textile floorings with silicone sprays after shampooing does assist in repelling water and preventing water based stains.

Figure 6.5
Duct covers not flush with carpeted floors can lead to tripping. While it is possible to firmly clamp the cut edges of carpets, there will be times when carpets and duct covers (and any other protrusions offered by inspection covers, matwell frames etc) will be at different levels. The same sort of risk of tripping will be encountered at doorways fitted with divider strips, especially where screws fixing the strips have worked loose

Figure 6.4
A sample of carpet tested for abrasion resistance from the BRE carpet testing programme of 1970

Figure 6.6
These carpet tiles have shrunk because of migration of plasticiser from the PVC backing into the PVC tiles beneath. The degree of shrinkage is shown relative to a small coin – about 5%. See also Figure 6.17 which shows the effect on PVC tiles

Thermal properties
Thermal conductivity of carpet: 0.055 W/mK.

Warmth to touch
There is usually no problem with textiles.

Fire
Although there is normally no requirement for fire protection of textile floor coverings in most situations in buildings, there is an increasing tendency for products to be tested to BS 4790(80). Where used within firefighting shafts, textile floorings are subject to special requirements(70).

Suitability for underfloor heating
Textiles will have an insulating effect on the flooring substrate, and the effects of heating should always be considered.

Sound insulation
Sound absorption is usually very good. Airborne sound insulation depends largely on the mass of the structural floor. Certain kinds of textile flooring can be stuck directly to a non-floating floor to achieve adequate impact sound insulation or can be used on a floating floor.

Durability
Life expectancy will be about 10 years for natural fibres and up to 20 years for man-made fibres, depending on traffic conditions. The backing or underlay has an important influence on the overall life of a product.

Risk of attack by insects (moths and beetles) which used to be a significant problem before the middle of the twentieth century, has been greatly reduced by pretreatments. Cleaner building environments have also kept down insect populations.

By far the most significant problem presented by textile floorings, however, is plasticiser migration from plasticised PVC backings. Plasticisers are used to soften PVC. Where the plasticised PVC is in contact with another material made from a polymer which also uses plasticisers, the plasticisers can move from one material to the other. The problem can occur where a textile flooring with a plasticised backing is stuck down with tape or adhesive also containing plasticiser, or is stuck down over a PVC flooring (Figure 6.6).

Plasticiser migration can lead to shrinkage or embrittlement, or to softening of adhesives. If the level of plasticiser in carpet backing is greater than in PVC flooring, this can lead to expansion of the PVC. Where PVC backed textile floorings are to be stuck down, it is therefore important that all other floorings containing plasticisers are removed.

Maintenance
In heavily trafficked areas, woven carpet will need renewal of stained and worn areas after three to four years of life even though regular cleaning may have been carried out. Vacuum cleaning on its own will not remove all grit embedded in the backing, and this is a major cause of wear.

Flocked carpets with a non-absorbent pile offer considerable advantages where frequent spillage and wet cleaning are likely.

Work on site

Storage and handling of materials
Carpet tiles and rolls should be conditioned at 18 °C for 24 hours on site before laying.

Restrictions due to weather conditions
Temperature affects the performance of adhesives and therefore manufacturers' instructions should be followed.

Workmanship
Full information is given in BS 5325(215) which should be followed.

Inspection

The problems to look for are:
- ◊ rucking
- ◊ excessive wear
- ◊ shocks from static electricity (usually man-made fibres)
- ◊ dampness
- ◊ floorings lacking adhesion
- ◊ natural materials not mothproofed (small holes in covering)

Chapter 6.2 Linoleum

The early history of linoleum has been described in the introductory chapter.

Linoleum in the early 1950s was by far the most commonly used floor covering (*Principles of modern building*[6]), even retaining some of its popularity today. However, since the 1950s it has progressively lost market share to the newer materials such as PVC (see Chapter 6.4).

Characteristic details

Basic structure

Linoleum is prepared from a mixture of linseed oil, resins, cork and wood flour, traditionally adhering to a hessian backing, though other materials have been used as backings in more recent years. Heat treatment then follows to create a tough material which is consistent throughout its structure.

If it is made with coarsely granulated cork instead of fine cork and wood flour, the result is known as cork carpet (which is no longer available).

Linoleum was formerly available in a range of thicknesses from 2 to 6.7 mm, with cork carpet up to 8 mm. (Battleship lino was a popular term used – though not by manufacturers – to describe the thicker grades.) The thicknesses available from stock are 2.4 and 3.2 mm though other thicknesses may be available to special order.

Linoleum can be laid with butt joints or it can be solvent welded to form a virtually continuous surface.

The appropriate standard is BS 6826[205].

Detailing

Normally abutments are covered with skirtings or a cover mould. The thinner products can be formed as coves by using formers to which the linoleum is stuck.

Linoleum is vulnerable to scuffing in areas of heavy traffic such as at thresholds; some protection, perhaps in the shape of a timber moulding or aluminium extrusion, may be needed.

Main performance requirements and defects

Appearance and reflectivity

Linoleum used to be available in a very wide range of colours and patterns: plain, jaspé, granite, marble and moiré, as well as many inlaid designs. The range available today is very much reduced from what was available in former years, and old designs may not be available for replacement purposes.

Light colours are available, giving excellent reflectivity. Light shades give reflectances of around 0.45, medium shades around 0.25, and dark shades around 0.10.

Choice of materials for substrate

Linoleum floorings have been laid successfully on a wide variety of substrates. Although the material is flexible, it performs best on a firm base. Suspended timber floors need to be in good condition, with no cupping and with nail heads punched below the surface. Timber floors susceptible to dampness also need to be well ventilated since the effect of linoleum coverings is to reduce ventilation. Several floors have been inspected by BRE where rot has ensued after laying this relatively impermeable covering over timber boards. Examples of loose laid linoleum floors have performed as well as the material stuck down, though, with old floors with wide gaps between the boards, billowing in strong winds has been known. Sticking down only at the perimeter of sheets generally has not proved successful.

If a timber floor is not level, for example boards are wearing or cupping, the surface should be covered with a suitable board or fabricated underlayment (eg hardboard or plywood).

Linoleum over a cement screed with a suitable DPM has usually given good service; indeed this is the most common arrangement to be seen. Linoleum over mastic asphalt has also proved successful. Attempts to create a surface DPM by painting on several layers of bitumen before laying the linoleum have not proved successful.

Magnesite can form a suitable substrate provided the floor has an effective DPM (see Chapter 5.8).

Adhesives used for sticking down linoleum have been of many varieties: vegetable and casein glues, lignin pastes, gum spirit adhesives, latex cements, bitumen rubber, and tar emulsions and bitumen rubber solutions in petroleum solvents. Not all these materials would now conform with health and safety legislation, or indeed be available. The adhesives most often used are resin alcohol and gum spirit. BS 5442:Part 1[112] applies.

Case study

Cutting and indentation of a linoleum floor in a public library

The 4.5 mm thick linoleum flooring in the public library of a local authority in the south east of England, which had been in use for eight years, was found to be so damaged by wear that it was no longer satisfactory for its purpose.

In the ground floor lending library, heavy foot traffic had led to cutting and indentation, particularly in the areas around the shelves. In the first floor reference library there was similar damage, and there was further damage caused by readers' use and misuse of chairs. In the lower ground floor offices, again there was damage of a similar nature, but also some from very heavy furniture. In the basement, where use was less frequent, damage was minimal. Apart from the surface cutting and indentation, the linoleum was in good condition. Insufficient stoving in manufacture which used to be a common problem with the thicker grades, may have been a contributory cause.

Linoleum gave the required degree of quietness for the building's use and, therefore, was advised as being the most satisfactory replacement for the damaged areas. Harder linoleum, of 3.2 mm thickness, would allow replacement without necessitating the removal of the original adhesive layer, a proprietary levelling compound taking up any difference in thickness.

The possibility of using textile flooring (needlefelt) in the reference library was considered, but linoleum was still thought to be the best surface available for the building's purpose.

Case study

Failure of linoleum to adhere to a floor screed in a hostel

Failure of linoleum to adhere to a floor screed in a hostel block was examined by BRE investigators. Lifting and bubbling of the linoleum was considered to have been caused by insufficient or no transference of the gum spirit to the hessian backing. The building owner was advised that most of the linoleum could be taken up and relaid with new adhesive without removing the old adhesive which had stuck well to the sand and cement screed.

Strength and stability

Standard linoleum may not be suitable for use in situations where there is a risk of impact loads from sharp objects and wheeled traffic. Linoleum was found to be very susceptible to indentation from stiletto heels during the 1960s. Harder grades were then produced to overcome the problem. A toughened form is also available in which the surface is reinforced giving improved resistance to point loads.

Dimensional stability, deflections etc

Generally, the indentation resistance of linoleum is fair, and it has a good rate of recovery. It also has sufficient resilience to follow normal moisture and thermal movements in cementitious substrates, even when fully bonded.

However, rippling of linoleum caused by movements in the substrate is a particular problem.

Wear resistance

Good.

Slip resistance

Good, although wax polishes should not be allowed to build up since surfaces can then become slippery. Linoleum can also be slippery when damp.

Coefficients of friction are:

- dry > 0.7
- wet 0.2 to < 0.35

Control of dampness and condensation

Linoleum is susceptible to degradation in the presence of water rising from below, or where spilt in quantity and remaining on the surface.

Thermal properties

Thermal conductivity of linoleum: 0.22 W/mK.

Warmth to touch

Good.

Fire

Combustible, though this is normally of little consequence unless used in escape routes in buildings where a fire certificate is required (see Chapter 1.6).

Although the material can be marked by cigarette burns, normally any scorching can be removed with care using a fine abrasive.

Figure 6.7
This linoleum tile floor shows evidence of upwards migration of the adhesive which has been affected by moisture – the prelude to curling and de-bonding

Suitability for underfloor heating

Suitable, though there is a small insulating effect from the cork content.

Sound insulation

Sound absorption is usually very good. Airborne sound insulation depends largely on the mass of the structural floor. Linoleum flooring can be used on a floating floor.

Durability

Life expectancy will be around 15 to 25 years depending on traffic.

The most common cause of lifting is construction moisture in the substrate (Figure 6.7). Other causes are insufficient adhesive or lack of adhesion due to late placing of the linoleum onto the adhesive (ie when it is too dry).

New linoleum is more susceptible to damage than older installations. New linoleum hardens slowly by oxidation of the oils.

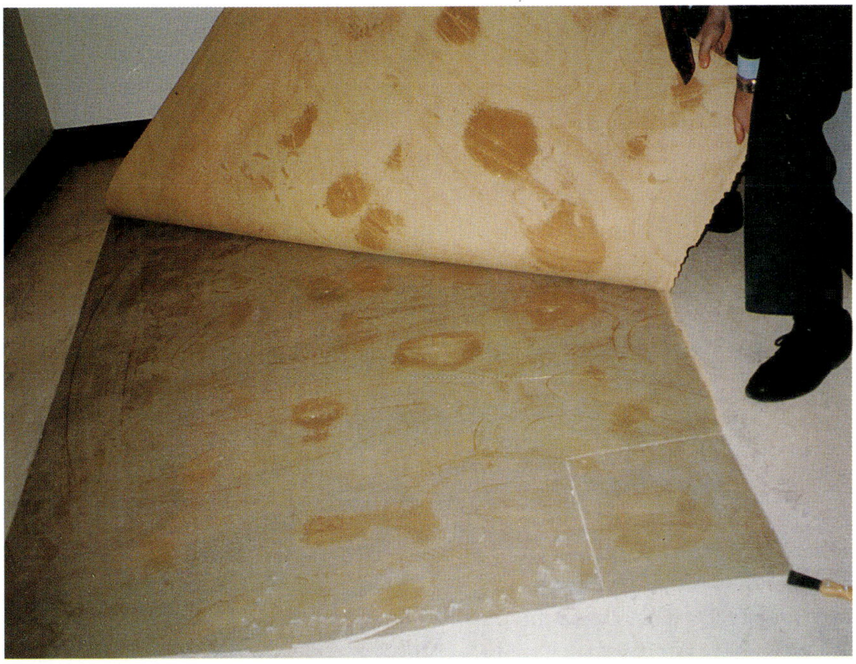

Figure 6.8
An example of linoleum having been laid off before it has set

Maintenance

Infrequent washing with a wet mop and a mild detergent is the preferred method of maintenance; scrubbing or abrasive cleaners, and particularly highly alkaline cleaners, must be avoided. Thin layers of either solvent or emulsion wax polishes are beneficial but should not be allowed to build up. White spirit, used sparingly, may be needed to remove deposits. Oil based seals will discolour linoleum.

Improved polishes have become available (eg dry-sheens, buffables etc). BS 6263:Part 2[114] applies.

Work on site

Workmanship

Laying linoleum is a skilled job. In particular, it needs to be conditioned and loose laid for a minimum of 24 hours – preferably a few days – before fixing so that it can adjust to ambient conditions. Less time needs to elapse when the linoleum is to be stuck down since the adhesive to some extent restrains the material, but it is important that the laying criteria – conditioning the linoleum, spreading the adhesive, rolling the linoleum and re-rolling after 30 minutes (BS 8203[35]) – are correctly observed (Figure 6.8).

Care should be taken that the 'bite marks'† induced during stoving are either removed before laying, or that methods are used that will ensure the material lies flat (ie without bite marks showing) after laying.

Adhesives used in laying linoleum should be those recommended by the manufacturer. The most common in use are the resin alcohols. The linoleum must be laid before the adhesive skins. Occasionally SBR (styrene-butadiene rubber) emulsions are used.

Thorough rolling with a 68 kg roller is essential.

† A 'bite mark' is a deformation formed where the linoleum passes round the bottom roller in the stoving part of the manufacturing process. The section which passes over the top roller is cut away. If the bite mark is not removed before or during laying, a hump is left in the finished surface – the 'memory effect'.

Linoleum is one of a number of materials that is susceptible to the memory effect, a phenomenon in which a material returns to an earlier form after an attempt has been made to reshape it. Paperhangers know the problem when they try to flatten a roll of wallpaper, but, when the ends of the roll are released, it returns to its original rolled up form.

Inspection

The problems to look for are:
- rucking
- excessive wear
- lifting and stretching
- floorings lacking adhesion
- ridging
- linoleum breaking down in localised areas
- surfaces of linoleum roughening
- materials discolouring or light shades yellowing

Chapter 6.3 Cork

Cork, in tile or sheet (carpet) form, provides a flooring which has been used for many years, taking advantage of its warmth and resilience. However, there are negative properties too, particularly its susceptibility to wear if not suitably protected and to damaging cleaning methods.

Characteristic details

Basic structure

Cork tiles used to be made by moulding granules of cork under heat and pressure, using the natural resins in the cork as a binding agent, although the addition of resin has permitted manufacture at much lower temperatures. Tiles range in thickness from 3 to 8 mm. The thinner tiles are usually to be found with square edges; thicker tiles are occasionally tongued-and-grooved.

Tiles faced and backed with clear PVC have become available, and these give improved abrasion resistance and cleaning properties.

Main performance requirements and defects

Appearance and reflectivity

Cork tiles are available in a range of natural shades of brown.

Cork in tile or sheet form gives reflectances of around 0.20, depending on exact shade.

Choice of materials for substrate

Virtually any substrate is suitable but it must be dry and not likely to become damp.

Strength and stability

See the same section in Chapter 6.2.

Dimensional stability, deflections etc

Cork has only fair resistance to indentation though there is usually some recovery when the load is removed. The vinyl faced tiles will give better performance.

See also the same section in Chapter 6.2.

Wear resistance

Different densities of the material are available, ranging from around 400 kg/m^2 for domestic use to heavy contract at around 500 kg/m^2 or more. It can be surfaced with PVC for use in areas experiencing heavy traffic, or, for less heavily trafficked areas, site coated with polyurethane or oleoresinous seals.

Slip resistance

Cork normally has good non-slip properties, though somewhat less so in damp conditions.

Coefficients of friction are:

- dry > 0.7
- wet 0.3 to > 0.5

Control of dampness and condensation

The material should not be used on floors without an effective DPM.

Thermal properties

Thermal conductivity of cork: 0.085 W/mK.

Warmth to touch

Cork is one of the warmest types of flooring available.

Figure 6.9
Cork flooring, which has not been sealed, showing ingrained dirt from foot traffic

Fire

Combustible.

Suitability for underfloor heating

Cork is not very suitable for use over underfloor heating because of its thermal insulation properties. The thinner tiles may prove to be satisfactory, though, if the temperature at the surface of the substrate does not exceed 27 °C.

Sound insulation

Sound absorption is usually very good. Airborne sound insulation depends largely on the mass of the structural floor. Certain kinds of cork flooring can be stuck directly to a non-floating floor to achieve adequate impact sound insulation, or can be used over a floating floor.

Durability
Cork is not very hard wearing, and can be poor when inappropriately specified and maintained.

Maintenance
Cork will eventually break down under inappropriate use of water for cleaning purposes, leading to ingrained dirt which becomes impossible to remove.

The tiles swell if wetted for long periods and should therefore not be used in wet areas.

Work on site

Storage and handling of materials
Conditioning on site for at least 48 hours is needed before laying. BS 8203[35] is relevant.

Inspection

The problems to look for are:

- ◊ materials rucking
- ◊ excessive wear
- ◊ cover moulds not provided at doorways and positions of greatest wear
- ◊ lifting and stretching
- ◊ floorings lacking adhesion
- ◊ ridging
- ◊ cork breaking down in localised areas
- ◊ heads of nails in boards not punched home
- ◊ surfaces of cork roughening
- ◊ materials discolouring or light shades yellowing

Chapter 6.4 PVC flexible

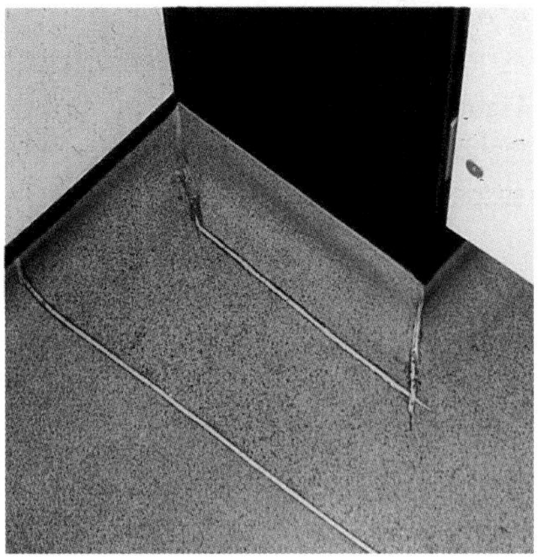

Figure 6.10
An integral cove – showing welding of indifferent quality – for PVC safety flooring in a wet area

The development of flexible PVC flooring – frequently referred to as vinyl – has been one of rapid growth since the 1950s. It is appropriate for use in a wide variety of situations and can be obtained in a wide variety of forms including those with resilient backing. Some grades can be used in heavy wear situations and special grades are produced with enhanced slip resistance.

Characteristic details

Basic structure

The tiles and sheets consist of PVC resins, plasticisers, stabilisers and extenders, forming a rubber-like material which varies in consistency between hard and soft. PVC cannot carry large quantities of fillers without losing its rubber-like properties. The material is calendered to thicknesses varying from 1 to 3 mm, formed into multi-layer sheets as appropriate, and left as sheets or cut to size as tiles. Some floorings of PVC are fabric backed, others cellular PVC backed and yet others unbacked. The cellular PVC backed, which are also known as cushioned vinyl, have superior performance with respect to impact sound insulation and warmth to touch, although having correspondingly increased vulnerability to surface damage from cuts and impacts.

Only the softer materials, Type A (see the next column), can be obtained as sheet.

There are two types of backed flexible PVC flooring covered by British Standards †:

- needleloom felt-backed flooring to BS 5085:Part 1[216]
- cellular PVC-backed to BS 5085: Part 2[217]

Both types are manufactured by spreading soft PVC on the back of the material.

There are two types of unbacked flexible PVC flooring covered by BS 3261:Part 1[203] †:

- Type A – fully flexible
- Type B – less fully flexible, but more flexible than those to BS 3260[218] (see Chapter 6.5)

Type A is available in sheets 2 m wide or occasionally in tile form while Type B is available only in tile

† The standards covering flexible PVC materials – BS 5085: Part 1 and Part 2, and BS 3261:Part 1 – are due to be replaced by BS EN 649, 650, 651, 652 and 653.

Figure 6.11
Rippling in PVC flooring generated by a crack in the sand and cement screed

form. (Type B is insufficiently flexible to roll up.)

Flexible PVC can be laid with butt joints, or it can be hot air welded to form a virtually continuous surface. With some materials joints can be solvent welded.

Main performance requirements and defects

Appearance and reflectivity

As already noted, unbacked flexible PVC is available in roll or tile form and backed material usually in roll form only. Since most of the basic materials are light in colour, a wide variety of clean bright colours and marbled effects is available. Seams can be welded, giving an almost unbroken surface. Welding rods can be found occasionally in colours that will provided contrasting effects to tiles.

Some variety in appearance is possible with the backed kinds by embossing the final surface during manufacture.

Light shades give reflectances of around 0.45, medium shades around 0.25, and dark shades around 0.10.

Choice of materials for substrate

PVC flooring can be laid on a wide variety of substrates. The most common substrate will be a screed or concrete. Although this type of PVC is relatively flexible, the flexibility may not be sufficient to overcome major movements at cracks and joints in the substrate (Figure 6.11). It is too flexible to bridge cavities (Figure 6.12).

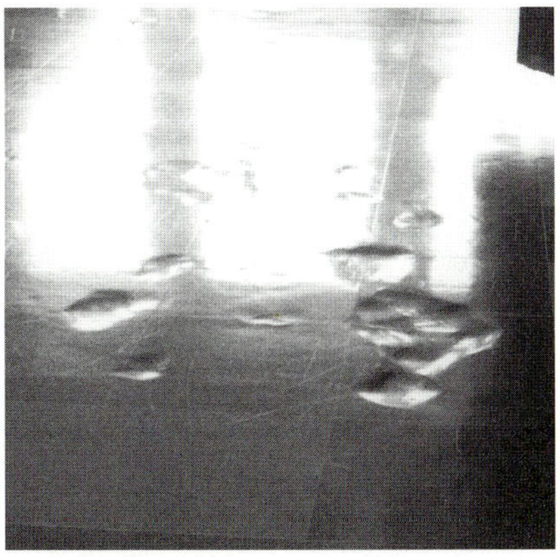

Figure 6.12
These elephants' footprints in the PVC tiles are an indication of a screed which has collapsed

Timber boarded floors will not usually cause flexible PVC flooring to crack, though any cupping will show and movements in the timber may result in rippling (Figure 6.13). Prefabricated timber underlays (eg plywood or hardboard) may be needed. Adequate ventilation to prevent rot will be vital.

Existing floors of ceramic tiles may provide a suitable base if a levelling screed or latex underlayment is applied (see Chapter 4.3). PVC flooring has also been laid successfully on magnesite.

Although there have been successful applications when PVC has been laid directly on mastic asphalt, there is a risk of failure due to the plasticiser migrating into the asphalt and softening it (see the first case study on page 222 and also the same section in Chapter 5.7). A latex underlayment, at least 3 mm thick, should be applied to mastic asphalt before laying PVC flooring.

Strength and stability

The resistance of unbacked PVC flooring to indentation is good, though the material is soft. The backed type, though, can be cut by furniture feet.

Dimensional stability, deflections etc

Coefficient of linear thermal expansion per °C: 40 to 70 × 10^{-6}.

Reversible moisture movement: 0%.

Flexible PVC shrinks for three main reasons:

- calendering effects (ie the memory effect)
- plasticiser migration into unsuitable adhesives
- removal of plasticiser by unsuitable cleaning materials

Any one of these phenomena or any combination of them can lead to excessive shrinkage; for example up to 25 mm in a 2 m sheet. The shrinkage forces are sufficient to split welded seams because these are the weakest points (Figure 6.14).

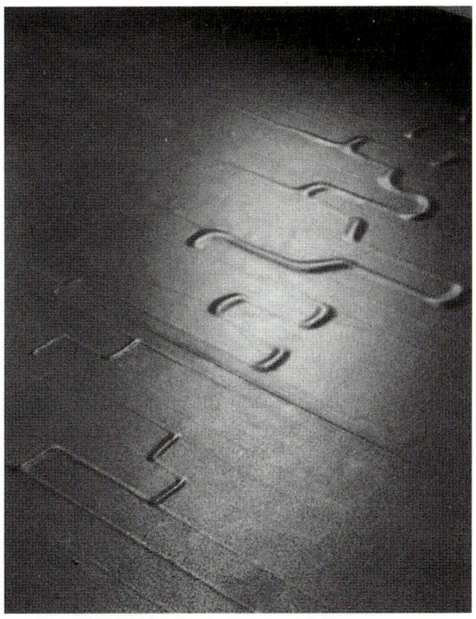

Figure 6.13
Rippling in PVC flooring laid over a suspended timber strip floor

Figure 6.14
Failure of a weld in PVC flooring due to shrinkage

Case study

Indentation and shrinkage of PVC tiles on an asphalt underlay in a factory

At a West Country factory, and associated office and canteen building, indentation of flexible PVC floor tiles had occurred where they had been laid over a mastic asphalt base. Another problem was shrinkage of the tiles in some places with gaps appearing between them.

A number of areas were examined: some where heavy traffic had occurred, others where there had been no traffic, others where furniture on castors had caused heavy indentation, and yet others adjacent to walls and duct covers. In places there had been no adhesive between the tiles and their base, but generally they were stuck down with the same adhesive (styrene-butadiene rubber emulsion). Usually there was a concrete base, but in areas over service ducts no concrete subfloor existed. Site examination had been supplemented by laboratory examination, and by tests on mastic asphalt and PVC tiles in combination.

A number of reasons for the corruption of the flooring were considered. The laboratory tests ruled out the adhesive as a possible cause. Although the tests provided no immediate proof, the most likely cause of the problem was the migration of plasticiser from the PVC tiles to the asphalt, causing softening of the asphalt. The problem was exacerbated by heavy weights and movements on the surface, such as from castors on furniture, which, in some places, had caused a mixing of the asphalt, adhesive and PVC. The shrinkage of the tiles in some places confirmed the likelihood of the plasticiser being the culprit.

The suggested solution was that in areas where little or no damage had occurred – the areas of lowest traffic – no action should be taken at that time, although softening of the asphalt might continue. In areas of maximum deformation, the PVC should be overlaid with needlefelt tiles of at least 6 mm thickness, without adhesive, to provide resilience and disguise the indentations. Ultimately all areas of PVC tile flooring would need to be replaced.

Wear resistance

Normally the wear resistance of these materials is between good and excellent.

Slip resistance

Very good in dry situations, but the smoother types can be very slippery when wet.

Coefficients of friction are, for ordinary PVC:

- dry > 0.7
- wet 0.2 to 0.3

and for PVC incorporating carborundum or aluminium oxide (special slip resistant grades):

- dry > 0.7
- wet 0.3 to > 0.45

Control of dampness and condensation

None of these PVC flooring materials should be used in groundbearing floors without an effective DPM. As the case study at right indicates, rising damp can be a problem with this type of finish.

Thermal properties

Thermal conductivity of flexible PVC flooring: 0.40 W/mK.

Warmth to touch

The material itself is not warm to touch. However the backed materials can show an improvement, depending on the insulation value of the backing.

Fire

Combustible. Electrically conducting grades are available for specialised applications in areas where flammable gases might be ignited by sparking.

All types of PVC flooring are very vulnerable to cigarette burns: the material can melt and the damaged area then collects detritus. There is no suitable remedy short of replacing the damaged tile or welding in a patch in sheet floorings. However, colour matching a new patch with old flooring may be a problem.

Suitability for underfloor heating

Unbacked PVC can be used where temperatures at the surface of the substrate do not exceed 27 °C. Special grades of adhesive should be specified for these conditions because of the risk of enhancing plasticiser migration. Backed PVC is not recommended because of the insulating effect of the backing which will reduce the effectiveness of the heating.

Case study

Entrapped moisture leading to rippling in PVC flooring in a hospital

The cause of rippling in the PVC flooring of a hospital ward was investigated by BRE Advisory Service. The flooring had been laid on hardboard over a concrete base. From the early 1970s the PVC flooring had provided an impervious membrane, but, due to the omission of a DPM in the original floor construction, moisture rising from the ground was being trapped beneath the flooring. The cause of rippling in the flooring was attributed to expansion of the hardboard underlay as a result of moisture absorption; this caused ripples in the flooring over the joints between hardboard sheets. Remedial measures recommended included:

- removing the PVC flooring and hardboard
- allowing the floor construction to dry out
- providing a permeable floor covering, such as needled nylon carpet if this was acceptable from a hygiene standpoint, which would allow water rising into the concrete to evaporate from the top surface

Alternative measures included:

- removing the PVC flooring and hardboard
- laying a DPM on top of the slab
- laying a 50 mm thick sand and cement screed
- when dry, laying new PVC flooring

or:

- removing the PVC flooring and hardboard
- laying mastic asphalt to BS 1076 (subsequently withdrawn and replaced by BS 6925[170]) on top of the concrete slab, with 3 mm of levelling compound over the asphalt to prevent plasticiser migration
- laying new PVC flooring

Since the case study was investigated, surface DPMs have become available which would allow the flooring to be relaid without raising the level of the finished floor.

Safety

Vinyl chloride, the base chemical from which PVC is made, is a hazardous substance; occupational exposure to it is controlled by EC directive and a maximum exposure limit set by the UK Health and Safety Executive. PVC resin is always contaminated with vinyl chloride monomer (VCM) due to the kinetics of the additional polymerisation in the manufacturing process. Although the subsequent stripping and recovery processes considerably reduce the level of VCM contained in the polymer, as much as 100 parts per million can be found in the resin after drying. PVC products are widely used in buildings and therefore VCM is potentially widespread in indoor environments. VCM is considered to be a problem only during the manufacturing process and, being in minute quantities, does not apply to PVC flooring when installed.

Sound insulation

Felt backed vinyl will give a degree of protection against impact sound. It has often been used in this way over separating floors having sufficient mass to provide airborne sound insulation.

Durability

Life expectancy can be 15 to 30 years for a good quality PVC flooring. PVC does not age as rubber does and should retain much of its resilience. If PVC materials are not well adhered to the substrate, though, they can stretch under traffic and bubbles can form which will become torn in use.

Even so, much of the workload of BRE Advisory Service is concerned with faults in PVC flooring. This probably reflects the considerable market share of the product. The case studies aside and on page 224 give a variety of different circumstances leading to failure.

Maintenance

PVC is relatively easy to clean using only water and neutral detergents, so making it possible to maintain a satisfactory appearance. Nevertheless, some types of cleaning agents can give problems, mainly

Case study

Bubbling and blistering of PVC flooring caused by excessive moisture in the substrate – a very common problem

An investigation was undertaken into bubbling and blistering of PVC flooring in a health service building in the West Midlands.

The floor structure consisted of a hardcore bed overlaid with 25 mm polystyrene slabs, a 300 µm (1200 gauge) polyethylene DPM, 150 mm powerfloated concrete slab and, finally, flexible PVC flooring or unbacked textiles (carpets).

The concrete base had been completed some three months before the roof. Laying the flooring began some three months after the roof was installed and was completed in two months. The building was not occupied until some six months after that. Soon after occupation, bubbling of the PVC flooring was reported.

The flooring was found to have bubbled in all the rooms inspected. The bubbles varied in size from 30 to 300 mm in diameter; some were elongated, and others were in the form of rings around the legs of furniture.

Some rapid estimates were made of moisture conditions under the flooring using electrical moisture meters. These meters, which rely on electrical conductivity, are not very accurate, but readings on the 'arbitrary' scale (1–100) can be used as an approximate guide. (In these circumstances, the sharp electrodes of the meter are pushed through the floorings into the surface of the base.) Over 95% of the readings obtained were between 80 and 100% RH (relative humidity), showing the underside to be quite wet.

In the corners of two rooms, the PVC flooring was peeled back and recently calibrated hygrometers were sealed to the surface of the concrete, following the method described in BS 8203[35]; in such situations equilibrium readings are normally reached in about one to two hours. After one hour the readings obtained were 88–91% RH. For acceptable conditions BS 8203 requires the readings to be 75% or less.

When the flooring was pulled up (with relative ease), it was found that the adhesive had re-emulsified and had not developed its full strength.

The BRE investigator concluded that there was no evidence to indicate that rising damp, condensation or spillages had played any role in the failure. For that amount of moisture to have come from the ground would have meant that the DPM was missing over the whole area of the slab, and it was independently confirmed that it had been laid. As the PVC had been welded at the seams, and in the absence of defective seams, any water from above could only have penetrated at the perimeters. However there was little difference between moisture contents at perimeters or edges and the centres of rooms.

It was reported that measurements of the moisture conditions in the base had been made before flooring was laid, but there was no information on the method used. For a 150 mm thick concrete slab, BS 8203 would require fully insulated hygrometers to be in position for at least three days, but readings taken with an electrical meter on such a thick slab would still have been misleading.

The base beneath other areas of the slab covered with unbacked textiles was dry, showing that the bases in these areas had dried out.

The investigator concluded, therefore, that excess construction water and rainwater in the slab had not been allowed to dry sufficiently before the flooring had been laid – a conclusion supported by the three months lapse between installing the concrete slab and completion of the roof. This had resulted in the adhesive losing strength, allowing the flooring to bubble. Further bubbling was likely to occur, so all areas of the affected flooring would have had to be removed.

There were two possible solutions to the problem:

- leave the exposed slab to dry out and then re-lay with new flooring
- provide a surface DPM and then re-lay with new flooring

Since the first solution, involving an out-of-use period of several months, would not be practicable in a working building, the second solution was preferred. Surface DPMs based on epoxy resin have a good track record in these conditions. The surface of the concrete would have had to be cleaned, and a 3 mm layer of latex poured over the slab followed by the DPM and the new flooring.

Case study

Moisture induced expansion causing rippling of PVC flooring in a factory

Anti-static PVC flooring, laid in a factory using an acrylic adhesive over a 200 mm thick concrete base constructed by the long strip method in bays 5 m wide, was rippling over construction joints, bubbling over copper strip intersections and lipping and tenting at some welded joints.

At the time of construction, movement joints had been sawn in the strips at intervals coinciding with column bases and filled with a type of mastic. The slab was allowed to dry for 8 months. Immediately after laying the flooring, the faint outline of the construction joints could be seen underneath the PVC flooring. Some two to three days after the heating had been switched on, ripples began to appear.

Taking the thermal expansion of concrete as 10×10^{-6} per °C, it can be shown that for a 5 °C rise in temperature, expansion of 0.25 mm will occur over a 5 m wide strip. Closing of a joint by 0.25 mm is sufficient to produce a sizeable ripple in flooring laid over the joint. The moisture content of the base was also found to be in excess of that recommended by BS 8203 (ie in excess of 75% RH when measured with a hygrometer). Insufficient drying time had been allowed for such a thick base.

Rippling had occurred because construction joints had closed up since the flooring was laid. Movement was induced by thermal and moisture changes.

Bubbles had formed immediately over copper strip crossovers because of moisture induced expansion of the PVC. The PVC over these areas had been left unbonded. Trucking over these unbonded areas led to stretching of the PVC. The bond was also broken above areas of surface laitance.

Moisture induced expansion of the PVC had led to lipping and tenting at some tile joints.

Since most of the floor was in good condition, careful repair of affected areas was recommended. Tiles containing ripples were to be cut out and the construction joints filled with epoxy resin, the object being to prevent any shrinkage cracks closing up after the new flooring was laid. Replacement of the flooring should have been delayed for as long as possible to allow further drying out of the slab.

shrinkage. This has been attributed mainly to the effects of stress relaxation in sheets or tiles, or to the migration of plasticisers into certain adhesives. Problems with shrinkage of PVC floorings may be due also to the action of cleaning fluids, and so discretion and care is necessary in selecting and using fluids.

As a result of research at BRE in the 1970s, it was found that gel cleaners containing pine oil or pine oil substitutes could cause excessive shrinkage of PVC floorings by removing plasticisers.

Highly alkaline cleaners should not be used as they can react with plasticisers.

Case study

Migration of plasticiser from PVC flooring into cleaning fluids at a college of education

The PVC flooring at a college of further education had shrunk but investigators could not find a single cause of the problem. The evidence suggested a number of factors were involved including the loss of plasticiser resulting from the use of a gel cleaner. A further possible factor, the migration of plasticiser from the PVC flooring into an SBR modified screed, had not been seen before. As well as recommending repairs to the flooring, guidance was given to minimise migration of plasticisers into cleaning fluids.

Case study

Rippling and blistering of PVC flooring over construction joints in the floor of a laundry

BRE Advisory Service investigated the rippling of both sheet and tile PVC finishes in the newly built laundry of a public transport undertaking. The rippling had occurred almost exclusively over construction joints. Occasional blistering was also evident during the inspection. Because of the danger of further rippling occurring when the laundry was in use, it was recommended that the joints should be exposed and filled with a **non-compressible** material prior to carrying out repairs to the flooring. This would prevent the joints from closing and, hence, producing further ripples. However, due to the pressure to carry out repairs prior to the laundry being commissioned, the repairs could be carried out without first cutting and filling the joints. Preferential wear would then be bound to occur on rippled tiles, but the tiles could be replaced either when this happened or sooner if the ripples were considered to be dangerous.

Work on site

Storage and handling of materials

BS 8203 is relevant.

Case study

Defective welding of joints between PVC tiles in a hospital

Welded joints in anti-static PVC tile flooring laid in operating theatres at a district general hospital were found to be loose or very weakly bonded. There was little or no sign of excessive shrinkage of the tiles in any of the theatres inspected. The welds which failed had come loose on both sides of the joint or were weakly bonded to one edge only. Though there was some loss of bond between the tiles and the screed, this could have been the result of water penetration at loose or broken welds. Cleaning procedures appear to have been carried out sensibly with due regard to the possible effects on the flooring of the machines and materials used. The investigators concluded, in the absence of any other explanation, that the initial welding process did not produce the strong durable joints that might have been expected with this type of flooring.

Case study

Failure of PVC sheet flooring due to contamination by food products

Adhesion between PVC sheet flooring and its base had failed in a food factory resulting in tears and splits at the welded joints between the sheets. This culminated in water penetration of the flooring; clearly, PVC flooring was not sufficiently robust for this location and purpose. Primarily because of the risk of contamination by sugar and, to a lesser extent, by lactic acid, corrosion resistant flooring employing ceramic tiles of sufficient thickness to resist impact, combined with suitable bedding and jointing materials, was recommended as being the only satisfactory flooring system for the particular use of the factory area concerned. (More suitable materials, including resin floorings, have become available since this investigation was undertaken in the 1960s.)

Inspection

The problems to look for are:

- ◊ rucking
- ◊ profiles of adhesive layers or disposition showing through the tiles (Figure 6.15)

Figure 6.15
Undulations in a flexible PVC floor caused by inexpert laying of tiles with a notched adhesive spreader

- ◊ excessive wear
- ◊ cover moulds and thresholds not provided at doorways or positions of greatest wear
- ◊ floorings lifting and stretching (Figure 6.16)

Figure 6.16
Failure in printed PVC strip flooring due to moisture rising from below

- ◊ ridging
- ◊ floorings breaking down in localised areas
- ◊ heads of nails in boarding not punched home
- ◊ slipperiness
- ◊ floorings showing unacceptable indentation
- ◊ floorings showing shrinkage
- ◊ floorings showing expansion (Figure 6.17)

Figure 6.17
A PVC tile floor which had been overlaid with a carpet tile incorporating a PVC backing. Plasticiser from the backing of the carpet tile has migrated downwards into the PVC flooring causing it to soften and expand before buckling. See also Figure 6.6 which shows the effect on carpet tiles

- ◊ flooring breaking up at joints
- ◊ poor welding

Case study

Failure of welded joints compounded by inappropriate cleaning methods

BRE Advisory Service was called in to investigate the strength of the welded seams of PVC flooring which had failed in a retail store. The investigator reasoned that the failure of flooring had two origins:

- the installers of the flooring used a welding technique which had not produced a sufficiently strong bond to resist the service conditions
- shrinkage caused by cleaning fluids

At the time of the first examination, the Advisory Service knew of no recognised method of test for assessing the strength of welded PVC seams. Since then the CSTB (Centre Scientifique et Technique du Bâtiment – the French equivalent of BRE) has issued a method of test. In a second examination, samples of the PVC flooring were removed from the store and tested according to the CSTB method. The results showed that the strength of the welded seams did not meet the required standard.

Further failure was being caused by shrinkage of the PVC flooring due to loss of plasticiser following penetration by cleaning fluids at the broken welds. The only longterm recommendation that could be given was to renew the flooring. Additional advice was given to the client on a cleaning procedure which would minimise the harmful effects caused by the action of cleaning fluids on PVC flooring.

Case study

Loss of plasticiser from PVC tiles in laboratories and unsuitable cleaning practices

The floors in the laboratories and storerooms of an industrial company's training centre in south west London had shown a variety of problems including gapping, lifting and buckling. The floors consisted of proprietary PVC tiles laid with a synthetic resin emulsion adhesive over a dense sand and cement screed on a structural concrete base. In areas subjected to few or no wet processes – offices, canteen etc – tiles had curled and shrunk. In the laboratories and other places where water was always present, tiles had become detached or had shrunk and lifted at the edges.

The causes of the problems were varied. The plasticiser had probably been leached out of the PVC tiles by using an alkaline degreasing agent. A polish stripper of acidic nature, used for removing layers of polish containing dirt, could also have caused various distortions. Water penetration in the PVC floor coverings of laboratories, kitchens and cloakrooms was further evident. In the drier areas, the slight defects were more acceptable and future problems were avoidable by using a more suitable cleaning regime.

However, in the wet areas removing the existing floorings was the only solution as the specified materials were obviously inadequate for the job. Suggested alternatives for replacement were:

- ceramic tiles or some sort of terrazzo (although this would raise the level of the floor by as much as 40 mm, which might not be acceptable)
- vinyl asbestos tiles or vinyl sheeting welded at the edges to ensure that moisture did not penetrate.
- a thermosetting resin floor

In all cases the underfloor would have needed to be totally dried out before resurfacing.

Chapter 6.5 PVC semi-flexible or vinyl (asbestos)

This type of tile was introduced in the UK in 1954, offering superior performance to the thermoplastic tiles then in common use.

Prior to 1983 they were called vinyl asbestos tiles but in that year the name was changed to semi-flexible PVC tiles. (At the same time as the name change, asbestos was removed as a constituent.) These semi-flexible tiles were manufactured by the same process as and in similar sizes to thermoplastic tiles (see Chapter 6.7). The binder was PVC with low levels of plasticisers. The resin being white, pigmentation was easier and light coloured tiles were readily made. They had similar properties to thermoplastic tiles but had superior fat, oil and abrasion resistance, and were slightly more flexible.

Characteristic details

Basic structure

Older tiles manufactured to BS 3260[218] may contain asbestos fibres, though this material has now been discontinued in tiles of UK manufacture.

They have been available in a range of thicknesses, usually 1.5, 2.0 and 2.5 mm.

The relevant standard is EN 654[219] (formerly BS 3260).

Main performance requirements and defects

Appearance and reflectivity

The tiles were originally made to 9 × 9 inches (228 × 228 mm) by ⅛ inch thick (3.2 mm), although a few other sizes and thicknesses were also produced. Tiles are now normally 300 × 300 mm, available in a wide variety of colours.

Cream coloured tiles give reflectances of around 0.45. Marbled and darker shades give around 0.25.

Choice of materials for substrate

Substrates must be firm and dry. The most common bases to be found will be cementitious screeds or concrete. Timber floors will need a sheathing board to prevent joints showing.

Strength and stability

Large static loads placed in one position for long periods could cause indentation. Standard tests set limits for allowable indentation.

Dimensional stability, deflections etc

Coefficient of linear thermal expansion per °C: 40 to 70 × 10^{-6}.

Reversible moisture movements can occur but are insignificant and are not normally a problem.

Wear resistance

These materials have good abrasion resistance but not as good as the more flexible PVCs (see Chapter 6.4).

Slip resistance

Slip resistance is good when dry, fair when wet, but poor if worn to a very smooth texture.

Coefficients of friction are:

- dry > 0.7
- wet 0.2 to 0.35

Case study

Loss of plasticiser from vinyl asbestos tiles in a hospital

BRE Advisory Service was called in to advise on the causes of failure of vinyl asbestos tiles at a general hospital. Large gaps between the floor tiles were found throughout the hospital. In this particular case it was not possible to positively identify the cause of the large gaps, but shrinkage due to plasticiser migration into adhesives and loss of plasticiser into cleaning fluids were not ruled out. No remedial measures have been available to close the gaps between tiles, although some diminishing of gap widths can be expected in heavily trafficked areas from the calendering effect of wheeled traffic.

The only satisfactory longterm solution was to replace the flooring in the affected areas with materials that would avoid the problems of plasticiser migration and loss.

Case study

Moisture in the concrete slab beneath vinyl asbestos flooring in a hospital

Examination of vinyl asbestos floor tiles which had detached from their substrates in the floors of a hospital, and taking sample drillings at various depths throughout the concrete floor slabs, indicated that detachment was due to entrapped moisture in the slabs. The high levels of moisture were caused, first, by flooding and, then, by surface wetting of the floor and possibly rising damp. Even a solvent bitumen adhesive applied when the flooring was laid would not have overcome the problems of the very wet substrate. Remedial measures recommended included removing the tiles and applying a surface DPM to the existing floor, finished with a new sheet PVC covering.

Figures 6.18
The tiles shown in these photographs have expanded slightly. This has resulted in tenting at the tile edges (top), the primary cause being a wet substrate leading to re-emulsification of the water based adhesive. Heavy, wheeled traffic has then stretched the flooring, making the tenting worse. It has also caused multiple fractures at right angles to the direction of travel, and one of the tents has collapsed (bottom)

Control of dampness and condensation

When stuck down with a solvent bitumen adhesive to an initially dry concrete base, these PVC tiles have been moderately tolerant of damp conditions if the concrete subsequently became wet. For this reason it was common to lay these floorings on bases without DPMs. However, they should not be considered to act as a DPM.

On very wet sites, water rising through the concrete can bring soluble salts with it. This water rising through the tile joints will, on evaporation, deposit the salts as a white crystalline material. The salts are mainly sodium and potassium carbonate which tend to creep towards the centres of tiles from the edges, producing white bands up to 25 mm wide along the tile joints. This is often called window framing.

The modern water-and-bitumen based adhesives are not so tolerant of damp substrates as the solvent based bitumen adhesives that have been used in the past.

Thermal properties

Thermal conductivity of semi-flexible PVC flooring: 0.40 W/mK.

Warmth to touch

Poor.

Fire

Combustible. PVC vinyl tiles are particularly susceptible to cigarette burns.

See also the same section in Chapter 6.4

Suitability for underfloor heating

The tiles can be used where temperatures at the surface of the substrate do not exceed 27 °C.

Sound insulation

Sound absorption is poor. Airborne sound insulation depends largely on the mass of the structural floor. Semi-flexible PVC can be used over a floating floor for impact sound resistance.

Durability

Life expectancy will be around 15 to 30 years for a good quality flooring. However, where the substrate is wet, earlier failure may occur (Figure 6.18).

Maintenance

Certain cleaning fluids may lead to the loss of plasticiser and, consequently, to gaps opening

Case study

White crystals on vinyl asbestos tiles in local authority housing

Flooring laid by a local authority in its housing over a period of three years had shown evidence of surface breakdown. The layers comprised a concrete slab, a cold applied liquid membrane, a further layer of concrete and a topping of 2.5 mm vinyl asbestos tiles applied with adhesive. Examination of damage was undertaken at a number of sites. Tiles had broken down at the edges and white crystals had appeared in the areas of breakdown. The crystals were found to be largely of sodium carbonate, no chlorides having been found.

The corruption of the floors was confined generally to areas which would normally have been less well heated. Areas in the most used parts of the houses, particularly those covered with rugs and carpets, were not affected. Examination of the sites suggested that the breakdown was the result of a physical process (based on moisture) and not a chemical one. Various possibilities for the origins of the moisture were considered.

There was no evidence that the moisture had come from washing the surface of the tiles and none that suggested that alkaline cleaning materials could have been responsible. There was a possibility that, in some cases, moisture could have come from normal condensation within the houses or entered as rain or rising damp via cavity walls. The most likely source, however, was water retained in the original concrete slab from the time of construction. Where the original slab had been laid before the walls were built, as was known in some cases, the extra time for drying may not have been sufficient because the uncovered slabs were subject to rainfall until the buildings were eventually roofed over.

As a guide for future practice, a sufficiently long time should be allowed for the original concrete slab and any subsequent screeds to dry out, or the placing of an intervening membrane of mastic asphalt. As a remedy at the time, the affected tiles should all have been removed, the damp areas thoroughly dried out, and a screed consisting of a blend of epoxy resin and pitch applied. New tiles could then be laid with little fear of the problem recurring.

between tiles. Manufacturers should be consulted – gel cleaners, particularly, may need to be avoided.

Work on site

Access, safety etc

Bitumen adhesives for semi-flexible PVC tiles traditionally contained solvent naphtha. Under COSHH (Control of Substances Hazardous to Health Regulations 1988), these adhesives are being phased out and replaced with water based adhesives. BS 8203[35] is relevant.

Inspection

The problems to look for are:

◊ floorings lifting and stretching
◊ slipperiness
◊ unacceptable indentations
◊ shrinkage
◊ expansion in underlying layers
◊ surfaces of floorings breaking down
◊ DPMs absent or defective

Chapter 6.6 Rubber

The coating of fabrics with rubber for use as floorings first took place in the middle of the nineteenth century. These coatings were comparatively thin – since rubber was a relatively expensive material at that time – and it was not until the wider availability of rubber, coupled with the invention of vulcanising, that sheet and tile flooring using this material became economic.

Characteristic details

Basic structure

Rubber flooring is available in sheet or tile form, in varying degrees of hardness, and with various forms of surface profiles to enhance slip resistance (Figure 6.19). It may also be found laminated to a sponge backing, particularly for use in blocks of flats where it has been used to obtain improved impact sound insulation in separating floors. The appropriate standard is BS 1711[220].

A wide range of material specifications has been available comprising natural and synthetic rubbers with various fillers, pigments and curing agents. At one time even anti-static rubber was available! (BS 3187[221])

Figure 6.19
Ribbed rubber flooring in a school

Main performance requirements and defects

Appearance and reflectivity

Sheet and tile rubber flooring is available in a wide range of plain or marbled colours. Light shades give reflectances of around 0.45, medium shades around 0.25, and dark shades around 0.10.

Choice of materials for substrate

Rubber flooring can normally be stuck to a flat screed using a contact adhesive or other recommended proprietary adhesive. Some very thick tiles, with an undercut (Figure 6.20 on page 230) or studded profile to the back, are designed to be laid in a wet cementitious grout which fully bonds the rubber to the screed or power floated slab; this type of rubber flooring is unaffected by moisture in the base. It can also be laid satisfactorily on metal plate floors or on timber boarded floors covered with a suitable underlay.

Where smooth backed rubber flooring is to be stuck to groundbearing concrete floors with adhesive, there must be an effective DPM.

Strength and stability

The resilience of rubber flooring can be exploited for a variety of applications; for example for indoor sports.

Dimensional stability, deflections etc

Coefficient of linear thermal expansion: rubber tends to have good dimensional stability, though if it becomes detached from the substrate it will stretch under the action of traffic.

Reversible moisture movement: negligible for most rubber products, but some filled rubbers can expand a little under the effects of moisture.

Wear resistance

Very good – at least as good as high quality PVC.

Figure 6.20
Undercut dovetail studded rubber sheeting having failed to bond. The bedding mortar should have been forced into the dovetails

Case study

Detachment of rubber tiles in a bus station caused by lack of control over adhesive

A number of rubber floor tiles had detached from their floor base at a bus station and were missing; some were loose and others were blistering. It was found also that, in areas near open entrances, the base was wet and tiles were losing bond for the usual reason associated with dampness and this type of flooring – the adhesive had been adversely affected by moisture. In some apparently dry areas, tiles had bubbled.

The most likely cause of the detachment and bubbling was that the impact adhesive had been exposed to the air for too long, allowing it to dry and lose some or all of its adhesion qualities, before placing the tiles. Subsequent traffic had stretched the loose rubber.

Case study

Detachment of rubber flooring in an office building

Inspection of rubber flooring in an office building, particularly at entrance doorways, but also at internal doorways, showed that large areas of the floor screed were damp; other areas laid with the rubber flooring showed signs of blistering. The blisters contained water under pressure suggesting that osmosis was the cause of these problems (see the feature panel in Chapter 1.9).

However, the flooring within the revolving door was also showing signs of detachment. The cause was identified as insufficient protection from driving rain leading to large quantities of water accumulating on the floor of the drum and percolating through the joints in the rubber flooring. This clearly was not due to osmosis.

The remedial measures recommended included removing the rubber flooring, and improving the dampproofing and drainage arrangements of the revolving door drum floor. As an alternative the possibility of providing canopies and glazed screens at the entrance doors to deflect the driving rain was considered. Since the floor slab was very wet, it was also suggested that floor finishes such as ceramic or terrazzo tiles would be more able resist the conditions.

The symptoms of blistering of rubber flooring caused by osmosis can sometimes be alleviated by drilling a small hole through the blisters to relieve the pressure.

Slip resistance

Rubber flooring is unsuitable for areas connected with cooking, washing or laundering, or in spaces near entrance doors. It is very poor when smooth and wet unless ribbed or studded.

Coefficients of friction are, for smooth rubber:

- dry > 0.7
- wet 0.2 to 0.3

Control of dampness and condensation

Although rubber as a material is largely unaffected by dampness, the same cannot be said of adhesives used to stick it to screeds or other substrates. Defects caused by dampness rising from below usually take the form of blisters or bubbles which can be distorted or stretched by traffic. It is essential, therefore, that rubber flooring which is to be stuck down is protected from rising damp by an effective DPM.

Since rubber tiles and sheeting are relatively thin, and their thermal insulation value is low, floors covered with these materials are just as likely to be affected by condensation as most other types of floorings.

Thermal properties

Thermal conductivity of rubber flooring: 0.40 W/mK.

Warmth to touch

Rubber flooring is too thin to add substantially to the thermal characteristics of the substrate.

Fire

Rubber flooring offers low flame spread and low smoke emission.

Suitability for underfloor heating

It can be used where temperatures at the surface of the substrate do not exceed 27 °C, provided the adhesive is unaffected.

Sound insulation

Sound absorption is usually very good. Airborne sound insulation depends largely on the mass of the structural floor. Backed rubber flooring can be stuck directly to a non-floating floor to achieve adequate impact sound insulation, or it can be used over a floating floor.

Durability

The lifespans of natural rubbers normally exceed 10 years after which they oxidise and become brittle. Failure will occur much earlier, though, if the substrate is wet (Figure 6.21).

Maintenance

Solvent based polishes can cause softening of the rubber which, in turn, leads to stretching and detachment from the substrate.

Figure 6.21
Residual construction moisture in the slab has caused this rubber flooring to detach

Work on site

Safety
Rubber solution adhesives for rubber flooring have in the past contained ketones. Under COSHH regulations, these adhesives have been phased out and replaced with water based adhesives.

Storage and handling of materials
BS 8203[35] is relevant.

Workmanship
Until 1987 rubber floors should have been laid to BS CP 203[222], and since then to BS 8203.

Inspection

The problems to look for are:

- ◊ floorings lacking adhesion and bubbling
- ◊ floorings lifting
- ◊ slipperiness
- ◊ floorings blistering
- ◊ surfaces of floorings breaking down
- ◊ floorings stretching

Chapter 6.7 Thermoplastic

Thermoplastic tiles came into general use around 1947. They were originally developed from asphalt tiles in the USA in the 1920s. Some floors were being laid with imported tiles in 1939, with manufacture starting in the UK in 1947. Production quickly rose to nine million square metres per annum to meet the needs of the post 1939–45 war building boom and the tiles were laid extensively in houses, schools and offices.

Characteristic details

Basic structure

Thermoplastic tiles were made by heating a mixture of thermoplastic resin binders with fluxes, asbestos fibres, powdered mineral filler and pigments and passing the mixture through heated rollers. Tiles were cut from the hot rolled material.

The relevant standard is BS 2592[223]. A number of different formulations comply with the standard.

Main performance requirements and defects

Appearance and reflectivity

The main resins used were based on coal tar or petroleum distillates, and as a result, tiles were dark in colour. Lighter colours were produced with coumarone-indene and gilsonite resins. The majority of tiles produced were black or dark brown with lighter colour marbling. They were mostly manufactured to a 9 × 9 inches size, but some to 6 × 6 or 12 × 12 inches. Two thicknesses were used, ⅛ inch in domestic applications, and $^3/_{16}$ inch in shops, offices and schools.

The dark colours have poor reflectivity at around 0.10.

Figure 6.22
A very early (1950s) thermoplastic tile floor

Choice of materials for substrate

Thermoplastic tiles can be accepted by a range of substrate materials which suffer little or no thermal and moisture movement, and little deflection, but the most usual are the cement based screeds and concrete. The tiles were usually laid with a cutback bitumen adhesive.

The defects commonly occurring during the early days of using thermoplastic tiles were nearly all related to the substrate; they included:

- lack of adhesion between screed and base concrete leading to curling of the screed and fracturing of the tiles
- dusting of the screed making adhesion difficult
- uneven surfaces
- irregularities at joints with other floorings.

Some thermoplastic tile floors are known to have been laid on timber decks, with totally unsatisfactory results. The tiles are too brittle to be laid successfully on bases which are not very rigid.

Strength and stability

The tiles have very little transverse strength; indeed, they may be broken by hand pressure at room temperatures before laying unless care is taken. Impact resistance is poor, especially where bedding has been uneven. Due to this brittleness, they are warmed on site before being placed in position.

It is essential that the subfloor to which the tiles are to be applied is flat as the tiles will fracture or deform where loads bear at or between high spots.

Figure 6.23
Total failure of a thermoplastic tile floor caused by moisture rising from below

Dimensional stability, deflections etc

Coefficient of linear thermal expansion per °C: 40 to 70 × 10^{-6}.

Thermoplastic tiles expand slightly when wetted, but this is usually insufficient to disrupt the surface.

Wear resistance

Wear resistance has been fair in domestic situations, though some more intensively used floors have shown signs of premature wear if the thinner tile was used. All exposed edges needed protection with a strip of wood, metal or plastics.

Slip resistance

Slip resistance is good when dry, but can be slippery in the wet if worn smooth.

Coefficients of friction are:

- dry > 0.7
- wet 0.2 to 0.35

Control of dampness and condensation

Most thermoplastic tile floors were laid on solid groundbearing floors without the benefit of DPMs. When stuck down with a solvent bitumen adhesive to an initially dry concrete base, the tiles have been moderately tolerant of damp conditions if the concrete later became wet. However, neither tiles nor adhesive could be considered to act as a DPM (Figure 6.23).

On very wet sites, water rising through the concrete can bring soluble salts with it. Without a DPM, the water rising through the tile joints, on evaporation, will deposit these salts as a white crystalline material along the joints between the tiles. (This phenomenon was also observed in the case study on vinyl asbestos tiles in local authority housing in Chapter 6.5.) The salts are mainly sodium, and potassium carbonate which tend to creep inwards from the tile edges to produce white bands up to 25 mm wide along the tile joints ('window framing', Figure 6.24).

These salt deposits can be removed by careful cleaning. Sometimes the evaporating solution is absorbed by the tiles. Crystallisation pressures set up within the tiles are often sufficient to cause delamination and, ultimately, powdering of the edges of the tiles.

Thermoplastic tile installations have been in use for more than 25 years although, over this time, householders may have covered them with loose laid carpets or printed sheet PVC flooring. Often this has worked well, but sometimes problems have arisen because plasticiser from the more flexible PVC flooring or from PVC backed carpets has migrated down into the thermoplastic tiling causing softening and expansion. The carpet or flexible PVC will then often exhibit shrinkage. The effect tends to be worse if the new flooring is stuck down to the existing thermoplastic tiles. Problems have also arisen where moisture coming through the tile joints has built up beneath the new flooring.

Thermal properties

The tiles are unsuitable for use below boilers or other heating appliances.

Thermal conductivity of thermoplastic flooring: no information is available but probably around 0.5 W/mK.

Warmth to touch
Poor.

Fire
Combustible.

Figure 6.24
Typical 'window framing' on a thermoplastic tile floor caused by soluble salts efflorescing from the base

Suitability for underfloor heating
Successful installations have been known where temperatures at the surface of the substrate do not exceed 26 °C.

Sound insulation
Resiliency is poor. Sound absorption is also poor. Airborne sound insulation depends largely on the mass of the structural floor. Thermoplastic tiles can be used over a floating floor but performance depends on the overall thickness of the screed.

Durability
The tiles are brittle at ordinary temperatures so they will not accommodate irregularities in the substrate. The tiles will withstand light trolleys but will tend to crack under heavy wheeled traffic.

Most tiles have not been resistant to oils or greases although special grades have been available with enhanced resistance.

Maintenance
Washing with a detergent and warm water is normally sufficient to remove dirt. However, stains may require treatment with a very fine abrasive. Solvents should not be used, since they will soften the tiles. By the same token, only water based emulsions of wax should be used for polishing.

Work on site

Safety
Bitumen adhesives for thermoplastic tiles traditionally contained solvent naphtha. Because of COSHH regulations, these adhesives are now being phased out and replaced with water based adhesives which are not as tolerant to moisture in the substrate.

Storage and handling of materials
Tiles should be stored at a minimum temperature of 18 °C for 24 hours before laying, and the temperature should not fall below this figure during the laying operation (BS 8203[35]).

Workmanship
Thermoplastic tiles should be warmed before laying, but taking care not to overheat them owing to the risk of blistering. BS 8203 recommends that thermoplastic tile floorings are not covered by similar flooring. Existing thermoplastic floorings should be removed and the subfloor brought up to the required standards to receive new floorings.

Inspection

The problems to look for are:
- ◊ floorings lifting, cracking or indenting
- ◊ DPMs defective or absent leading to moisture collecting on undersides of floors
- ◊ slipperiness
- ◊ expansion of tiles
- ◊ surfaces of floorings breaking down
- ◊ excess construction water in bases

Chapter 7 Jointed hard finishes

Figure 7.1
A ceramic tile floor at the medieval Byland Abbey

All the floorings dealt with in this chapter tend to be both hard and durable. They are commonly specified for use in public access areas such as concourses.

The materials included are ceramic in various forms; for instance ceramic tile (Figure 7.1), brick, concrete flag, stone and its derivatives including terrazzo tile, composition block (mainly in the form of a cementitious binder with sawdust and wood flour) and metal.

Movement joints in hard finishes are extremely important and the recommendations in standards should be strictly observed.

Toughened sheet glass of appropriate thickness is occasionally used as a flooring for special effects such as concealed lighting within the floor. Supports are similar in character to those used in platform floors described in Chapter 2.6, though additional measures will be needed in mountings to enable the glass to safely accommodate impacts. The glass does not retain its gloss surface for very long. However, there is probably insufficient use of this material to justify a separate chapter; BRE has never been asked for advice or to carry out any investigation of glass flooring.

Chapter 7.1 Ceramic tiles and brick paviors

Ceramic floorings, in the form of tiles and bricks, have been used for several millennia – the introductory chapter referred to Roman and medieval examples, but they go back much further than Roman times. The survival of examples from the earliest times testifies to their longevity, given appropriate conditions of use (Figure 7.2).

There are, however, several important provisos on their installation, and the rest of this chapter (including case studies) draws attention to the most important of these.

Characteristic details

Basic structure

Ceramic floorings – tiles, quarries, faience, paviors and mosaics – are available in great variety. There are divergent opinions in the building industry on the precise definitions applied to the names of these fired clay flooring units. However, for the purpose of this book, the following descriptions are offered.

The two main classes of fired clay tiles are the quarry (Figure 7.3) and the ceramic tile with the latter available in semi-vitreous and vitreous bodies.

Quarries are normally made thicker than tiles, sometimes with less accuracy, by the extrusion or moulding process, whereas tiles are usually made by compaction or pressing a drier mix. The term quarry comes from the French word *carreau*, a square stone. It is believed originally to have applied only to a square shaped stone obtained in a range of sizes from a quarry. Later quarry came to describe a 9 × 9 × 1¼ inches ceramic unit, moulded with a wettish clay, fired at relatively low temperatures and left unglazed. Later still the term was extended to cover a semi-vitrified or fully vitrified unit of a similar size fired at a higher temperature (eg up to red heat).

Figure 7.3
Quarries laid with a cementitious joint

Until the 1980s, 'quarry' was used without the appendage 'tile', but in recent years the term quarry tile has come to mean units of similar size to ceramic tiles – say 150 × 150 × 12 mm or 100 × 100 × 12 mm – but made by the extrusion or moulding process rather than by pressing semi-dry clay bodies. (Quarry tile, as a term, is tautological.) With the introduction of a European dimension to standards, however, the term quarry has now largely fallen into disuse, and they have all become known as ceramic tiles.

Clay tiles for flooring were also referred to in BS 1286[224] as quarry tiles.

Clay bricks used for flooring are usually referred to as paving bricks or paviors (Figure 7.4). They can be laid in a mortar bed with narrow or wide joints, usually on a concrete base.

Figure 7.2
Brick paviors more than 200 years old. The flooring is worn and stained, but users tend to make allowances for features that are very old provided they are not actually dangerous

Figure 7.4
Brick paving laid to falls in a dairy

Fired clay finishes have been used extensively in industrial premises where clean, dust-free environments were necessary, as in switch and motor rooms, and in food processing areas. Where the very heaviest impacts and wear, combined with heat, are experienced – as in foundries – the high strength red or blue paviors have been used. Brick-sized paviors frequently are found laid on edge, though they are to be also found laid flat in areas which experience lower impacts.

Laying techniques were covered by BS CP 202[225] until 1989; since then by BS 5385:Part 3[226] and Part 4[227].

Problems have arisen in the past from differential movements between ceramic flooring and its substrate. A common preventive measure was to lay the units on a thin separating layer of sand or building paper but this has now been superseded by the semi-dry thick bed method. Calcium sulfate screeds need to be primed with an appropriate primer before laying ceramic tiles and must not be used in wet areas.

Detailing

Abutments, upstands and drainage channels in areas subjected to spillages should normally be formed in the same material as the remainder of the flooring (Figure 7.5).

Main performance requirements and defects

Appearance and reflectivity

Ceramic tiles are available in an enormous variety of sizes and colours (Figure 7.6), and it is not practical to list them. However, most manufacturers produce a range of special sizes to fit awkward spaces, and to form coves and upstands. The larger the tile, the more likely is it to distort slightly in the firing process, and therefore the more difficult to lay to an acceptable finish. The advantages of fewer joints need to be balanced with the disadvantages of warping in the kiln. Modern production techniques have mainly overcome the problem of warping and large tiles, fired to high temperatures and as little as 8 mm thick, are readily available.

When the tiles are subject to wide dimensional deviations in manufacture, it should be remembered that the wider the joint, the less apparent will be these variations in size when the floor is laid. This needs to be balanced against the greater cost of high performance jointing materials used in the wider joints.

Encaustic tiles are made by incising decorative hollows in the unfired body, and filling with slip

Figure 7.5
Ceramic tiles forming a gully in a dairy

Figure 7.6
A Victorian ceramic tile floor

Figure 7.7
Making encaustic tiles (Photograph by permission of BT Harrison)

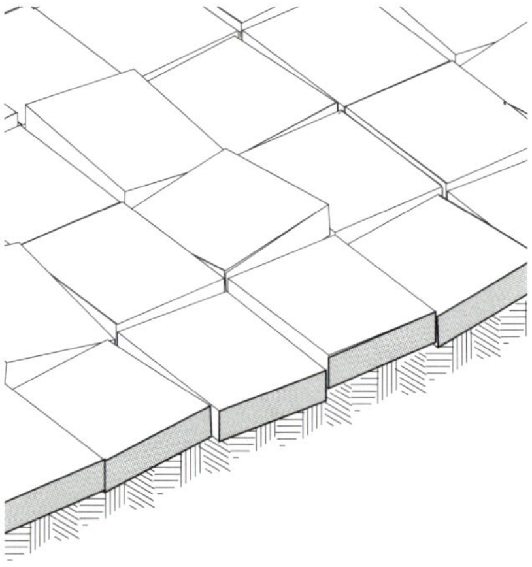

Figure 7.8
Floors of quarries may not be durable if laid directly onto prepared ground instead of onto a concrete slab

clays (Figure 7.7). They provide a very wide range of patterns and have been made for several centuries.

Red and brown quarries give reflectances of around 0.10, and modern glazed new encaustic tiles around 0.5.

Paviors are normally available in standard brick size, but with reduced thickness of 50 mm. Thicker ones are available.

Choice of materials for substrates and joints

Substrates and beddings

Tiles are normally laid on concrete bases or screeds. If they are laid on timber bases, great care will be needed to accommodate their brittle characteristics; either the floor will deflect, tending to break or spall the arrises at the joints, or the sheathing will move, tending to open at joints and removing support for the tiles.

Many methods have been used over the years to fix tiles to substrates. The following have been used for bedding onto a solid base:

1 wet cement mortar bonded to the base
2 wet cement mortar laid on a separating layer of building paper or polyethylene
3 wet cement mortar laid over a compacted layer of dry sand
4 semi-dry mortar, 40–70 mm thick, bonded or partially bonded or laid on a separating layer
5 bedding mixtures made from bitumen emulsion and sand, 6–7 mm thick
6 cement or organic based proprietary adhesives, either thin bed (up to 3 mm) or thick bed (about 6 mm)
7 thermosetting adhesives used for chemical resistance applications

With methods 1–3, the mortar bed was usually between 10 and 15 mm thick. Methods 1, 4 and 6 are the ones most commonly used, and 7 is used for special applications.

Quarries and brick paviors can be found laid on sand or ash beds, or even upon consolidated earth (Figure 7.8).

Joints

Joint width is usually between 3 and 6 mm but varies from butted to about 12 mm. Joints are filled with grout, usually based on cement, with fillers, admixtures and pigments. However, in particularly onerous service conditions, surveyors may sometimes encounter a number of different jointing compounds. A few observations follow on the jointing materials that have been used for this type of flooring.

- Silicate cements are acid resistant but not resistant to alkalis.
- Sulfur cements are made from molten sulfur to which sands, pitch, tar, resins or gums have been added. These cements are acid and alkali resisting but prone to attack by vegetable oils and some solvents. Since sulfur changes its form at 95.5 °C, it can be broken down where very high temperatures are met in service or where very hot water is used for cleaning.
- Rubber latex cements are produced from high alumina cement (or OPC), sand and rubber latex. The mixture adheres strongly to paviors and ceramic tiles giving a good waterproof joint. Under attack from acids, the rubber content swells to maintain a watertight joint.
- Phenol-formaldehyde jointing compounds are made from resin cements with inert fillers and set by catalytic action. They are acid resisting but are attacked by strong alkalis.
- Cashew nut resin cements are combined with inert fillers and set by catalytic action. They are acid resisting and moderately alkali resisting, but are attacked by organic solvents.
- Furan resin cements are combined with inert fillers and set by catalytic action. They are acid and alkali resisting, and organic solvent resisting. When laying at lower temperatures, there could have been a tendency to add too much filler in an attempt to accelerate the set. However, this could lead to failure from overfilling (Figure 7.9).

- Epoxy and polyester resin cements give good waterproof joints since adhesion to paviors and tiles is good. Epoxy grouts have largely displaced jointing materials previously used in chemical resistance applications.

Strength and stability

Thin tiles will be less able to resist impact damage unless they are fully bonded.

In specifying thin tiles of comparatively large superficial area, a balance has to be struck between initial cost and impact resistance. The thinner tiles may cost less but are more vulnerable to impact damage than thicker tiles, body for body. If for any reason a thinner tile becomes unbonded (due to differential shrinkage in substrates or beddings for example) then it becomes vulnerable to damage and cracking from impacts or heavy loads (Figure 7.10). It is optimistic to believe that large areas of tiling can be laid without some tiles losing bond. Most poorly bonded tiles will become detached within 6 months to a year of being laid. Subsequently, fewer tiles will become unbonded.

Loosening or debonding of tiles from the substrate is not necessarily the result of bad workmanship. In a number of cases examined by BRE the cause was just as likely to be due to drying shrinkage or moisture expansion of the tiles.

Brick paviors instead of tiles are used where impacts are expected to be severe. Paviors thicker than the normal 50 mm can be obtained, and, if necessary, improved impact resistance can be obtained by laying the paviors on edge.

Dimensional stability, deflections etc

Coefficient of linear thermal expansion per °C: 4 to 6×10^{-6}.

Large areas of tiling need movement joints. BS 5385:Part 3 provides guidance on the spacing of joints in ceramic floor tiling, but encourages designers to calculate the specification for jointing in relation to the design. The data needed for this are provided in the standard. The standard also provides a number of recommended joint design details. BS 5385:Part 4 gives requirements for more onerous service conditions including swimming pools.

As with other rigid floorings, particular attention should be paid in design to restraints arising from containment – for example by perimeter walls, steps, plant machinery bases and columns – and the appropriate allowances made, otherwise failure can occur (Figure 7.11 on page 240).

Figure 7.10
Ceramic tiles which have cracked because they have been laid on a timber floor

Figure 7.9
Failure of furan resin joints in a ceramic tile floor

All porous ceramic bodies run the risk of moisture expansion. If the body is glazed, the glaze may craze as it is weak in tension. To avoid this it is normal for the glaze to be put into compression during manufacture. Provided the compression is not all removed by subsequent moisture expansion of the body, the glaze will not craze. Most expansion occurs within hours of leaving the kiln, and after several months the expansion reduces to a very small amount. Most tiles with low water absorption characteristics suffer very low longterm moisture expansion.

Wear resistance

Abrasion resistance is good to very good, especially for the harder vitrified or semi-vitrified products fired at high temperatures. Impact resistance is normally fair to excellent. Resistance to longterm indentation is excellent.

Where paviors, quarries and tiles are laid herringbone pattern at 45° across the main lines of traffic, they can offer improved resistance to wear from wheeled traffic. This is because impacts on protruding edges of tiles are glancing rather than direct. Clay tiles which are slightly proud of their neighbours are prone to chipping of the arrises.

Ceramic tiles are normally tested for surface hardness by a Mohs test to BS 6431:Part 13[228]. Normally the grout is the weakest point in a tile system; if not of good quality, it can wear away leaving tile edges proud and vulnerable to damage.

Case study

Tiles exploding from the floor of a swimming pool

In a 10-years-old swimming pool the temperature of the water was raised from 24 to 29 °C. Subsequently, on the first occasion that the water level was lowered, the imported ceramic tiles on the base of the pool exploded into the air. No movement joints had been provided. Moreover, it was calculated that the rise in temperature would have created an induced stress of about 0.35 N/mm^2, resulting in a bow of about 6 mm over about 2 m which was only restrained by the weight of water on the base of the pool. By the combination of lowering the water level (which released the pressure on the tiling) and expansive forces in the tiling (which induced it to debond from its bedding), it was not surprising that tiles exploded. The provision of movement joints would have obviated the problem.

Figure 7.11
Arching of ceramic tiles in a floor

Slip resistance

Depending on their characteristics, the slip resistance of the surfaces of smooth tiles is fair to good in dry conditions, except when polished. They can be slippery when wet. They can be particularly slippery when both polished and wet. For this reason, textured tiles are specified for the surrounds to swimming pools. With textured tiles, which offer improved grip, a panelled finish is easier to clean than a ribbed one. Tiles are also available with a carborundum finish which may be useful in kitchens and food preparation areas.

Coefficients of friction are, for smooth tiles:

- dry > 0.6
- wet 0.2 to 0.3

for clay tiles with a carborundum finish:

- dry > 0.6
- wet 0.3 to 0.5

Control of dampness and condensation

Ceramic tiles are almost totally unaffected by moisture from above or below. However, the property of water absorption gives a good indication of other properties of tiles such as durability. The old standard (BS 1286) gave the following values for water absorption:

- Class 1 quarry 6%
- Class 2 quarry 10%
- fully vitrified ceramic 0.3%
- ordinary vitrified ceramic 4%

Under the new standard, BS 6431: Parts 1–23[229] †, for which there are also equivalent European standards, tiles are classified according to their method of manufacture: A, extruded; B, dust pressed; and C cast. Each of these classes is divided into four groups according to their water absorption characteristics: < 3%, 3–6%, 6–10% and > 10%.

† BS 6431 was being revised at the time of publication. Also, the intention is that an ISO standard will be issued in due course.

Paviors and tiles should be laid to break joint across any falls; that is to say there should be no continuous joint down which water or effluents can drain.

Floors will sweat in unheated buildings under certain weather conditions and therefore may become slippery.

Swimming pools

Ceramic tiles are the finish most often used for swimming pools. Nevertheless it is in swimming pools that some of the more onerous service conditions are encountered. Additional precautions (described in BS 5385:Part 4) are therefore necessary to achieve good performance.

Adhesives suitable for use in conditions of continuous immersion in pool water and in pool surrounds are usually selected from BS 5980[230]:

- cementitious adhesives of Type 1 (not suitable in aggressive water conditions)
- dispersion cement adhesives of Type 3 (with slightly improved resistance to chemicals)
- resin adhesives of Type 5 (which give good performance in aggressive waters)

The appropriate choice will hinge largely on pool housekeeping decisions affecting acidity and dissolved salts in the water; for example cementitious adhesives and grouts should not be used unless the pH of the water is carefully controlled. Materials used in movement joints will also need careful selection to suit the service conditions.

See Clauses 13.2.5–7 of BS 5385: Part 4.

Ceramic tiles will not act as a tanking layer, even when jointed with epoxy mortars which give performance superior to cementitious mortars and grouts.

Thermal properties

Thermal conductivity of tiles or paviors: depends on density but probably around 1.8 W/mK.

Warmth to touch

Cold.

Fire

Non-combustible.

Suitability for underfloor heating

Very suitable, though the bedding may not always be so.

Sound insulation

Ceramic tiles can be used over a floating floor to achieve adequate impact sound insulation. Performance in respect of airborne sound insulation depends on overall thickness of the screed.

Durability

Life expectancy will be about 50 to 65 years for most tiles – with brick paviors about the same – depending naturally on the severity of service conditions. Non-vitrified fired clay tiles may have shorter lives.

A problem which has occurred on a significant scale in the past has been ceramic tiles which become detached from their bedding layer and lift to form an arch (Figure 7.11). This phenomenon has been recorded within a few months of laying and up to 50 years after laying. Using a separating layer, or the **correct** use of the semi-dry thick bed method, has almost eliminated this problem but it can still be seen,

Case study

No movement joints in the ceramic tile floor of a boiler house

BRE investigated cracking in and detachment of the ceramic tile floor in the boiler house of a district general hospital, finding that the cause of the failure was shrinkage of the screed for which no provision had been made in the design of the floor. The absence of movement joints and, possibly, longterm moisture expansion of the tiles contributed to the failure. BRE recommended that the remaining tiles in the boiler house, together with the screed which was heavily contaminated with salts, be removed and the floor relaid using the separating layer method. The expense of cutting movement joints in the floor, and the accompanying disruption, need to be balanced against the possible longterm failure due to moisture expansion of the tiles.

Figure 7.12

Ceramic tiles round a swimming pool, ribbed for enhanced slip resistance

Case study

Failure of ceramic tile flooring at a swimming pool

Areas of ceramic floor tiles around a public swimming pool had become detached from their base, including one area that had arched and was completely disrupted (Figure 7.13). Other areas in the changing rooms and the spectators' area had become similarly detached, and there had been considerable loss of tiling grout. The tiles had been laid on fully bonded screeds of 40 to 50 mm thickness. Proprietary grouts were used within the pool itself, but site mixed sand and cement grouts had been used elsewhere.

About one year after the swimming pool had been completed, and before the problems with the tiled areas became apparent, the main pool had been emptied for some minor repairs. Some time after the work on the repairs had finished, an area of tiling in the spectators' area became detached and arched. Part of this area had been laid over a suspended concrete floor. The maximum gap beneath the arched tiles was 10 mm. (It was possible for a person to stand on the arched tiles without the arch collapsing.) Several square metres of tiles to the floor in the splash pool also were defective and many adjacent tiled areas had detached.

Some of the affected tiles were taken up and the undersides examined. The adhesive beneath was still wet and, when scratched with a penknife, appeared to be soft. Most of the adhesive stayed on the screed. Serrated trowel marks could easily be seen in the adhesive bedding and it was clear that the tiles had not been solidly bedded. It was estimated that individual tiles were in contact with the adhesive for only 50 to 80% of their area. Lime scale was found on the faces of some tiles near to grouted joints, especially on tiles close to the arched areas. Examination of the screed showed it to be of excellent quality and fully bonded to the base.

Movement joints had been formed in the tiling at positions around the splash pool where the tiling changed direction. As far as could be seen, most of these joints also went right through the screed. However, the perimeter movement joint in the surround at the shallow end of the pool did not go through the tiling and screed, and some of the tiles were touching. Silicone sealant had been applied to the re-entrant corner to make it look like a movement joint.

The jointing grout to the tiling was badly eroded in most areas, and was down by over 3 mm in the main circulation areas. Probing with a penknife showed the grout to be very soft and like a mush in some places where it could easily be scraped out to its full depth.

Figure 7.13
Arched and collapsed tiles in a swimming pool surround

Samples of the glazed tiles were examined in the laboratory under a microscope and no crazing or cracking was found. There were, however, some patches of calcium carbonate on the surface of the tiles, caused by the free lime in cementitious material reacting with pool water leaking through the grout. The progressive loss of free lime leads ultimately to a loss of strength, and the rate of loss depends on the chemistry of the pool water together with the degree of access to the lime. In this case, the arching had given free access.

Lifting and arching of ceramic tiles is a well known phenomenon which is normally caused either by shrinkage of the base or screed, or by moisture expansion of the tiles, or by a combination of both. Detachment and arching of tiling can be prevented by carefully following the recommendations made in the various Parts of BS 5385. For swimming pools these are to be found in Section 7 of Part 4.

In this case it was concluded that the loss of bond between the tiles and the background structure, and the resulting arching, had been caused by the compressive forces set up by the drying shrinkage of the substrate combined with some expansion of the tiles. The following factors also contributed:

- the tiles in the pool were not solidly bedded
- some of the movement joints were not correctly formed
- insufficient drying time had been allowed between the various stages of construction

Loss of jointing grout from the floor tiling had been caused by the use of soft water for washing down, combined with a weak grouting mortar.

Shrinkage of concrete and screeds, and expansion of tiles, decrease with time; it was recommended, then, that undamaged tiles could be refixed with little likelihood that the problem would recur. Where replacement tiles were required there might be some expansion, but this would be relatively small, obviating the need for new movement joints. When replacing the tiles, it would have been essential that they be fully bedded with at least 90% of the tile base in contact with the adhesive, and any gaps in the adhesive to be well spaced out. The pool should not have been filled for three weeks following the repairs.

The white calcium carbonate deposits could have been removed by dilute hydrochloric acid, treating small areas at a time; the operative would have had to wear protective clothing, gloves and goggles.

particularly where unsuitable bedding has been used. (The case studies on swimming pools, on page 240 and on the opposite page, describe the problem in greater detail.)

The chief mechanisms of failure are as follows:

- drying shrinkage of the base concrete, breaking the bond between tile and mortar bed. This may be exacerbated by longterm moisture expansion of the tiles (Figure 7.14)
- differential thermal movements between base and tiles
- adhesive failure

Maintenance

Modern ceramic tiles are relatively easy to clean and can provide very hygienic surfaces (*The cleaning of ceramic tiles*[231]). Brick paviors may not be so easy to clean, depending on the surface textures.

A floor of brick paviors is straightforward to repair, provided matching paviors are obtainable. Defective paviors can be cut out with a carborundum disc and the joints cleaned off with a bolster, matching the original pointing if the repair is to be inconspicuous.

Case study

Impact damage to ceramic tiles in a factory

A ceramic tile floor in an East Anglian factory had suffered from breakages in various parts. The 150 x 150 mm tiles had been fixed with a 20 mm thick sand and cement mortar over a 65–70 mm thick dense sand and cement screed, with a layer of building paper sandwiched between the screed and the structural floor. In some places the floor sounded hollow where the tiles had become detached from their mortar bed. Tiles had broken near to the soft bituminous movement joints.

The most likely cause of the broken tiles was compression forced by shrinkage of the screed layer. This could have been prevented by placing the layer of building paper in the correct position (ie between the screed and the mortar) or by allowing the screed to shrink before laying the tiles. The separating paper layer was of no use where it had been laid.

By the time of the investigation it was likely that all shrinkage of the screed which could take place had done so, therefore repair of the existing damage was all that was required. Careful removal of the broken tiles and replacement with new ones over a very thin layer of adhesive was suggested. Where the edges of tiles had chipped at the movement joints, metal angle sections fixed into the bedding should have been sufficient to protect them from further damage.

Case study

Arching of tiles at a technical college caused by an unsuitable bedding mix

BRE Advisory Service was called in to investigate the failure of ceramic tile floors at a technical college. The ceramic tile floor laid in a teaching kitchen had failed by becoming hollow and arching as a result of shrinkage in the sand and cement bedding and the concrete base on which the tiles were laid. The failure arose because the bedding mix did not conform to the requirements of the semi-dry method specified: the mix was too strong and too wet, and hence had high drying shrinkage. Re-laying the tiles using a proprietary thin bed adhesive was recommended.

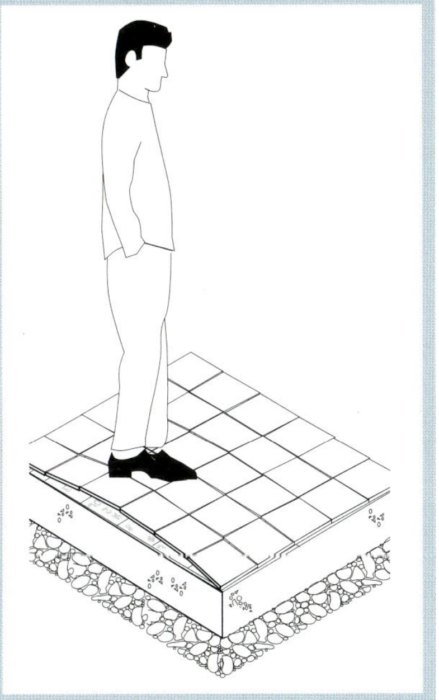

Figure 7.15
Tile floorings which have arched may still support full body weight

Figure 7.14
Failure of a slip resistant ceramic tile floor in a swimming pool: the tiles have arched due to longterm moisture expansion

Case study

Impact damage to ceramic tile flooring at a shopping centre

Breaking and loosening of ceramic floor tiles at two stores within a shopping complex appeared to be due to deficiencies in both the tiling and the semi-dry screeds beneath. While the tiles were sufficiently dense and the screeds sufficiently well compacted to sustain the static loads imposed on them, the tiles were too thin and the screeds proved inadequate to resist impact damage when trolleys were offloaded or when the floors were subjected to forklift truck traffic. It was recommended that, following repairs to screeds and tiles, there should be a reduction in the loads carried by trucks and trolleys. As a further measure, the thin ceramic tiles could have been replaced by a stronger and thicker type of tile.

Ceramic tiled surfaces can be repaired with comparative ease by replacing the tiles.

Medieval tiles need special consideration and further information on their conservation is available in *Architectural ceramics: their history, manufacture and conservation*[232].

Work on site

Storage and handling of materials

Section 11.1 of BS 8000:Part 11[233] is relevant.

Restrictions due to weather conditions

Cementitious beddings will need to be protected from frost during curing. The likelihood of damage is rare in normal buildings, but concourses open to the weather will be at risk.

Workmanship

A common problem which leads to failure is that installers tend to confuse the bedding methods in common use; for example when using the semi-dry method, too wet a mix is prepared which leads to excessive shrinkage.

A clean and well prepared surface, free of contamination with oils and free of surface laitance, is required to achieve the best performance when bedding mortar is applied directly onto the building carcass. A well prepared surface is perhaps of less crucial significance where there is a separating layer.

The laying of paviors and ceramic tiles is a highly skilled job and should never be attempted by unskilled labour. Cut tiles should be avoided wherever possible by employing special sizes produced by the manufacturer. Where cut tiles are unavoidable, they should be used at the tops of any falls and never at the bottom, and oriented so that water drains away from the cut edge.

The larger the superficial areas of tiles, the more difficult it is to get even bedding, especially with the stiffer cementitious mortars. When fixing thin tiles of large area by the thick bed method there may be a tendency for operatives not to compact the bed as thoroughly as when laying thicker tiles of smaller area. If, to achieve levelness, thin tiles have to be beaten hard with rubber hammers, breakages will occur. Levelness can be more easily attained without breaking tiles if the bedding is not fully compacted. This will lead, though, to lower strength in the bedding than if it were fully compacted with, again, increased risk of breakage of the tiles.

Full bedding of tiles means that at least 90% of the undersides of the tiles should be in contact with the bedding or adhesive. Any voids should be widely spaced.

Poor quality grouting has been a feature of many cases investigated by BRE. It has been seen as non-adhesion to the sides of the tiles, allowing the grout to be removed in strips, or of intrinsic material deficiencies such as inappropriate choice of mix and low strength. Some grouts have been of such poor quality that they had been removed by cleaning brushes.

The low strength of grouting can be attributed to:

- incorrect mix constituents or proportions
- poor mixing of materials
- poor compaction
- air bubbles in the mix
- incompletely filled joints

Further aspects of workmanship are covered in Section 11.1 of BS 8000: Part 11. BS 5385:Part 3 and Part 4 also apply.

Inspection

The problems to look for are:

- ◊ tiles cracking
- ◊ tiles loosening
- ◊ longterm moisture expansion in tiles
- ◊ floors arching
- ◊ movement joints absent
- ◊ slipperiness
- ◊ arrises of tiles chipping
- ◊ grouting worn away
- ◊ jointing materials breaking down
- ◊ floors in unheated buildings suffering condensation, especially where open to the atmosphere (nothing can be done to prevent this)
- ◊ cementitious grouting friable

Chapter 7.2 Concrete flags

Concrete flags for flooring purposes are those normally used for external pavings, frequently in smaller sizes.

Flags can be seen laid in the foyers of blocks of flats, offices, factories and workshops (Figure 7.16), providing robust flooring at very low prices.

Characteristic details

Basic structure

Although concrete paving flags have been used for a great number of years, they have seen a considerable increase in popularity since the 1970s, not just for external use. When used internally, concrete flags offer many advantages compared with in situ material, including being manufactured to a consistent quality and cured under controlled conditions. Consistent quality is achieved by hydraulic pressing and vacuum de-watering of the mix. BS 7263:Part 1[(234)] is relevant

The normal method of laying concrete flags is to fully bed them in mortar on a concrete base which has been well wetted to reduce suction. It is possible to find flags lifting through shrinkage of a relatively newly laid base, so adequate time should be allowed before bedding them.

Precast concrete 'rafts', approximately 1 m square and weighing upwards of 1 tonne, were used in the 1950s and 1960s on a considerable scale in heavy industrial workshops and foundries, and some may still be in use. These rafts, reinforced top and bottom, were edged with steel angles and laid on a 50 mm gravel bed.

Main performance requirements and defects

Appearance and reflectivity

Well used flags give around 0.25.

Choice of materials for substrates and joints

Normally flags are laid on a concrete base which should be clean. Old bases may need to be scabbled. Flags are jointed with cementitious grout to which bonding agents are sometimes added.

Strength and stability

See Chapter 1.1 and the same section in Chapter 4.1.

Dimensional stability, deflections etc

Coefficient of linear thermal expansion per °C: 10 to 13 × 10^{-6}.

Reversible moisture movement: 0.02–0.06%.

Wear resistance

Abrasion resistance varies between good and excellent, impact resistance is normally fair to good, and longterm indentation very good.

Slip resistance

The slip resistance of surfaces is fair to good, depending on characteristics. Flags are available with carborundum or other non-slip finishes. Ripple finish is slip resistant, even when wet.

Coefficients of friction are, for smooth finish:

- dry > 0.5
- wet 0.2 to 0.3

for carborundum finish:

- dry > 0.6
- wet 0.3 to 0.5

Figure 7.16
Concrete flags in the entrance hall of an office block

Control of dampness and condensation

Although unaffected by dampness, floors laid with concrete flags will be at much the same risk of condensation as other similar floors of concrete.

See also the same section in Chapter 7.1.

Thermal properties

Thermal conductivity of concrete flags: depends on density but probably around 1.8 W/mK.

Warmth to touch

Cold.

Fire

Non-combustible.

Suitability for underfloor heating

Suitable.

Sound insulation

Sound absorption is poor. Concrete flags can be used in a floating floor to achieve adequate impact sound insulation but performance in respect of airborne sound insulation depends on overall thickness of the flags and the bedding.

Durability

See the same section in Chapter 2.2.

Maintenance

See the same section in Chapter 2.2.

Work on site

There is no standard for laying these products internally but the recommendations given in BS 5385: Part 5[235] provide a good guide, particularly with the semi-dry method.

Storage and handling of materials

See the same section in Chapter 2.2.

Restrictions due to weather conditions

See the same section in Chapter 2.2.

Workmanship

The flags are laid by forcing them into the bedding mortar. The joints are pointed up afterwards with a similar, perhaps wetter, mix. If a proprietary bonding agent has been used in the mix, cleaning mortar off exposed surfaces immediately is necessary to avoid discoloration.

The laid floor should be allowed to dry slowly. To enable it to cure in hot weather, covering it will be necessary.

Inspection

The problems to look for are:

- ◊ flags cracking and loosening
- ◊ bedding suffering drying shrinkage
- ◊ substrates not cleaned or scabbled before laying
- ◊ slipperiness
- ◊ arrises of flags chipping
- ◊ jointing materials breaking down
- ◊ floors in unheated buildings suffering condensation, especially where open to the atmosphere (nothing can be done to prevent this)
- ◊ cementitious grouting friable
- ◊ poor compaction of grouting

Chapter 7.3 Natural stone

Natural stone flooring has existed for at least 3,000 years. The range of stones used has been very wide indeed, and the performance obtained has been extremely variable.

This chapter includes floors of granite, marble (Figure 7.17), travertine, limestone, sandstone, slate and quartzite.

Characteristic details

Basic structure

Most stone floorings have been in the form of large thick flags (originally called, confusingly, quarries – see the basic structure section in Chapter 7.1) laid either on ash beds or spanning between timber joists. The more exotic stones such as marble have been sawn into thin slabs (eg of 12 or 20 mm thickness) and bedded on mortar on a concrete base. Cobbles and setts have also been used in particular areas, whether for particular effect or simply to achieve cheap hard wearing surfaces (Figure 7.18); they were often laid on the bare earth.

Surveyors may also encounter flooring tiles made from natural marble aggregates or other stone aggregates commonly bound with polyester resins instead of Portland cement; they are included in this chapter. Such resin bound tiles have many of the characteristics of natural marble, their performance depending on the quality and relative amounts of marble in the matrix.

Main performance requirements and defects

Appearance and reflectivity

The colours and textures of natural stone finishes are numerous (Figures 7.19 and 7.20 on page 248). Even slate is available for floorings in a range of surface textures and colours. Although most marbles will polish without the aid of waxes, the effect soon wears off in heavily trafficked areas; a fine matt finish is often preferred, therefore, as it retains its appearance for longer.

Figure 7.18
A cobbled floor in a stable block

Figure 7.17
A highly polished marble floor in an office complex

Light shades give reflectances of around 0.45, medium shades around 0.25, and dark shades around 0.10.

Choice of materials for substrates and joints

Stone paving has been laid on bare earth, though such floors are often damp. Normal bases of concrete are satisfactory. See Chapter 3.1 for suitable forms of construction. Flags are commonly jointed with cementitious grouts to which bonding agents are sometimes added.

Strength and stability

See the same section in Chapter 7.1.

Figure 7.19
A traditional chequerboard pattern marble floor

Dimensional stability, deflections etc

Coefficient of linear thermal expansion per °C:

- sandstone 7 to 12 × 10^{-6}
- limestone 3 to 4 × 10^{-6}
- slate 9 to 11 × 10^{-6}
- marble 4 to 6 × 10^{-6}

Reversible moisture movement:

- sandstone 0.07%, but variable
- limestone 0.01%
- slate, negligible

Deviations in the surfaces of natural stone floorings after installation are likely to be due to movements in the subfloors on which they are laid. The remedies for these irregularities have been discussed in Chapters 3.1 and 3.2.

Figure 7.20
Travertine flooring in a shopping mall

Figure 7.21
These stone staircase treads in a Brecon farmhouse dating from the seventeenth century are in remarkably good shape after more than 300 years of wear. The walls and inner parts of the treads have been whitewashed to enhance lighting levels (Photograph by permission of D Thomas)

Wear resistance

Stone floorings have a wide range of resistance to wear. Granite and quartzite are excellent for wear resistance whereas limestone and travertine are less good.

Concern has been frequently expressed about the rates of wear on stone floors in UK heritage buildings. As long ago as the 1920s, the Building Research Station was carrying out abrasion tests on samples of stone flooring.

The results of a more recent programme of monitoring undertaken by BRE in four historic buildings floored with natural stone show that some floors are wearing at rates of around 10 mm per 100 years (Figure 7.22). The abrasion resistance of samples of some existing floor stones and of possible replacements have been measured. These results indicate that replacements can be found which will be more resistant to abrasion than those already in place. Most preventive actions are within the control of the authorities responsible for historic buildings; for

Figure 7.22
Marble ledgerstones, as this one in a heritage building, are susceptible to wear and need protection

Thermal properties
Thermal conductivity of stone flooring: depends on density but probably around 1.8 W/mK.

Warmth to touch
Cold to the touch.

Fire
Non-combustible.

Suitability for underfloor heating
Suitable.

Sound insulation
Sound absorption is poor. Natural stone and slate can be used over a floating floor to achieve adequate impact sound insulation but performance in respect of airborne sound insulation depends on overall thickness and mass of the floor.

Durability
Several instances of brown staining on Carrara marbles with grey veining have been seen (Figure 7.23). A common factor has been a significant amount of moisture in the substrate, either from the construction process or from flooding. It appears that some component, probably iron from the pyrites in the marble, is being drawn to the surface. The problem seems to be worse with heated floors.

Prevention, rather than cure, is the preferred solution. The substrates must be kept dry by effective DPMs.

example by the provision of adequate barrier matting at entrances. The damage and disfigurement which can occur as a result of the accumulation of grit and combustion particles, and staining and efflorescence from the movement of water, may be tackled less easily. Unfortunately, the most effective way to reduce damage is to prevent access to the floors by visitors and to prevent diffusion of soluble salts from the ground – anything less than these measures must result in some damage and, in the longterm, erosion of the surface with consequent loss of artistic and historical information.

Figure 7.23
This Carrara marble floor has suffered from staining, components of the marble being brought to the surface by flooding

Slip resistance
Slip resistance is fair to good in dry conditions; some stones can be very slippery when waxed. Most smooth stones will be slippery when wet; granite is the worst.

Coefficients of friction are:
- dry >0.5
- wet 0.2 to 0.45

Control of dampness and condensation
If dampness is being transmitted to floorings from the substrates, soluble salt crystallisation deposited at the surface of softer stones can lead to their disruption. There could also be staining, particularly of the lighter coloured marbles.

Figure 7.24
Contour scaling on a York stone flagged floor in a shopping mall caused by moisture transporting soluble salts up through the stone to crystallise under the surface

Case study

Cracking in the finish of a stone floor of a bank due to shrinkage of the substrate

Cracking of the stone tile flooring in a banking hall was found to have been initiated by shrinkage of a very good quality screed to which the tile bedding had bonded well in some places, not so well in others. Investigators reasonably assumed that many of the original tiles had not cracked simply because they were fortuitously less well bonded to the mortar bedding and the bedding to the screed. Only replacement of the cracked tiles was therefore recommended.

Case study

Hollowness under areas of marble paving in a bank

Although the method used to lay the paving was observed to be sound, there was some evidence that not all the slabs were fully bedded – they sounded hollow when tapped. However, it was considered unlikely that the normal foot traffic expected in a building of this kind would cause any extensive breakdown of the paving. Nevertheless, some remedial action, including replacement of affected slabs, would be warranted in areas subjected to heavy wheeled maintenance traffic.

Case study

Defective appearance of floor tiles in a retail store

BRE Advisory Service was asked to investigate the defects and generally poor appearance of a marble and resin tile floor at a retail store. It was found that there was extensive chipping, cracking, crazing and pitting of the tiles. The chipping and cracking was attributed to instability in the screed and plywood base on which the tiles were laid. Some apparent crazing and pitting had most likely been caused by latent defects in the tiles and some of the chipping to mechanical damage.

A problem which sometimes occurs with slabs of travertine (or terrazzo and resin marbles, and ordinary marble for that matter also) is that the edges of the tiles erode (Figure 7.25). Lack of bedding at the edges of the tiles leads to tilting caused by traffic. Hard grouting which does not completely fill the joint then exerts sideways shear forces on the tile arris, leading to spalling. The amount of movement required to cause this spalling in a brittle material is not very great (Figure 7.26). If the tiles were solidly bedded, right up to their edges, the problem would not arise.

Maintenance

Seals are not normally necessary on natural stone floorings. Cleaning should be with neutral detergents only.

Most stones and marbles can be stained by oils and greases which will be difficult to remove. Poulticing with a whiting paste may help.

Work on site

Section 6 of BS 5385:Part 5 is relevant.

Restrictions due to weather conditions

Stone floorings should not be laid until the building is weathertight.

Workmanship

One of the most important considerations in laying stone floorings is to ensure that they are fully bedded: at least 90% of each unit should be in contact with the bedding. Any voids should be spread out. It is also important that joints are grouted to their full depth.

Repairs to worn or broken stone flooring and staircases are problematical. So far as flooring is concerned, the stone is probably best left alone until replacement of the whole floor surface becomes necessary. Inversion and rebedding

Figure 7.26
Measuring deflections across joints in flooring; deflections are often caused by heavy cleaning machines

Figure 7.25
Spalling of travertine flooring slabs at a joint. Scale is indicated by the 50 pence coin

of the flags or slabs may be possible, but unlikely as the surface texture of inverted units will probably be unacceptably different from the original finish.

Piecing-in of small tablets of stone should be possible to renew broken nosings or arrises of stairs. The broken area should be cut back to a rectangular shape (Figure 7.27); a slight undercut to the side edges be beneficial. Bedding should be in a proprietary resin. Cement mortars should not be used in this situation.

Considerable care needs to be taken when recaulking balusters into mortises in the treads that the stone is not spalled.

Inspection

The problems to look for are:

- ◊ slabs cracking
- ◊ slabs loosening
- ◊ slipperiness
- ◊ arrises chipping
- ◊ slabs or jointing materials stained
- ◊ floors in unheated buildings suffering condensation, especially where open to the atmosphere (nothing can be done to prevent this)
- ◊ cementitious grouting friable

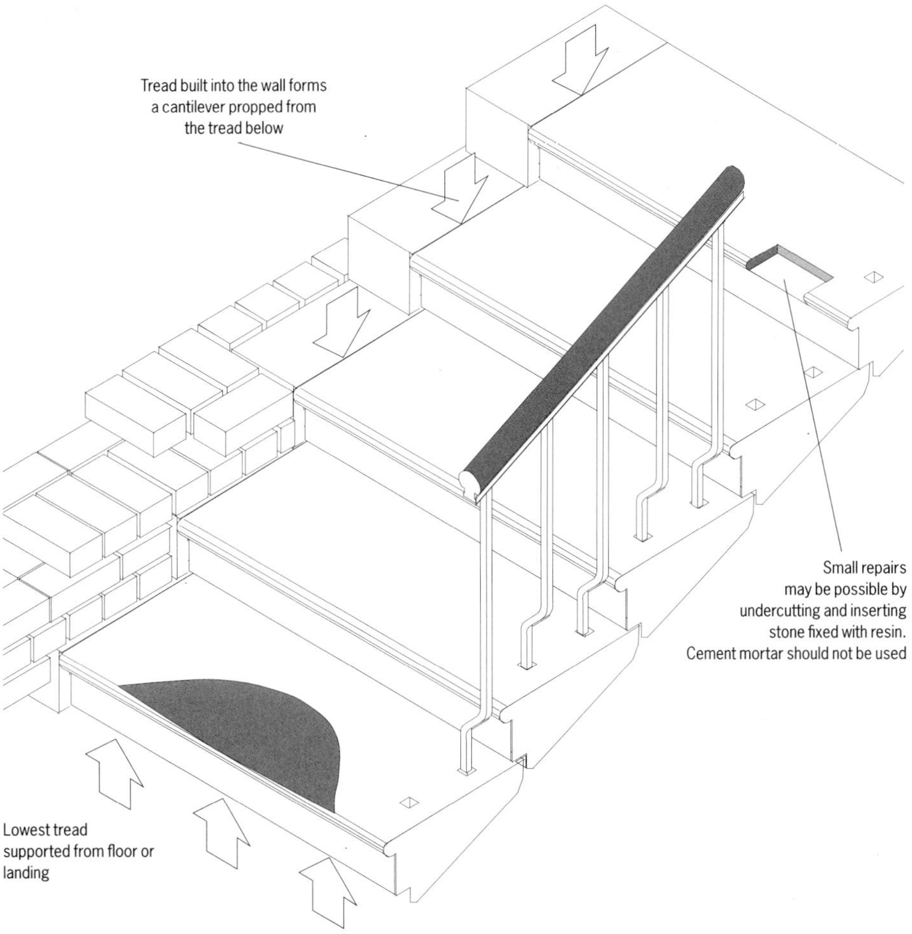

Figure 7.27
A cantilevered staircase in natural stone

Chapter 7.4 Terrazzo tiles

Figure 7.28
Hard wearing terrazzo tiles in the conservatory of a public house

Case study

Staining due to dirt accumulation
BRE Advisory Service was called in to advise on dark staining of the terrazzo tile floor finish at an industrial location. The staining had occurred along the sides of tiles in an unusually random manner. The investigator concluded that the problem was due to dirt accumulating on the surfaces of the tiles, particularly at edges. It was recommended, in particular, that the affected tiles should be reground to a clean surface and the whole floor area sealed to protect it from further soiling.

Terrazzo tiles have been used in considerable quantities where a decorative, robust and easily cleaned flooring is needed (Figure 7.28). They are now generally preferred to in situ terrazzo because of their lower vulnerability to cracking and their greater durability and assured quality.

Characteristic details

Basic structure
Terrazzo tiles consist of similar mixes to in situ terrazzo; namely crushed marble in a cementitious matrix. Consequently, floorings of these tiles share many characteristics with in situ mixes (see Chapter 5.4). The most common size of tile is 300 × 300 × 28 mm: the upper half of the 28 mm thickness being terrazzo and the base being cement mortar. Overall thickness tends to vary with plan size. Hydraulic pressure is used to obtain a dense mix and the surface of the tile is ground after curing.

Main performance requirements and defects

Appearance and reflectivity
Light shades give reflectances of around 0.45, medium shades around 0.25, and dark shades around 0.10.

Staining is a potential problem.

Choice of materials for substrates and joints
Tiles should be laid on a clean concrete base. Old bases may need to be scabbled. Often the joints are formed with strips of brass or plastics to offer improved performance when subjected to movements.

Strength and stability
See the same section in Chapter 7.1.

Figure 7.29
Terrazzo tiles which have broken up under trolley traffic because they were not fully bedded

Dimensional stability, deflections etc

Coefficient of linear thermal expansion per °C: 10 to 13 × 10^{-6}.

Reversible moisture movement: 0.02–0.06%.

Movements in terrazzo tile flooring are governed largely by the behaviour of the material in the tiles and their bedding.

Wear resistance

Resistance to wear depends mostly on the hardness of the aggregates used in the manufacture of the tiles (Figure 7.30). Soft marbles are particularly susceptible to damage from small heels, and many examples will be seen of wear in the aggregate while the cementitious matrix remains unaffected (Figure 7.31).

Slip resistance

Terrazzo tiles have good slip resistance when dry but are exceptionally slippery when wax polished. Smooth tiles can be very slippery when wet. Polish should not be applied to surfaces adjacent to terrazzo since it can be transferred by traffic. Slip resistant nosing is necessary on terrazzo tile stairs.

Coefficients of friction are, for smooth finish:

- dry > 0.5
- wet 0.2 to 0.3

for carborundum finish:

- dry > 0.6
- wet 0.3 to 0.5

Control of dampness and condensation

Terrazzo tiles are unaffected by water spillages, but they should still be laid on slabs which have an effective DPM. Although the tiles are unaffected by moisture, moisture rising from below can lead to staining of tile surfaces.

Thermal properties

Terrazzo tiles have little thermal insulation value.

Thermal conductivity of terrazzo tiles: no information is available but probably around 1.0 to 1.8 W/mK.

Warmth to touch

Cold.

Fire

Non-combustible.

Suitability for underfloor heating

Suitable.

Sound insulation

Sound absorption is poor. Terrazzo tiles can be used over a floating floor to achieve adequate impact sound insulation but performance for airborne sound insulation depends on overall thickness and mass of the floor.

Case study

Slipperiness of the floor finish in a public building

The slip resistance of terrazzo tile flooring in a public building was investigated by BRE Advisory Service using a skid tester. The results of the tests indicated that in dry conditions the coefficient of friction between shoe materials and the terrazzo flooring was close to or in excess of the recognised safe value of 0.4. Tests made on wetted terrazzo in the entrance hall of the building indicated that some form of barrier matting was needed to inhibit the amount of moisture carried into the building by footwear.

Figure 7.30
Precast terrazzo slabs on a staircase in an office block showing 30 years of wear

Figure 7.31
Wear in a terrazzo tile floor caused by heavy foot traffic. The soft aggregate in the material wears away and the remaining matrix then picks up soiling

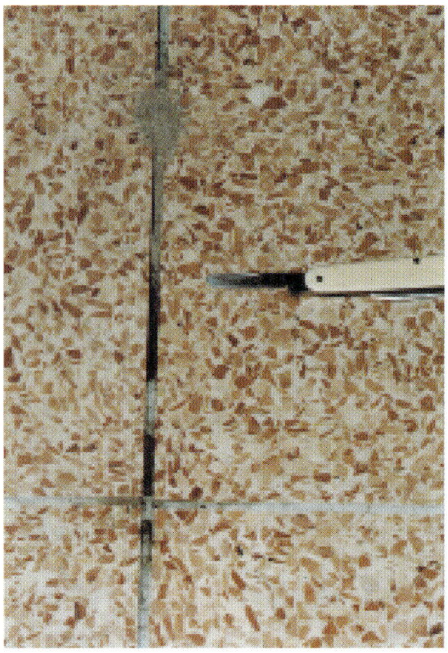

Figure 7.32
Poor quality jointing of terrazzo tiles

Figure 7.33
A diagonal crack in terrazzo tiles replicates a joint in the substrate. This is commonly known as reflective cracking

Case study

Thin tiles and poor quality bedding for flooring in a shopping centre

The combination of too thin a tile and poor mortar bedding was found to be the reason why the terrazzo tile flooring at a shopping centre had cracked under the loads imposed by foot traffic and the machine used to clean the flooring. Replacement with thicker tiles, well bedded, over the whole of the floor area was recommended.

Case study

Pitting of dark coloured terrazzo tiles in a shopping centre

The surface of black terrazzo tile flooring at a shopping centre had pitted because a component of the tiles, a soft loosely compacted mixture of fine aggregate and cement, was being removed during the floor cleaning operations. A BRE investigator recommended that the tiles were carefully inspected, any soft material being removed from surfaces and the resulting holes repaired with an epoxy based mortar – a SBR (styrene-butadiene rubber) or acrylic resin cement based mortar of the same colour. If after two years the amount of pitting that occurred in the terrazzo tiles still caused concern, consideration should be given to the replacement of the tiles.

Durability

Because this material is inherently durable, it is often used in heavily trafficked areas such as shopping malls and station concourses. To a large degree, however, durability depends on adequacy of the bedding and the completeness of filling the joints (Figure 7.32). Terrazzo tiles can suffer from spalling (see the same section in Chapter 7.3).

Life expectancy will be between 50 and 65 years for a good quality flooring.

All joints in substrates should be brought through the tiling, at least as simple contraction joints. Failure to do this may result in reflective cracking of the tiling (Figure 7.33).

Maintenance

The use of disinfectants containing phenols, which can stain some marble aggregates pink, should be avoided.

Figure 7.34
Contamination rising through the joints between terrazzo tiles as a result of flooding

Work on site

The appropriate standard is BS 5385: Part 5[235], Section 5.

Restrictions due to weather conditions

Terrazzo tile floorings should not be laid until the building is weathertight.

Workmanship

Aspects of workmanship are covered in BS 8000:Part 11: Section 11.1[233].

Inspection

The problems to look for are:

- ◊ staining
- ◊ wear resistance poor
- ◊ slipperiness
- ◊ finishes breaking down (eg pitting)
- ◊ tiles cracking
- ◊ tile floors in unheated buildings, or buildings open to the atmosphere, sweating (mostly unavoidable)

Figure 7.35

Replacement of broken tiles will very often cause mismatching problems. It may be less obvious to rebed the broken ones if spares are not available

Figure 7.36

Hand finishing terrazzo slabs at Waterloo International railway station

Chapter 7.5 Composition block

Many buildings have been fitted out with composition block floorings. Indeed, by the early 1980s the floors of many hundreds of sports halls alone were composition blocks (Figure 7.37). Although having a somewhat drab appearance, they normally produce a robust flooring provided they are fully bedded and laid with care.

These floors have been normally laid by specialist suppliers, and sanded and sealed after laying. However, some floors may be found laid using a patterned surface – for instance in imitation of cobble, tile or brick – which cannot be sanded.

Before the early 1970s, many of these floorings may have been laid without a DPM.

Figure 7.37
A composition block floor in a sports hall

Characteristic details

Basic structure

'Composition' flooring blocks – consisting of sawdust, sand and pigment bound with Portland cement or calcium sulfate, and impregnated with linseed oil – have been manufactured since the 1920s under several proprietary names of which the best known is probably Granwood. Plan sizes were around 150 × 50 mm in thicknesses of, approximately, between 6 and 12 mm with a dovetail-keyed undersurface. Unlike the wood blocks which they superficially resemble, they were laid bedded in cement mortar. Blocks were also available for staircase treads and risers, and skirtings. Since the late 1980s, a sprung composition block floor has been specially made for sports halls.

Main performance requirements and defects

Appearance and reflectivity

The blocks are available in a limited range of dark colours, mainly greys and browns, some imitating wood block. Some surfaces may be found in muted shades of orange, green and blue. Reflectances are around 0.10 or perhaps slightly more.

Choice of materials for substrates and joints

Cementitious bases are appropriate. Since the blocks are laid to abut, no jointing materials are required.

Strength and stability

Where fully bedded, they provide a robust floor for sports halls, refectories, classrooms etc, and even light industrial situations.

Dimensional stability, deflections etc

Measurements of the moisture sensitivity of composition block flooring were carried out by the Building Research Station during the 1939–45 war and in the early 1950s. These showed that although moisture expansion of unrestrained blocks was high, those stuck to a concrete slab using a 1:3 cement:sand bedding showed no sign of becoming loose. It was concluded then that, on sites not subjected to a head of water, they

Figure 7.38
A composition block floor: the bay joints in the heated screed have widened leading to lack of support and cracking of the blocks

could be laid without a DPM. Moisture content ranged from 2 to 10.4% when fully saturated.

Drying shrinkage has been measured in sample blocks as between 0.011 and 0.029% (Figure 7.38).

Wear resistance
Composition block flooring is suitable for light or medium traffic only. Flooring laid in the Mount Pleasant mail sorting office in 1925 had become so badly worn near the entrance and in doorways from the heavy trolleys that it had to be entirely replaced in 1938.

Slip resistance
Composition blocks become slippery when oiled and polished. No values are available.

Control of dampness and condensation
The flooring is not suitable for use in areas liable to dampness such as kitchens and bathrooms. It should be laid on an effective DPM, though not all early examples were.

Cases have arisen where existing floors laid without DPMs have been coated with polyurethane seals. Whereas the unsealed blocks would allow some dispersion of moisture rising from the ground, sealing prevents this. Consequently, where moisture has built up in sealed blocks, the blocks have softened and the seals have detached.

Thermal properties
Thermal conductivity of wood-and-cement blocks: no information is available but probably around 0.3 to 0.5 W/mK.

Warmth to touch
Fair, due to the sawdust content.

Fire
Combustible. Class 1 surface spread of flame is claimed.

Suitability for underfloor heating
Underfloor heating has been used successfully under composition block flooring up to relatively high temperatures although, normally, systems should not be designed to run at temperatures exceeding 28 °C. Early examples date from the 1930s. Indeed, there is some indication that the material was at first developed with underfloor heating in mind since the thermal movements of blocks were comparatively small. When laying on screeds containing heating elements, it is essential that sufficient drying time is allowed for screed shrinkage to fully take place before fixing blocks (see Table 1.3 in Chapter 1.4).

Case study

Opening of joints in composition block flooring due to insufficient drying time for the substrate

BRE Advisory Service investigated the cause of 'cracking' in a composition block covered floor in which the construction comprised a 150 mm thick in situ cast reinforced concrete slab and DPM, topped with a 85 mm concrete layer, insulation and screed. Electric heating cables were incorporated in the screed. The specification included laying the composition blocks on a sand and cement bedding layer.

On drying, shrinkage movement had caused the screed to crack and the cracks were reflected at the joints between floor blocks. The shrinkage was attributable mainly to insufficient drying time between laying the screed and the blocks. Movement within the screed had been aggravated when the electric underfloor heating was switched on. It was recommended that all the affected areas of floor blocks should be lifted and replaced using a cement based, thin bed adhesive; all open joints between blocks should be grouted; and on completion, the whole floor area should be sanded and resealed.

Case study

Cracking and hollowness in composition block flooring due to inadequate bedding

Following examination by consultants who had diagnosed breakup of the screed as the reason for cracking of composition blocks in the floor of a sports hall, BRE Advisory Service was asked to investigate the reason for the problem. The screed was tested with the BRE screed tester and found to be well finished and of good quality.

From examination of the blocks it became evident that the bedding mortar, which had been well spread on the screed, had not made good contact with the dovetails on the undersides of all the blocks when the flooring was laid, although most of the blocks lifted still had mortar well stuck to the outer ribs.

The BRE investigators found that the blocks were slightly hollow on their undersides. Though the hollowness amounted only to approximately 1 mm (Figure 7.40), the bedding mortar failed to make full contact with all the recesses between the ribs of the blocks. Where there was hollowness between the blocks and the bedding, some of the blocks had cracked and others had deformed (Figure 7.41). The damage was sufficiently widespread and serious for the composition block flooring to be taken up and replaced with new composition block flooring laid on the existing but cleaned base.

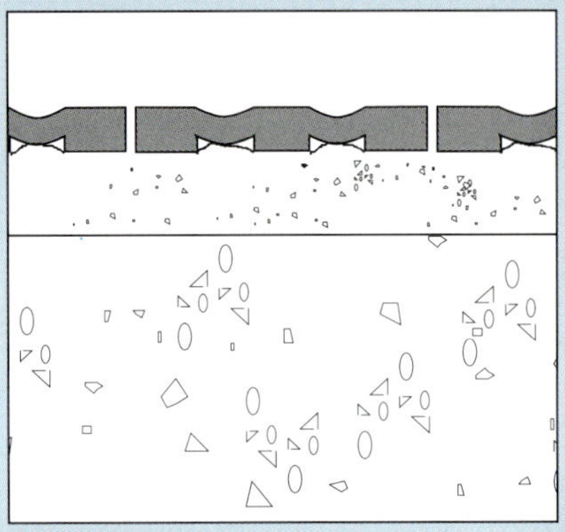

Figure 7.40
Where bedding mortar does not completely fill the recesses in the underside of the composition blocks, the flooring surface deforms in use

Figure 7.41
Indentation of composition block flooring by wheeled traffic. The problem had been initiated by mortar not being forced up into the dovetails of blocks when laying

Figure 7.39
Moisture rising from below has caused the seal to lift on this composition block flooring

Sound insulation

The flooring provides a relatively hard surface which does not absorb sound. Composition block can be used over a floating floor to achieve adequate impact sound insulation but performance in respect of airborne sound insulation depends on overall thickness of the screed.

Durability

Life expectancy will be about 40 years for a good quality flooring. However, there have been problems with joints between blocks opening up – some of this due to poor screeds, but some due to inadequate bedding of the mortar for the blocks and some where the blocks were continued over daywork joints. The 'cracking' normally arises early in the life of the flooring, say within a few months of laying. If the floor lasts more than five years it is normally good for another thirty, depending on there being no change in the service conditions.

There have been one or two cases where blocks have been fixed to asphalt screeds with adhesives containing organic solvents. The asphalt screeds have subsequently softened and led to hollowness and breakup of the flooring. The presence of solvents can be tested for in samples by thermal absorption gas chromatography.

Moisture rising through the structure of a composition block floor without a DPM, and collecting below the surface of the flooring which has been sealed with polyurethane, will cause softening and a reduction in the strength of the blocks (Figure 7.39). The blocks then fail due to the kneading action of wheeled traffic.

Maintenance

Composition block flooring will need sanding and resealing about 10 years after it was installed, or even earlier under adverse traffic conditions.

Work on site

The appropriate standard is BS 5385: Part 5[235], Section 7.

Restrictions due to weather conditions

Composition block flooring should not be laid until the building is weathertight.

Workmanship

Spreading mortar evenly and tamping in the blocks is vital if hollowness between the blocks and their beddings is to be avoided.

Inspection

The problems to look for are:

- ◊ blocks cracking
- ◊ blocks loosening
- ◊ slipperiness
- ◊ blocks softened by solvents in adhesives used on asphalt screeds
- ◊ DPMs absent where composition block floorings are being replaced by moisture sensitive floorings

Chapter 7.6 Metal

Figure 7.42
Steel 'anchor plate' floor plates with a modelled surface to enhance slip resistance

Floorings made of metals (whether ferrous or non-ferrous) tend to be used in very specific situations such as might be found in industrial plants where service conditions are onerous. They can take many different forms, almost all of a utilitarian rather than a decorative nature.

Characteristic details

Basic structure

There are three basic kinds of metal floorings designed to be supported on concrete or mastic asphalt bases.

- Plates or tiles set into a wet concrete base to provide a wearing surface. Steel is much more common than cast iron. Cast iron plates usually have feet carried on the base concrete. Inverted steel trays of this type are commonly called anchor plates (Figure 7.42).

Metal faced tiles have been used for a number of years where enhanced impact resistance has been needed. The tiles are held in close contact with the concrete by means of tabs or anchors punched through the top faces. Sometimes the wearing surfaces of the tiles carry a profiled pattern.

- Shallow trays of welded steel or cast iron are filled with fine concrete to provide a wearing surface, but with the upstands in metal providing a degree of armouring to the edge of the surface.
- Grids of open mesh steel or cast iron are buried in the wearing surface to give a high degree of armouring to increase the impact resistance of floors: the interstices can be filled with fine concrete or mastic asphalt. Up to the 1950s, these were frequently called dairy grids (Figure 7.43). As *Principles of modern building*[(6)] pointed out, these floors can be very noisy with hard-wheeled traffic.

In addition, as has already been described in Chapter 2.6, there are two kinds of metal decked suspended floors:

- inverted trays of steel, cast iron or aluminium alloy designed to span directly between steel joists
- grids of riveted or welded open steel diamond-patterned mesh (designed to be left open) spanning between steel joists

Figure 7.43
A 'dairy grid' floor

So far as free-spanning metal floors are concerned, these are usually of open mesh grids but can be of plate form in cast iron, mild steel, stainless steel or aluminium alloy. Steel and aluminium plate is usually impressed with a non-slip chequer pattern. They are normally made in panel form, carried on steel channel or angle, and used in industrial areas.

Main performance requirements and defects

Appearance and reflectivity

Rusting is probably unavoidable with ferrous metals unless stainless steels are used, but it is rarely a problem in the industrial situations where these floorings are normally used. Relatively polished surfaces can give high light reflectance, but painted surfaces in light shades give reflectances of around 0.45, medium shades around 0.25, and dark shades around 0.10. Slats and open grids are more complex, and no indicative values can be given.

Choice of materials for substrates

Plates and grids are usually laid on concrete.

Strength and stability

BRE have tested particular cast iron floor panels for strength. While they have been found to be generally safe, provided they were not showing defects in the central or edge ribs, nor excessive central deflections, there remains the possibility that some might not be safe under the design floor load of 5 kN/m^2. Defects originating in the casting process are not always apparent. Other metals, most of which can be more easily assessed for strength, are more consistent and predictable in behaviour.

Dimensional stability, deflections etc

Coefficient of linear thermal expansion per °C:

- cast iron 10×10^{-6}
- mild steel 12×10^{-6}
- aluminium 24×10^{-6}

Reversible moisture movement: all metals 0%.

Wear resistance

The abrasion resistance of soundly bedded metal tiles is very good, impact resistance very good, depending on thickness, and longterm indentation very good. Inadequately bedded tiles do not perform well. With slats and grids, little surface area is available to resist wear, and those routes subjected to heavy traffic will undoubtedly show wear in time, particularly if steel has been galvanised.

Slipperiness and safety

Slip resistance of surfaces is fair for tiles, but can be very good for specially cast metal (chequered or indented) surfaces (Figure 7.44). Where surfaces become worn smooth, they can become very slippery, and there is no easy way to restore good slip resistance short of coating with epoxy resin flooring. However, some surfaces retain a rough texture even when worn, and these will have good characteristics in both wet and dry conditions. Metal gridded surfaces can be very good.

When metal surfaces become oily, they can be very slippery.

Cast or extruded nosings for stair treads and landings can have non-slip additives or inserts.

There are very many metal floors in existence which have not had the benefit of electrical earthing protection (equipotential bonding). However, in today's safety conscious world, it may be worthwhile considering whether additional measures for existing metal floors are necessary. This topic was mentioned in Chapter 2.6 in relation to platform floors. For any metal tiled or gridded floors which are to remain in use or accessible, specialist advice from an electrical engineer should be sought.

Thermal properties

Thermal conductivity of aluminium: 160 W/mK.

Thermal conductivity of steel: 50 W/mK.

Figure 7.44
An aluminium plate floor (Photograph by permission of B T Harrison)

Warmth to touch
Very cold.

Fire
Aluminium has the advantage that it does not cause sparking when struck, though it has a low melting point. Steel and cast iron can spark when struck.

Suitability for underfloor heating
Very suitable if the geometry is appropriate and, in the case of electrical heating, all necessary measures are taken to guarantee safety.

Sound insulation
Sound absorption is poor. Metal tiles, however bedded, are unlikely to achieve adequate impact sound insulation, but performance in respect of airborne sound insulation depends on overall thickness and mass of the floor.

Durability
Steel trays which are incompletely filled with concrete can be easily deformed and lead to problems, as also can anchor plates which become detached. Where the grid is filled with concrete, the alkaline environment affords a degree of corrosion protection to the steel or cast iron. If concrete has been contaminated with chlorides, this can lead to premature deterioration. Steel trays filled with concrete provide an extremely robust finish for some of the heavier industrial situations; for example loading bays, dairies and breweries.

Maintenance
Loose plates relying for fixity on punched anchors are difficult to refix. Slippery metal plates usually require coating with resin floorings which adhere well to cleaned metal.

Work on site

These materials are not covered by standards and reliance must be placed on manufacturers' advice. Where grids are to be filled with cementitious materials, normal practice for mixing these materials applies (see Chapters 2 and 3)

Inspection

The problems to look for are:
- corrosion
- loosening of plates
- slipperiness
- earthing arrangements dangerous or unsatisfactory (an electrical engineer should be consulted)

Chapter 8 Timber and timber products

Timber has provided the majority of decking for suspended floors over the ages, mainly softwood but sometimes indigenous hardwoods, and in most cases providing both deck and finish. Since the 1939–45 war there has been a significant growth in the proportion of floorings formed from processed timber such as plywoods and chipboards. Both traditional and newly developed forms are included in this chapter.

There has been concern expressed about the profligate use of timber from non-renewable sources. This has meant that some of the hardwoods traditionally used for floorings have declined in usage, and other species have been investigated.

Figure 8.1
Wood strip flooring provides a high quality ambience to this museum gallery

Chapter 8.1 General

This chapter deals with topics which apply to most kinds of timber flooring, while later chapters deal with additional topics specifically relating to the different kinds of product.

Characteristic details

Basic structure

The species of timbers suitable for flooring are too numerous to mention in detail in this book. Nevertheless, the following references may be found useful:

- Timber for flooring, Forest Products Research Bulletin No 40[236]
- Flooring and joinery in new buildings, FPRL Technical Note No 12[237]
- Hardwoods for industrial flooring, FPRL Technical Note No 49[238]
- BS 8201[239], Tables 1–11

Small firm knots are normally acceptable in softwoods intended for use as the final flooring, but are not acceptable in hardwoods.

Detailing

Wood and wood based panel products are subjected to dimensional movement as moisture content changes, and usually provision must be made for this movement to take place without disrupting the flooring. Movement is normally accommodated at the perimeters of rooms where skirtings cover the edges. Vulnerable areas are at door thresholds and staircase landings.

Figure 8.2
Random width floorboarding, screwed and pelleted to the bridging joists, in a heritage building. A coin – a 50 pence piece, centre left – indicates the massive size of the boards

Main performance requirements and defects

Appearance and reflectivity

The species of timber is often selected for its appearance, provided durability can be predicted. There are too many species to permit meaningful comment on appearance.

The quality of the wood is crucial to the appearance of the finished floor, though, as a natural material, it is inevitable that there will be differences in grain and colour between units (Figure 8.2). Indeed, it can be argued that it is this variability which gives a wooden floor much of its appeal.

In general some of the paler coloured woods, such as maple and sycamore, tend to bleach slightly on exposure to ultraviolet light (sunlight). They may yellow too. Other woods, such as iroko, afrormosia and fir, tend to darken. Sapwoods and heartwoods tend to change colour at different rates.

It is possible to slow down the rates of colour change by incorporating filter media in any coatings, but specialist advice will be needed.

It should be remembered that resanding old wood floorings will often reveal a quite different colour from that to which building occupants have become accustomed.

Maple, birch and beech floorings give reflectances of around 0.35, light oak around 0.25 and dark hardwoods around 0.20.

Table 8.1
Thermal and moisture movements of timber flooring products

	Coefficient of linear thermal expansion °C x 10^{-6}	Reversible moisture movement %
Wood and wood laminates:		
softwoods	With grain 4–6 Across grain 30–70	Negligible with grain Across grain: 0.6–2.6 tangential 0.45–2.0 radial
hardwoods	With grain 4–6 Across grain 30–70	Negligible with grain Across grain: 0.8–4.0 tangential 0.5–2.5 radial
plywood	*	With grain 0.15–0.2 Across grain 0.2–0.3
blockboard and laminboard	*	With core 0.05–0.07 Across core 0.15–0.35
Wood chipboard and fibrous materials		(On length or width, values for thickness may be up to 30 times greater)
hardboard	*	0.3–0.35
medium hardboard	*	0.3–0.4
softboard	*	0.4
chipboard	*	0.35
woodwool	*	On length 0.15–0.3 On width 0.25–0.4

* No data available

Choice of materials for substrate
See Chapters 8.2–8.5.

Strength and stability
See Chapters 1.1 and 2.1 for a general discussion.

Dimensional stability, deflections etc
The section on detailing has already mentioned the need for provision of movement gaps which are essential with certain kinds of wood floorings. These gaps are usually about 10 mm and can be conveniently located under skirtings at room perimeters in domestic situations.

Amongst the species giving least movement in service are Douglas fir, larch, mahogany, makore, meranti, muhuhu, padauk, pine, and teak; and amongst those giving large movement are beech, gurjun, and ramin. A list of species and the amount of movement for each is given in BRE Technical Note 38[240]; also *Handbook of hardwoods*[241] and *A handbook of softwoods*[242].

The most vulnerable point in wood based boarded or panelled floor finishes laid over thermal insulation boards is at internal door thresholds. Unless extra support is provided, the floor will deflect under foot traffic. This has been seen many times in BRE site inspections. Housing Association Property Mutual reports that, in 1 in 6 cases they examined, the necessary support was lacking.

Wear resistance
Wear resistance depends on the species and structure of the timber. Although surface finishes, especially those based on polyurethanes, can slow down the rate of wear of a timber surface, they have only a marginal effect on hardwoods.

Abrasion resistance is fair to very good for unprotected hardwoods and poor to fair for unprotected softwoods; impact resistance is fair to very good for hardwoods and poor to fair for softwoods; and longterm indentation is fair to very good for hardwoods and fair for softwoods. The pattern of sawing can affect wear rates: quarter sawn stock, the most expensive, gives the better rates.

Some softwood blocks have traditionally been available with end grain exposed, purportedly giving improved wear and impact resistance. In reality, wear rates depend on the closeness of the grain.

Industrial situations offer the most severe wear conditions for wood floorings. FPRL Technical Note 49 recommends three species providing superior performance:

- rock maple
- Rhodesian teak
- East African olive

Slip resistance
Slip resistance of surfaces is very good for both hardwood and softwood blocks and strip when dry. It can be very poor if the surface is wax polished. On stair treads a slip resistant nosing is usually necessary.

Coefficients of friction are:

- dry > 0.5
- wet 0.2 to 0.4

Control of dampness and condensation
Woods are hygroscopic in nature, and they will move in sympathy with ambient conditions. When laid on a solid base, there should be an effective DPM (Figure 8.3 on page 266).

Thermal properties
Thermal conductivity of hardwoods: 0.15 W/mK.

Thermal conductivity of softwoods: 0.13 W/mK.

Timber is a naturally good insulator, but this advantage can be easily lost if gaps between floorboards permit draughts. Square edge boards are particularly bad in this respect and may need to be overlaid with a sheet material to control air movement. Where a crawl space gives access to the underside of the floorboards, underdrawing with a thermal insulation board may be feasible. The introduction of central heating may exacerbate the problem by causing shrinkage of boards.

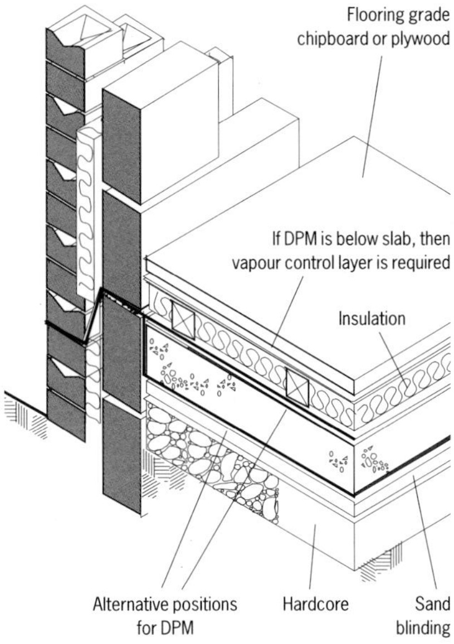

Figure 8.3
Wood floorings need the protection of a DPM

Warmth to touch
Most timbers are fairly warm to the touch.

Fire
Timber is combustible. Surface spread of flame for all timbers used in buildings is usually Class 3 to BS 476:Part 7[128].

Suitability for underfloor heating
Timber and timber based floorings are not the best of materials for use over underfloor heating owing to their thermal insulating properties and their propensity to dry out and shrink (unless moisture contents can be well controlled) when heating is on; also their ability to pick up moisture and expand when heating is off.

Sound insulation
Wood raft floating floors consisting of timber boards over battens laid over a resilient quilt on a concrete or timber subfloor will give good impact sound insulation. The performance of airborne sound insulation depends on overall thickness and mass of the floor and the number of joints. Therefore, panel products, with fewer joints per unit area, give slightly better performance than strip or board flooring.

Structural floors of wood, surfaced with boards or panel products, occasionally give rise to noise when trafficked; in fact the worst kind of floor is the floating floor described in the previous paragraph. Such noise usually emanates from the drying shrinkage of the wood opening joints slightly so that the adjacent edges of the boards rub on each other, or the undersides rub on the supporting joists. This is exacerbated if there is accompanying twisting of the units. Ringshanked nails or screws, which hold better than plain nails, will often offer improved resistance to unwanted noise. In addition, a strip of adhesive over the joists will help. There is not much that can be done to remedy noise from some existing floorings short of re-laying and glueing, though extra nailing or screwing of softwood boards in situ may help. One possible alternative which might be tried in narrow width secret-nailed floorings is to drill small holes in the joints over joists and inject a foaming adhesive to fill any gaps which have appeared between the joists and the boards.

Durability
Where wood floorings are maintained at moisture contents of less than 22%, fungal decay is most unlikely to occur.

Softwoods which have not been surface protected, when subjected to traffic, will rapidly deteriorate; particularly if joints open, foot traffic will breaks the arrises of boards, creating splinters. In attempts to toughen the surface, softwoods traditionally were treated with oils or varnish stains which could become tacky and pick up dirt, quickly darkening the floor. Since the 1960s more sophisticated materials have become available, particularly polyurethane seals which give vastly improved performance compared with oils and varnishes.

Hardwoods are more durable, though they still need sealing or 'feeding' to keep them in good condition.

Maintenance
Caustic cleaning agents should not be used on wooden floorings. Furthermore, using water to clean wood floorings should be kept to an absolute minimum, even if the flooring has been sealed.

Periodical stripping of old emulsion waxes will be necessary, and it may be necessary to use a mildly alkaline detergent. If waxes used have been of the solvent type, then stripping should be unnecessary. The thicker timber floorings are often resanded before resealing.

Care should be taken during and after sanding wood floors to avoid combustion of bagged dust or explosion of dust-laden atmospheres.

Work on site

Storage and handling of materials

Timber has a relatively high moisture movement across the grain, and consequently timber flooring should have a moisture content at installation appropriate to the future service conditions of the building in which it is fixed (see the workmanship section which follows). For all kinds of timber, and especially where it has been artificially seasoned as is usual for flooring grades, it is imperative that storage under cover is provided. It will also be beneficial if that accommodation is heated to maintain the moisture content levels to approximately those at the time of delivery.

Restrictions due to weather conditions

Timber flooring should be laid only when the building is weathertight; and, in winter conditions, blocks and strips only when the heating is on. Rainwater has frequently been seen to saturate timber sheeted floors during BRE site inspections. Not all floors affected were relaid.

Workmanship

Timber floorings should be laid at moisture contents not substantially different from what they will achieve in use (Table 8.2).

Timber floorings should not be laid directly on screeds or concrete bases, or directly over unventilated cavities until a reading of 75 to 80% relative humidity or less has been obtained by the hygrometer method (see Chapter 1.4).

Timber floors laid in winter, when the humidity in heated buildings tends to be low, may need larger than normal movement joints than floors laid in summer in order to accommodate greater expansion during the summer.

See also the appropriate sections of Chapters 8.1–8.5.

Inspection

The problems to look for are:

- ◊ wood of unacceptable quality (eg knots)
- ◊ timber species not appropriate to density of traffic in service
- ◊ slipperiness
- ◊ movement joints absent
- ◊ DPMs absent or defective
- ◊ nailing or glueing inadequate
- ◊ wood splitting from moisture shrinkage
- ◊ attack by rots or insects
- ◊ moisture movements excessive

Table 8.2

Target moisture contents of timber to be laid as flooring*

Unheated	15–19%
Intermittent heating	10–14%
Continuous heating	9–11%
1Underfloor heating	6–8%

* Further information can be obtained from BS 8201

Chapter 8.2 Board and strip

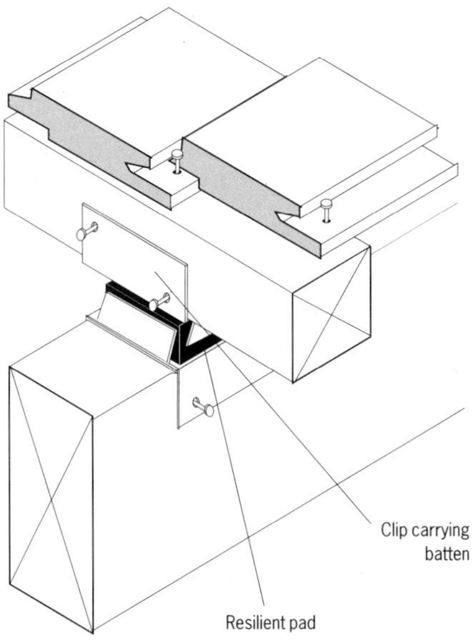

Figure 8.4
The principle of the sprung floor

Sawn, adzed or planed-and-rebated softwood boards were the first choice of specifiers to form the normal flooring for suspended timber floors until the development and use of wood chipboards during the early 1960s. Few boarded floors were designed to be left totally uncovered, although domestic room perimeters remaining exposed after standard sized carpets had been laid have often been stained and varnished.

This chapter examines some of the main performance attributes affecting the acceptability of board and strip floorings.

Boards are normally wider than 100 mm and strips less than 100 mm, but having similar thicknesses.

Characteristic details

Basic structure

In domestic construction, softwood tongued-and-grooved boarding has been favoured for suspended floors, with hardwood boards, traditionally of oak, used in other building types where wear is more severe. Fixing is invariably by nailing, often through the face of the board, although better class work is normally secret nailed through the tongues. The latter practice is much more labour intensive, for each board has to be cramped individually, rather than several boards at once. For the largest boards, screwing and pelleting is an alternative.

Wood strip tongued-and-grooved flooring is more commonly used over groundbearing floors, carried on battens which may be supported on resilient pads (Figure 8.4). This gives a semi-sprung floor, which is often specified for gymnasia and sports halls. Another type of semi-sprung floor is where the battens are loose laid on a sheet foam underlay.

Main performance requirements and defects

Appearance and reflectivity

See the same section in Chapter 8.1.

Choice of materials for substrate

Boards are normally laid to form the deck, whereas strip flooring is normally laid on battens on a concrete or dense screeded base. Where the substrate is solid, it is essential that it is dry, and a DPM should be provided.

Where the substrate is of timber, it helps normally to lay the flooring at 45° to the substrate boards so that the effects of shrinkage are minimised (BRE Digest 364[37]).

Strength and stability

Safe maximum spans for softwood tongued-and-grooved boards are given in the Building Regulations 1991 Approved Document A[116]. Maximum spans for hardwood boards will depend on the species, and further advice in this area may be sought from the Timber Research and Development Association or BRE.

Floorboarding may be insecure where it has been cut and lifted to install or maintain services. Large gaps may be found where boards have shrunk allowing excessive draughts, particularly through gaps in plain edge boards. Sloping floors may be due to settlement or simply have been poorly built. Badly cupped boards are common and will cause carpets to wear rapidly in heavily used areas. Projecting nail heads also disrupt and damage floor finishes.

See Chapter 1.1 for a general discussion.

Dimensional stability, deflections etc

Although most strip and board floors have been in use for many years, mistakes can still occur in relatively modern buildings. Timber is a hygroscopic material which will absorb moisture from the atmosphere; changes in relative humidity (RH) of the atmosphere, say in intermittently heated buildings, can cause dimensional changes in flooring which need to be accommodated.

The pattern of conversion (sawing) of softwoods also will have a considerable influence on the shape of any distortion following moisture movement, with quarter sawn boards being the most stable.

See the same section in Chapter 8.1.

Wear resistance

See the same section in Chapter 8.1.

Slip resistance

The slip resistance of timber board and strip can be very poor if wax polished. On stair treads a slip resistant nosing is necessary.

See also the same section in Chapter 8.1.

Control of dampness and condensation

Strip and board is adversely affected by dampness, with the risk of rot.

Thermal properties

Thermal conductivity of hardwoods: 0.15 W/mK.

Thermal conductivity of softwoods: 0.13 W/mK.

Warmth to touch

Good.

Case study

Buckling due to defective movement joints in a timber strip floor

BRE were asked to advise on the cause of buckling in a new two-layer timber strip floor laid over a groundbearing concrete base. The floor had begun to rise within one month of the building opening. At first the problem was thought to be due to the overenthusiastic use of water by the cleaners, but, although cleaning practices had immediately been modified, the floor did not return to its intended level state. The investigator was not able to find out what arrangements had been made for dampproofing the slab.

The investigator noted that the floor bulged upwards by several millimetres over an area of approximately 2 x 1 m, but the timber strips were still tight to each other. The floor could be made to spring up and down by applying full body weight. Measurements showed that the moisture content of the flooring was about 15%.

An examination was made of the design provision for movement at perimeter abutments. In general this was found to consist of a 10 to 20 mm gap which was either hidden under the skirting board or cover mould, or was filled with a cork strip. However, where metal fittings penetrated the covering, including duct covers and staircase cleats, no provision for movement had been made (Figure 8.5).

Figure 8.5

One edge of this timber flooring had been provided with a cork movement joint which is partly covered by a timber moulding. At right angles to this joint, against the steel cleat, no movement joint was provided. When the flooring expanded, the absence of the second movement joint caused the flooring to disrupt at the other side of the room

The investigator reported to the client that it was essential that timber floors were laid in accordance with BS 8201[239]. It is crucial to satisfactory future performance that allowance is made against **all** upstands for accommodating the movements caused by changes in moisture content of the material. In the particular case, a general provision had been made but had been overlooked in a few places.

It was recommended to the client that he arrange for gaps to be formed in situ round the fittings in the floor, and for the gaps later to be filled with cork. After allowing the flooring to recover from the buckling, no further problems should ensue.

Fire

Combustible. Although the material can be burned by cigarettes, normally any scorching can be removed with care using a fine abrasive.

Suitability for underfloor heating

See the same section in Chapter 8.1.

Sound insulation

The sound absorption of timber board and strip floors is poor. Wood raft floating floors consisting of boards or strip laid over battens on a resilient quilt on the concrete or timber subfloor will give improved impact sound insulation compared with finishes laid directly on the

Case study

Arching of timber strip flooring due to moisture induced expansion

A new floor had been designed as a covering of polyethylene backed timber strips laid on a 5 mm foam sheet on a layer of plastics coated building paper on a concrete slab base. However, two months after the slab was laid it was still insufficiently dry to lay the other elements, so two coats of a bituminous paint were applied and the floor laying went ahead. After a further eighteen months, one of the timber strips had arched up by 200 mm. Removal of the skirtings showed that the movement gap at the perimeter had been completely filled. A number of the strips were removed to allow the flooring to settle.

BRE was then called in to advise on the problem with the floor. By the time the investigators saw the floor the strip had returned to its correct shape. On peeling back the intervening layers the slab was found to be noticeably damp. Recently calibrated hygrometers in thermally insulated boxes were fixed to the surface of the concrete and read after one hour and sixteen hours; readings of 88 to 94% RH were obtained. From an inspection of drawings, it appeared that there was no effective DPM under part of the slab.

Arching of the timber floor was typical of that caused by moisture and there was no reason to suppose that any other factor was involved. Checks were made for plumbing leaks but none was found; it was also established that water had not been used for cleaning, and there had been no spillages.

There was no conclusive evidence that established the reason for the moisture found in the base, but the most likely explanation was that the slab had had insufficient time to dry before the flooring was laid. Although the building paper and plastics backing to the strips cannot be considered to have the qualities of a DPM, they would have acted as a limited moisture vapour check. Under these conditions the flooring could have existed in good condition for 18 months until the movement gap had been filled by expansion of the timber strips. Continued transfer of moisture from the slab into the timber strip would then have caused it to expand and arch. Moreover, the continued absence of an effective DPM would mean that the problem would worsen over time.

BRE recommended that the client should remove the strips and lay a surface DPM consisting of an epoxy resin before re-laying the strips.

deck or screed. Performance in respect of airborne sound insulation depends on overall thickness and the mass of the floor.

Durability

Life expectancies for hardwood and softwood board and strip floorings depend on many factors (eg species, traffic conditions and degree of protection). Provided design detailing and maintenance have been appropriate, traffic is at a low or moderate level, and there is no rot or insect attack, lifespans can be often measured in centuries rather than decades.

Maintenance

Stain and varnish, or stain and polish, was the traditional finish for flooring of wood boards. After years of neglect, refinishing by sanding and repolishing may become necessary, depending on the density of traffic (Figure 8.6). Resanding cannot, however, be continued indefinitely, since perhaps 2 to 3 mm of surface is removed each time, and the sanding on strip flooring will soon expose the tongues and nail heads. Current practice is to apply a polyurethane or oleoresinous seal after sanding.

Care should be taken during and after sanding wood floors to avoid combustion of bagged dust or explosion of dust-laden atmospheres.

Work on site

Storage and handling of materials

Boards and strip should be stored in dry conditions.

See also the same section in Chapter 8.1.

Restrictions due to weather conditions

Boards and strips should not be laid until the building is weathertight and dry. If this has not been done, there will usually be evidence of movement in the following months.

Workmanship

It is important that boards and strips are laid at moisture content levels consistent with ambient internal conditions. Boards which are laid too dry are very prone to expansion and disruption. For boards or strips laid on battens set on or in a screed,

Case study

Two cases of buckling in timber strip flooring: same problem, different causes

Two schools in a London borough had problems with buckling of timber strip flooring. In both establishments the flooring was timber strips on timber joists laid directly over concrete bases. The cause of the problems in each case was dampness in the timber joists. In one case, the flooring of a gymnasium had lasted seventeen years before showing the dampness (moisture content of up to 26%); in the other – its use was not specified – the problem (moisture content of up to 30%) had come to light after only two years.

In the gymnasium, a tar rubber emulsion had been applied to the concrete as a remedial measure after the problem first became apparent; also brass ventilation grilles had been inserted, set into the strip flooring. In the second case, tar rubber emulsion had been applied at the time of construction.

While the two cases presented the same problem, investigation revealed very different causes.

In the first case the cause was condensation below the timber strips. Installation of the ventilation grilles, without arrangements for exhausting air in the subfloor void to the external atmosphere, had only exacerbated this situation by introducing more moisture laden air. Under no circumstances should ventilation of the subfloor void be made to the inside of the building, even where ventilation to the outside is not feasible.

In the second case the moisture was probably a legacy of the original construction and not given enough time to evaporate.

The solutions of the two problems were also different. Ventilation of the void to the outside of the building (effectively to provide cross-ventilation) should have proved a satisfactory treatment for the first case. For the second the flooring strips needed to be removed to allow the concrete base and timber joists to dry out to a satisfactory level of moisture content before replacing the strips. Suitable arrangements for ventilating the subfloor void, as with the first case, should have been ensured.

it is essential that the latter is dry; that is less than 75 to 80% RH established by the hygrometer method (see Chapter 1.4).

For best quality work, softwood boards should be cramped in groups of not more than four boards when face nailed, or individually secret nailed. All header joints should bear on joists or battens, and header joints on a single joist should be separated by at least two board widths.

See also the same section in Chapter 8.1.

Inspection

In addition to the items listed in the inspection section at the end of Chapter 8.1, the problems to look for are:

◊ spans of boards exceeding those permitted under building regulations
◊ boards (inadequately protected before fixing) showing blue stain
◊ traps cut into decks not supported by joists and noggings
◊ plain edge boarding showing gaps
◊ squeaking between boards or strips
◊ sanding marks not removed
◊ movement joints absent
◊ arching and disruption

Figure 8.6
Vigorous sanding on these 100-years-old floorboards has exposed the nail heads

Case study

Distortion in timber strip flooring in a university building

Wooden strip flooring at a university in the Midlands had suffered from distortion to varying degrees in various parts of a building. There was some shrinkage and some expansion, with less distortion adjacent to areas of textile flooring. The floor structure generally consisted of 12 inches thick hollow ceramic pots set in reinforced concrete, topped with wooden battens, infilled with a sand and cement screed and covered with thick strips of gurjun.

The timber used as the floor covering, gurjun, was capable of absorbing considerable amounts of water, the moisture content varying from 12 to 20%. This had obviously caused the distortion. The most likely source of the moisture was the structural base where originally there would have been large quantities of water which had not been given enough time to evaporate.

The suggested treatment for the floors was to lift the wooden strips and allow the base to evaporate over a period of time. This would, of course, have made the floors unusable. Alternatively, a damp proof layer, consisting of two coats of epoxy resin, could have been applied, although there would be a problem in assuring the continuity of the DPM. A further alternative, of introducing a flow of dry warm air to the area to reduce the moisture in the floor structure, was considered not to be feasible.

Chapter 8.3 Block

Wood block floors can be found in very many species of timber and laid in a variety of patterns – herringbone and basket being the most popular. End grain blocks are comparatively rare but can be seen occasionally on staircase treads. Appropriate practices can maintain or even improve the appearance of a floor for many years, and blocks have the advantage that they can be sanded to restore the original surface after many years of wear.

Characteristic details

Basic structure

Wood blocks for flooring have been manufactured in many different sizes, with many different jointing techniques used: from tongued-and-grooved to metal tongues or wood dowels. The earliest wood blocks were relatively thick at up to 40 mm; but as time went by there was a progressive reduction in thickness, especially for those blocks designed to be laid close butted rather than being tongued-and-grooved. For the very hardest wearing surfaces, end grain blocks were frequently specified.

Wood blocks were traditionally laid on a concrete base without the benefit of a DPM since it was assumed that the adhesive or bedding medium (hot pitch or hot bitumen) would act as a DPM. This was sometimes wishful thinking on the part of the specifier although it was usually considered to be an effective way of bedding the blocks. Modern block floors are stuck with a bitumen rubber emulsion, an effective DPM being provided separately in the floor structure.

Figure 8.7
Moisture expansion has caused this wood block flooring in a school to fail

Main performance requirements and defects

Appearance and reflectivity

See the same section in Chapter 8.1.

Choice of materials for substrate

Block floorings are normally laid on a solid or suspended base of screeded concrete. Since the blocks are laid directly on the surface of the screed, it is important that it be level. As already mentioned, it was not formerly the practice to provide block floors with a DPM when they were bedded in hot bitumen. It is now obligatory, of course, to provide new floors with a DPM.

Strength and stability

See Chapter 1.1 for a general discussion.

Dimensional stability, deflections etc

Wood blocks which are intended to provide a superior quality floor finish are more likely to be well packaged, and to receive greater care in storage and during construction, and so be kept dry. Therefore moisture induced size changes in service, although in principle reversible, are likely to be expansive changes. In wood block flooring these occur across the grain and thus in the plane of the floor; expansive size changes can be considerable.

If insufficient provision is made for expansion at perimeters, substantial upward displacement may occur, either as localised ridging or tenting or as uniform bowing over the entire area (Figure 8.7).

Figure 8.8
Rafting of wood blocks

BS 8201[239] suggests providing an expansion gap of 10 to 12 mm at perimeters, but, for any areas larger than domestic room size, it recommends that sufficient space be provided between successive panels as well as at perimeters.

However, shrinkage can also occur. If a seal is applied too early to the surface of a wood block floor, and the seal sticks the blocks together, any subsequent shrinkage of the blocks can be expected to result in the formation of wide gaps. This phenomenon is known as rafting (Figure 8.8).

Some woods are subject to much greater moisture movement than others. Maple, for example, has a comparatively large movement.

See also the same section in Chapter 8.1.

Wear resistance

Resistance to wear depends on the species of timber.

See also the same section in Chapter 8.1.

Slip resistance

Slip resistance can be very poor if the surface is wax polished. On stair treads a slip resistant nosing is necessary.

Control of dampness and condensation

See the same section in Chapter 8.1.

Thermal properties

Thermal conductivity of hardwoods: 0.15 W/mK.

Thermal conductivity of softwoods: 0.13 W/mK.

See also the same section in Chapter 8.1.

Warmth to touch

Wood block floors provide a surface which is warm to the touch.

Fire

Wood block floors are combustible, but this is relevant only in certain specific situations such as firefighting areas or protected escape routes.

Although the material can be

Case study

Rafting of new wood block flooring in an old building

Gaps of up to 5 mm in 2 m had appeared in a new wood block floor which had been treated with two coats of an oleoresinous seal. The blocks had been laid on an intermediate floor during refurbishment of an old building. The new blocks had been stored on site for a period of one month before laying, though a check on their moisture content had not been carried out. The blocks were laid during the winter season, with the heating on. Some six to nine months later the blocks had rafted. For an explanation of rafting, see the section of this chapter on dimensional stability and BS CP 209:Part 1[243].

A BRE investigator noted during a site visit that, in many areas of the flooring, the seal had penetrated between adjacent blocks. A gap of some 3 to 5 mm had appeared at the perimeter of the floor which suggested that the flooring had been pulled inwards at the perimeter.

The investigator reported that the failure was typical of rafting as described in BS CP 209: Part 1. Although this standard has been withdrawn, it is not entirely outdated; for instance it describes precautions to be taken in applying a seal on new wood block or mosaic flooring. The reason for this cautious approach is that the seal can penetrate between blocks and bind them together so that the effects of any shrinkage are concentrated at a few large gaps.

It was clear that the blocks had been laid at a moisture content in excess of those produced by relative humidities found in service. This defect need not have occurred if the provisions of BS CP 209:Part 1 (described in the section on maintenance) had been followed. The code also recommends that either the floor should not be sealed for the first six to twelve months to allow movement in the flooring while the moisture content of the blocks reaches equilibrium, or that the blocks are pretreated before sealing so that they do not stick together before the moisture content reaches equilibrium. If there is any possibility that the building will be heated later to higher temperatures, pretreatment rather than delay in sealing is the better preventive measure.

All kinds of seals can produce this kind of failure, but it is much more common with plastics seals.

It was calculated that the gaps could have been produced by a change in moisture content of as little as 2%.

Reinstatement of the floor to an acceptable appearance would be difficult short of complete re-laying. If re-laying were to be considered impractical, then consideration might be given to replacing some of the blocks adjoining the largest gaps with blocks slightly oversize. If the latter course were to be adopted, the blocks should be at their maximum size at the time of laying if subsequent arching were to be prevented.

burned by cigarettes, normally any scorching can be removed with care using a fine abrasive.

Suitability for underfloor heating
See the same section in Chapter 8.1.

Sound insulation
Sound absorption is poor. Floating floors consisting of timber blocks laid over a boarded substrate on a resilient quilt on a concrete or timber subfloor will give improved impact sound insulation compared with finishes laid directly on a deck or screed. Performance in respect of airborne sound insulation depends on the overall thickness and mass of the floor.

Durability
Life expectancy will be very long for a good quality flooring in wood block, provided design detailing and maintenance has been appropriate, and surfaces kept dry and only rarely sanded.

End grain wood block floors at least 25 mm thick, but sometimes up to 75 mm, have been much used in the past for those floor finishes, primarily in industrial areas, which are liable to suffer very high wear rates. They are also to be found on staircases in older public buildings. Timber species suitable for various intensities of traffic are still available. Hardwood is sometimes used in industrial flooring.

Maintenance
Seals should not be applied to wood blocks until the flooring has reached moisture equilibrium with the ambient conditions. This may mean a delay of three to six months before sealing. If some form of seal to the surface is deemed necessary before that time, it is possible to use a special primer or a thin liquid wax before the blocks are sanded to prevent the seal from penetrating into the gaps between the blocks (BS CP 209:Part 1).

If rafting has occurred, this can normally be repaired by removing some blocks and replacing with slightly oversize blocks. Alternatively, a filler of matching colour may disguise the worst effects.

Wood block floors generally are easy to repair, though they may be subject to seasonal movements and open joints collect dirt which is difficult to remove. Where it is necessary to patch re-lay old blocks that have become loose, all the old adhesive must be cleaned off to avoid the risk of incompatibility with a newer adhesive.

Resanding and resealing will probably be necessary after about ten years, depending on the density of traffic. It will be self-evident that resanding cannot, however, be continued indefinitely, since perhaps up to 2–3 mm of surface is removed each time.

Care should be taken during and after sanding wood floors to avoid combustion of bagged dust or explosion of dust-laden atmospheres.

Work on site

Storage and handling of materials
See the same section in Chapter 8.1.

Restrictions due to weather conditions
Block floors should be allowed to reach equilibrium moisture content before sealing. Otherwise, differential width joints may open.

Workmanship
See the same section in Chapter 8.1.

Inspection

In addition to the items listed in the inspection section at the end of Chapter 8.1, the problems to look for are:
- ◊ rafting
- ◊ blocks loose
- ◊ pattern faults
- ◊ colour mismatching
- ◊ sanding marks not removed
- ◊ arching and disruption

Chapter 8.4 Parquet and mosaic

Parquet has occasionally been denigrated as the poor relation of wood blocks, but this totally ignores the increased flexibility in appearance afforded by the smaller units, and the truly remarkable and intricate inlaid patterns achieved by craftsmen of years gone by who were masters of their trade.

There is often inconsistency in the terms used to describe these parquet and mosaic floorings, but those used in BS 8201[239] have been adopted for this book.

Characteristic details

Basic structure

Parquet often has the appearance of wood blocks but the material is much thinner, down to a few millimetres, and is normally found glued and pinned to a boarded subfloor. Mosaic is simply the very small sized units of similar thickness to parquet, but prefabricated and stuck to a backing for laying as a tiled finish.

Main performance requirements and defects

Appearance and reflectivity

Many parquet floors were laid in a mixture of woods to give variety in appearance. Reflectivity depends both on species and maintenance regimes.

See also the same section in Chapter 8.1.

Choice of materials for substrate

Parquet and mosaic floorings are normally laid on a suspended base of screeded concrete or on a timber or board substrate (eg plywood). It is important that any screed is level and dry, and an effective DPM is essential. However, many old floors will be found laid on timber substrates or timber joisted floors and movement in these materials could easily disrupt the thin parquet.

Strength and stability

See Chapter 1.1 for a general discussion.

Dimensional stability, deflections etc

Mosaic and parquet floorings behave similarly to wood blocks as regards movements and rafting (Figure 8.9).

See also the same section in Chapters 8.1 and 8.3.

Wear resistance

Resistance to wear depends entirely on the timber species and its thickness. There have been a number of panel products available which comprise a thin parquet veneer – much thinner than normal parquet – as a wearing surface on a backing of ply. These products have not survived long in heavy traffic conditions.

See also the same section in Chapter 8.1.

Slip resistance

Slip resistance can be very poor if surfaces are wax polished. On stair treads a slip resistant nosing is necessary.

Control of dampness and condensation

See the same section in Chapter 8.1.

Figure 8.9
Rafting of finger mosaic parquet flooring

Case study

Disruption of a felt backed mosaic flooring in a lecture theatre

A felt backed hardwood mosaic floor in a lecture theatre had started lifting. The flooring had been laid over a screed in the winter months when the building heating was on. Some two months after the heating had been switched off for the summer season, the floor began to lift.

Tests were carried out using hygrometers sealed to the surface of the screed which showed that construction moisture remaining in the slab and screed before laying the floor had contributed little to the problem. The flooring in the affected areas was eased but the problem recurred. It was decided then to seek BRE's advice.

When outside temperatures are low, the humidity inside heated buildings is also low. During the summer months, when buildings tend not to be heated, the internal relative humidity is usually found to be high. It is therefore not unusual to find that wood floors laid during the winter months expand during the summer, and, if insufficient joints or gaps have been allowed for during laying, this expansion results in lifting. Conversely, floors laid in summer often show gaps between the blocks or panels during the winter.

BRE advised that the problem could be cured by providing wider expansion joints, and filling them with cork.

Thermal properties

Thermal conductivity of hardwoods: 0.15 W/mK.

Thermal conductivity of softwoods: 0.13 W/mK.

Warmth to touch

Good.

Fire

Combustible. Although the material can be burned by cigarettes, normally any scorching can be removed with care using a fine abrasive.

Suitability for underfloor heating

See the same section in Chapter 8.1.

Sound insulation

Sound absorption is poor. Floating floors consisting of parquet or mosaic laid over a boarded substrate over a resilient quilt on a concrete or timber subfloor will give improved impact sound insulation compared with finishes laid directly on a deck or screed. Performance in respect of airborne sound insulation depends on the overall thickness and mass of the floor.

Durability

See the same section in Chapter 8.1.

Maintenance

Maintenance is the same as for other kinds of wood, but it should be remembered that parquet and mosaic is much thinner than strip and block, and does not provide sufficient depth to permit more than the minimum of sanding.

Care should be taken during and after sanding wood floors to avoid combustion of bagged dust or explosion of dust-laden atmospheres.

Work on site

Storage and handling of materials

See the same section in Chapter 8.1.

Restrictions due to weather conditions

See the same section in Chapter 8.1.

Workmanship

See the same section in Chapter 8.1.

Inspection

In addition to the items listed in the inspection section at the end of Chapter 8.1, the problems to look for are:

- ◊ rafting
- ◊ pattern faults
- ◊ colour mismatching
- ◊ squeaking between units of flooring
- ◊ sanding marks not removed
- ◊ arching and disruption

Chapter 8.5 Panel products

Chipboard, plywood, and other wood based panel products are not normally specified as finished floorings; nevertheless they are often used in their unprotected state as deckings for suspended timber floors and for raised access floors, and as overlays and substrates for covering with other floorings after completion of the building contract. They can be subjected therefore to wear and tear during the completion of building operations. Occasionally hardboard or chipboard may be used as a final wearing surface for very lightly trafficked areas where appearance is not at a premium.

Characteristic details

Basic structure

Table 8.3 shows European standards† expected to apply to panel products, and the equivalent British Standards.

BRE Digest 323[247] is relevant.

The wood content of most of these materials is relatively high (eg 87% for chipboards). They do not, however, behave similarly to products entirely of natural wood (eg as to dimensional stability) and must be detailed accordingly.

† It is intended that European plywood grades will be introduced from July 1997, and those for chipboard, OSB and cement bonded particleboard from January 1998.

Main performance requirements and defects

Appearance and reflectivity

See the same section in Chapter 8.1.

Choice of materials for substrate

Plywood to BS 6566:Part 7[248] and Part 8[249]

For domestic floors:

- MR bond or better and G durability or better for dry use
- CBR bond or better and M or better for risk of wetting
- Structural grade plywood must be used as listed in BS 5268: Part 2[119]

Table 8.3
European standards (EN) and their British Standard equivalents

EN 300	Oriented strand boards (OSBs) – definitions, classifications and specifications		BS 5669:Part 3[244]
EN 312	Particleboards – specifications		BS 5669:Part 2[180]
	Part 1	General requirements for all board types	–
	Part 2	Requirements for general purpose boards for use in dry conditions (P2 boards)	C1
	Part 3	Requirements for boards for use in interior fitments (including furniture) in dry conditions	C1A
	Part 4	Requirements for loadbearing boards for use in dry conditions (P4 boards)	C2
	Part 5	Requirements for loadbearing boards for use in humid conditions (P5 boards)	C4(M)
	Part 6	Requirements for heavy duty loadbearing boards for use in dry conditions (P6 boards)	–
	Part 7	Requirements for heavy duty loadbearing boards for use in humid conditions (P7 boards)	C5
EN 622	Fibreboards – specifications		BS 1142[245]
	Part 1	General requirements	
	Part 2	Requirements for hardboards	
	Part 3	Requirements for medium boards	
	Part 4	Requirements for softboards	
	Part 5	Requirements for dry process boards (MDF)	
EN 634	Cement bonded particleboards – specifications		BS 5669:Part 4[246]
	Part 1	General requirements	
	Part 2	Requirements for OPC bonded particleboards for use in dry, humid and exterior conditions (Type T2 boards)	
EN 636	Plywood – specifications		
	Part 1	Plywood for use in dry conditions (for flooring)	
	Part 2	Plywood for use in humid conditions (for flooring)	
	Part 3	Plywood for use in exterior conditions	

For non-domestic floors:

- MR bond or better and G durability or better for dry use
- WBP bond or better and M or better for risk of wetting
- Structural grade plywood must be used as listed in BS 5268:Part 2. See also BRE Digest 394[250]

Chipboard to BS 5669:Part 5[251]
For all floors as decks or overlays, C4(M) and C5 in domestic situations.

C2 was widely used in domestic floorings prior to the 1990s, but it was suitable only for dry conditions.

BS 5669:Part 5 provides information on panel thicknesses required, and on methods of laying and fixing.

Figure 8.10
This polyethylene sheeting has failed to protect the chipboard from saturation

Oriented strand board (OSB)
British Board of Agrément certificates apply.

BS 5669:Part 3 refers to the specification for the material and for the methods of laying and fixing. BRE Digest 400[252] also applies.

OPC bonded particleboard
This can be used as decking in domestic situations and as overlays in non-domestic situations; also for floating or raised access floors.

BS 5669:Part 5 applies. See also BRE Information Paper IP 14/92[253].

Figure 8.11
The results of wetting are evident long after the chipboard flooring has dried out

Fibre building board to BS 1142
This building board can only be used as an overlay. S grade should not be used where there is risk of wetting.

BS 5669:Part 5 applies. See also BRE Information Paper IP 12/91[254].

BS 8203[35] applies to all panel products where used as an underlay for sheet and tile floorings.

Strength and stability
Chipboard, OSB, cement bonded particleboard and fibre building boards can be used as decks in suspended floors provided they satisfy the loading conditions.

Chipboard decking
Consumption of chipboard in the UK peaked in 1988 at 3.28 million cubic metres and in 1990 was still 2.75 million cubic metres. About 40% of the UK consumption goes directly into the construction industry where it is widely used as floor decking. Currently there are six grades, each developed with a specific range of properties that best suit it for a particular application; three grades are normally used in flooring. It is vital that the correct grade is specified for a particular use (BS 5669:Part 5).

Problems have occasionally been observed where the incorrect thickness of chipboard has been used. For domestic floor loadings the thickness should be:

- 18 mm for joist spacings up to 450 mm
- 22 mm for joist spacings between 451 mm and 600 mm

For increased floor loadings with standard chipboards, the maximum spans will be smaller; for example for 2 kN/m^2 it will be 450 mm. Alternatively, non-British Standard special grades may be available which give increased spans.

Chipboard to C4(M) retains a high proportion of its initial strength after wetting. This property is due to the proportion of melamine in the composition. However, many situations will still exist where C2 board has been used and, for this material, the proportion of strength recovered after wetting is substantially lower. Where there is a risk of significant wetting, it is essential that the correct grade is used (Figures 8.10 and 8.11). Requirements for the protection of chipboard floors in potentially wet areas are given in BRE Defect Action Sheet 31[255]. Further information is given in BRE Information Paper IP 3/85[256], in BRE Digest 323 and in BS 5669: Part 2.

Dimensional stability, deflections etc

Reversible moisture movement of chipboard depends on relative humidity. The moisture content of boards direct from manufacture is around 6 to 9%, so expansion of boards when installed in the wetter conditions of sites will be a major problem.

Where tongued-and-grooved (T&G) chipboards are continuously supported by the sub-base or underlay of insulation, it is essential that all the T&G joints are glued continuously – spot glueing is insufficient. In BRE site surveys, many examples were found where there had been no glueing. It is now recommended for all situations.

Where decks such as chipboard are laid over compressible insulation at thresholds and other heavily loaded positions, the board should be fixed to battens supported by the subfloor (Figure 8.12).

Like all products based on timber, changes in the moisture condition of chipboard and OSB will produce changes in the dimensions. For example, a change in moisture content from 9 to 16% produces an expansion in the order of 7 mm in a typical room dimension of 3 m. It is essential to provide a gap at the perimeter of a floor to accommodate any likely moisture induced movement. For houses with rooms of normal size, a gap of not less than 10 mm is adequate (Figure 8.13, and Figure 8.14 on page 280).

Cement bonded particleboards have smaller movements in service than other particleboards, but a small allowance of 0.05 mm per metre run should still be made.

If chipboard, or insulation and chipboard, is to be laid on to a concrete base it is essential that the surface of the base is flat. The floor may oscillate up and down under dynamic loads (eg people walking across the floor) if boards are laid on an undulating base. A surface regularity of not more than 5 mm under a 3 m straight edge is acceptable. Bases which do not meet this standard can be levelled by applying a thin layer of sand and cement mortar which should be allowed to set, harden and dry out before proceeding with laying boards. Dry sand is not suitable for levelling bases as it can move around under the effects of pumping action induced by dynamic loads.

See also the same section in Chapter 8.1.

Wear resistance

When used as the wearing surface for light traffic, the abrasion resistance can be marginally improved by applying a polyurethane seal.

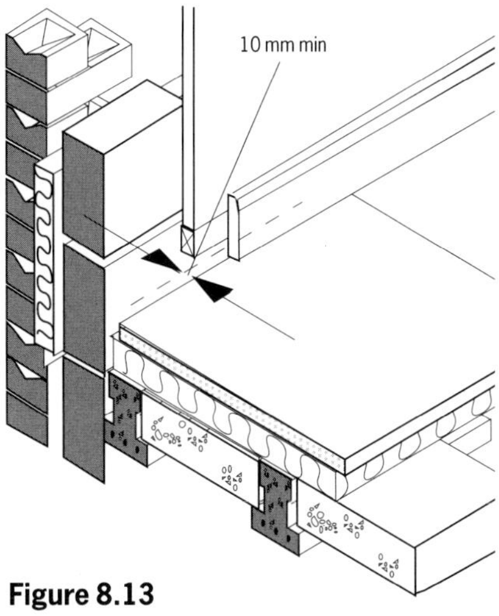

Figure 8.13
Edge detail of chipboard flooring

Figure 8.12
This chipboard has been laid on strips of polyurethane foam of insufficient loadbearing capacity, leading to collapse

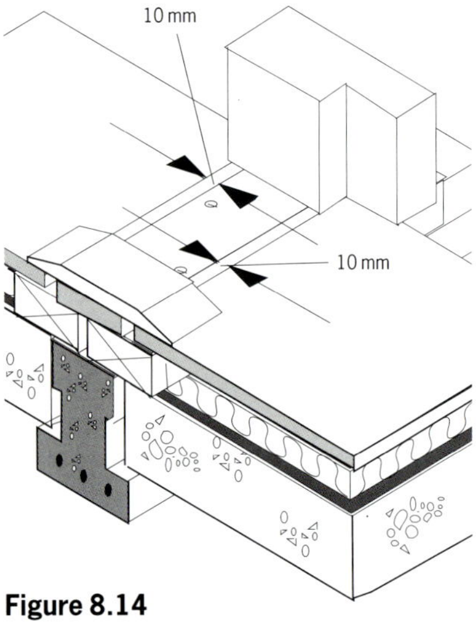

Figure 8.14
Extra support of the decking is needed at door thresholds

Case study

Cupping of chipboard in domestic floating floors

Early in 1993, BRE received a number of reports that chipboard in floating floors was cupping soon after installation even though it had been laid to agreed standards. The number of cases of distortion was relatively small compared with the total number of installations. The amount of cupping varied between 1 and 4 mm under a straight edge laid across a 600 mm wide panel; distortion of this magnitude is easily seen when the floor is covered in flexible PVC.

At first it was thought that the origins of the cupping were the combined effect of the impervious covering and heating, but the phenomenon has also been seen in three month old floors without coverings or heating. Subsequently the distortion was believed to be due to a moisture imbalance within the chipboard, but the precise conditions which cause such an imbalance are not known with any certainty.

Slip resistance

Slip resistance can be very poor if boards and panels are wax polished.

Control of dampness and condensation

See the section on dimensional stability earlier in the chapter.

Thermal properties

Thermal conductivity of hardwoods: 0.15 W/mK.

Thermal conductivity of softwoods: 0.13 W/mK.

Warmth to touch

Good.

Fire

Combustible. Although boards and panels can be burned by cigarettes, normally any scorch marks can be removed (to a greater or lesser degree) using a fine abrasive.

Suitability for underfloor heating

See the same section in Chapter 8.1.

Sound insulation

Floating floors consisting of panel products laid over a resilient quilt on the concrete or timber subfloor will give improved impact sound insulation compared with finishes laid directly on the deck or screed. The performance of insulation against airborne sound depends on the overall thickness and mass of the floor. Glueing of particleboards is important for avoiding squeaking caused by minute movements between adjoining boards.

Cement bonded particleboard of 18 mm thickness will offer, by itself, an airborne sound reduction of 33 dB.

Durability

Provided they are kept dry, particleboards should give a reasonable life as underlays.

Chipboard should normally be Type C4(M) or C5 to BS 5669:Part 2. Where chipboard is used in kitchens and bathrooms, it should be covered with a sheet flooring turned up at the perimeter and welded at the joints.

Work on site

Storage and handling of materials

Stacking boards and panels on level surfaces with appropriately spaced battens is important to avoid distortion. Boards should not be stored on edge. The height of stacked boards should be also considered in relation to the extra loadings they impose on floors.

The crucial requirements of site practice, so far as chipboard is concerned, are to ensure that boards are kept dry while in storage and during the construction process; that the correct grade of chipboard to BS 5669:Part 2 is used; and that any specified tanking, designed to protect chipboard floors in wet areas, is installed so as to provide complete protection against water spillages.

Restrictions due to weather conditions

See the same section in Chapter 8.1.

Workmanship

BS 8203 applies where these products are used as underlays. However, BS 5385:Part 3[226] covers

Case study

Fungal attack in chipboard decking under tiles

Fungal decay of chipboard flooring in a number of 10-years-old, two storey houses located on a sloping site was investigated by BRE Advisory Service. The construction of the houses comprised external loadbearing cavity brick walls, rendered externally and plastered directly internally. Floors throughout were of timber suspended construction with a vinyl tile finish.

The solums of the dwellings were below the level of adjacent ground and water tended to migrate into the underbuildings. The vinyl floor tiles provided an almost impermeable membrane and, due to the omission of damp resisting treatment to the solum and walls, moisture evaporating from the ground was being contained within the underbuilding. Ventilation had been inadequate to remove this moisture which had been absorbed by the timbers and chipboard resulting in fungal attack and loss of strength. Correction of the design errors and replacement of the floors were advised.

special provisions for fixing plywoods used as underlays for ceramic tiles. Particular attention should be paid to the following measures.

Chipboard should be conditioned by storage in the intended area of use for 24 hours prior to fixing in order to minimise subsequent movement. Boards should be installed at the following moisture contents:

- continuously heated buildings 7–9%
- intermittently heated buildings 9–12%
- unheated new buildings up to 15%

Ringshanked (or annular grooved) nails should be used for fixing particleboards in preference to plain shanked, and all cement bonded particleboards over 12 mm in thickness should be pre-drilled before nailing and punching at not more than 300 mm centres. Screws are an alternative and should be countersunk. It is not usual to pre-drill OSB and chipboard.

Preferably boards should be laid to break joint and all edges of square edged boards should be nogged.

All joints should be continuously glued to prevent squeaking in service. A PVA glue to BS EN 204[257] is satisfactory.

Where water services are enclosed in floors having chipboard decks, the Water Byelaws require properly formed openings with removable covers above every pipe at a change of direction to permit repair or removal of the pipe. This often means nogging the floor to accommodate traps of around 200 to 300 mm in each plan dimension.

Hardboard should be conditioned before use, either by storing in ambient conditions where the boards are of 4.8 or 6.4 mm thickness or where they are to be used in heated buildings, or by wetting where used elsewhere. Recommendations are contained in Clause 27.1.3 of BS 8201[239].

Hardboards used as overlays should be fixed to timber with staples or nails no closer than 12 mm to the edge. Nails and staples should be hammered flush to surfaces.

Inspection

In addition to the items listed in the inspection section at the end of Chapter 8.1, the problems to look for are:

- ◊ boards and panels sagging
- ◊ boards and panels buckling
- ◊ boards and panels loose or squeaking
- ◊ loss of strength and stiffness in chipboards
- ◊ chipboard flooring grades not used
- ◊ boards and panels not stored under the correct conditions
- ◊ perimeter gaps for expansion absent
- ◊ nogging, nailing and glueing of boards and panels inadequate
- ◊ moisture resistant boards and panels not used in potentially wet or humid areas
- ◊ movement joints at perimeters and all upstands not provided for
- ◊ chipboards for spans of incorrect thickness
- ◊ ringshanked nails or screws not used for fixing chipboards
- ◊ nails and staples not hammered flush
- ◊ nails and staples positioned too near to board edges

Where inspections are needed in cavities in floors, yet there is no possibility of opening up the construction, using a borescope (Figure 8.15) may provide a limited amount of information. The field of view is restricted and interpretation is sometimes difficult if the operator is unfamiliar with the type of construction under examination (Figure 8.16).

Figure 8.15
Using a borescope to examine the cavity under a chipboard floor

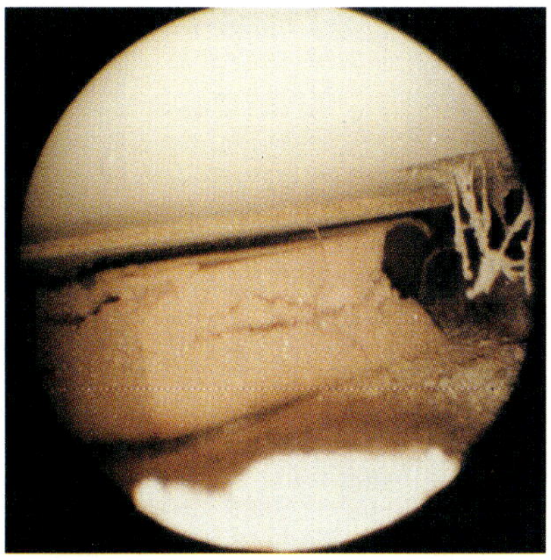

Figure 8.16
The view through the borescope of the construction in Figure 8.15, showing polyurethane strip partially crushed and disintegrating into a powder just visible at lower right. The joint between the boards was neither tongued-and-grooved nor glued. The board furthest from the camera has only partially recovered, leaving both a gap between it and the bearer, and a lip with the adjacent board. The cobweb to the right is collecting polyurethane powder

Alternatively the boards may be glued – for example with an SBR (styrene butadiene) emulsion.

Appendix A How to identify less recognisable floorings and their substrates

Most of the main characteristics of floorings are sufficiently distinctive to make identification easy. However there are a number of opportunities for confusion and the following notes may assist users.

There is often little problem with floorings: the problems arise with hidden layers beneath, including adhesives.

A relatively intractable problem is identification of whether or not a ground-bearing floor has a DPM. Apart from availability of the original drawings or specification, often the only sure way of finding out is to take a core from a relatively unimportant and unobtrusive area such as the floor within a cupboard. It is no use taking a moisture reading, for the moisture may come from sources other than rising damp such as cleaning water or condensation on uninsulated floors.

Magnesite can provide a source of confusion, for the range of fillers and pigments used can be quite different in appearance. Although magnesite is electrically conducting, a method of identification using electrical conductivity is unreliable since damp cement screeds can give similar readings. Magnesite is commonly pigmented brick red, sometimes straw yellow, and even occasionally marbled. It is always laid in situ and will usually contain sawdust. A laboratory test will always show large quantities of magnesium.

Another source of confusion is where a thin dense topping of sand and cement has been used over a lightweight screed. A blow with a sharp instrument will be needed to break through any topping.

The best way of identifying asphalt and pitchmastic is by smell when the materials are molten; laboratory testing is not straightforward and can be expensive. The identification process involves first experiencing the different smells given off by pitchmastic (eg when coal tar is sprayed on a road before laying tarmac) and asphalt (eg when bitumen is being applied to flat roofs). Then, at some later date when a problem of identification arises, burning a small piece of the material should enable it to be easily recognised. Burning these materials must not, of course, be done in situ, nor near the same or other materials which might be flammable.

Where identification of a species of wood proves difficult, a sample can be identified by BRE using the national collection.

Appendix B How to choose a flooring

The prime purpose of this book is to assist surveyors and others to assess and maintain existing floorings. However, it should be possible for specifiers to prepare, from the information given, an outline performance specification for floorings for particular applications. Naturally, however, the book cannot cover all possible applications in a building, and some tailoring or augmentation will be necessary for particular situations. A series of steps is involved in this process.

1 The first task of the specifier is precisely to identify the separate areas of the building having specific performance requirements for flooring. Particular attention should be paid to substrates and whether they may be subjected to deflection or movement, the relative traffic densities, the risk of and effects of impacts, and the existing and potential ambient conditions in each area of the building; also how much thickness can be allowed for the finish.

2 Next, the specifier should decide what categories to use in the performance specification. The main subject areas, roughly conforming to the functional headings used in this book, are:

- strength and stability
- dimensional stability
- energy conservation, thermal insulation, air penetration and ventilation
- control of dampness and condensation
- comfort and safety
- fire and resistance to high temperatures
- appearance and reflectivity
- sound insulation and quietness
- durability
- inspection and maintenance

3 The specifier will need to set the criteria of acceptable performance for each of the chosen categories (and component sub-categories). Some of the categories are easier to assess than others. For example, resistance to fire and the ability to control fire spread, slip resistance, surface hardness, impact resistance and thermal characteristics will be easier to deal with than colour and pattern, general appearance and quietness.

4 A life cycle cost plan should be established which takes into account initial cost, maintenance requirements, ease of repair and ease of removal for replacement. Warranties too are relevant.

5 The relative importance of the above list of functions must be decided with the client. Often it will be found that there are significant determinants of choice, especially with regard to appearance and hygiene.

6 Some assessment must be made of acceptable limits for site processes. For example, how long can laying moisture susceptible flooring be delayed waiting for screeds to dry?

7 Assembling and tabulating the information for comparison of the competing floorings with the performance requirements is the final step. The assessment might be made easier if the process included some system of scoring the performance attributes. With traditional solutions the risk of misrepresentation of likely performance is low, but newer solutions may justify third party certification. Part of the decision making process should include inspection of how similar materials have performed in similar situations.

At the end of the day specifiers and clients may have to accept that an entirely objective choice is not possible or that compromise will be required.

References and further or general reading

Each numbered reference below is shown only under the chapter in which it first appears in the text.

Chapter 0

[1] Harrison H W. Quality in new-build housing. *Building Research Establishment Information Paper* IP 3/93. Garston, Construction Research Communications Ltd, 1993

[2] Department of the Environment. *English house condition survey 1991*. London, The Stationery Office, 1993

[3] Welsh Office. *Welsh house condition survey 1993*. Cardiff, Welsh Office, 1993

[4] Scottish Homes. *Scottish house condition survey 1991. Survey report.* Edinburgh, Scottish Homes, 1993

[5] Northern Ireland Housing Executive. *Northern Ireland house condition survey 1991. First report of survey*. Belfast, Northern Ireland Housing Executive, 1993

[6] Building Research Station. *Principles of modern building*, Volume 2: Floors and roofs. London, The Stationery Office, 1961

[7] Building Research Establishment. *Assessing traditional housing for rehabilitation*. BRE Report. Garston, Construction Research Communications Ltd, 1990

[8] Building Research Establishment. *Surveyor's checklist for rehabilitation of traditional housing*. BRE Report. Garston, Construction Research Communications Ltd, 1990

[9] Building Research Establishment. *Quality in traditional housing*, Volume 2: An aid to design. BRE Report. London, The Stationery Office, 1982

[10] Building Research Establishment. *Quality in traditional housing*, Volume 3: An aid to site inspection. BRE Report. London, The Stationery Office, 1982

[11] Nixon P J. Floor heave in buildings due to the use of pyritic shales as fill material. *Chemistry and Industry*, 4 March 1978, 160–4

Further reading

Building Research Establishment. Domestic floors. *BRE Good Building Guide* GBG 28, Parts 1–5. Garston, Construction Research Communications Ltd, 1997

Chapter 1.1

[12] Pippard A J S and Chitty L. A study of the voussoir arch. *Building Research Station National Building Studies Research Paper* No 11. London, The Stationery Office, 1951

[13] Department of the Environment and The Welsh Office. *The Building Regulations 1991*. Statutory Instrument 1991 No 2768. London, The Stationery Office, 1991

[14] *The Building Standards (Scotland) Regulations 1990*. Statutory Instrument 1990 No 2179 (S 187). Edinburgh, The Stationery Office, 1990

[15] *The Building Regulations (Northern Ireland) 1994*. Statutory Rules of Northern Ireland 1994 No 243. Belfast, The Stationery Office, 1994

[16] British Standards Institution. Structural use of steelwork in building. *British Standard* BS 5950:Parts 1–9:1987–94. London, BSI, 1987–94

[17] British Standards Institution. Structural use of concrete. *British Standard* BS 8110:Parts 1–3:1985. London, BSI, 1985

[18] British Standards Institution. Structural use of timber. *British Standard* BS 5268:various Parts:1978–91. London, BSI, 1978–91

[19] British Standards Institution. Loading for buildings. Code of practice for dead and imposed loads. *British Standard* BS 6399: Part 1:1984. London, BSI, 1984

[20] British Standards Institution. Schedule of weights of building materials. *British Standard* BS 648:1964. London, BSI, 1964

[21] Armitage J S and Judge C J. *Floor loading in warehouses: a review*. Building Research Establishment Report. Garston, Construction Research Communications Ltd, 1987

[22] Building Research Establishment. Temporary support: assessing loads above openings in external walls. *BRE Good Building Guide* GBG 10. Garston, Construction Research Communications Ltd, 1992

[23] British Standards Institution. Structural design of low-rise buildings. Code of practice for stability, site investigation, foundations and ground floor slabs for housing. *British Standard* BS 8103:Part 1: 1995. London, BSI, 1995

[24] Building Research Establishment. Static load testing: concrete floor and roof structures within buildings. *BRE Digest* 402. Garston, Construction Research Communications Ltd, 1995

[25] European Union of Agrément (UEAtc). Directive for the Assessment of Floorings. *Method of Assessment and Test* No 2. Paris, UEAtc, 1970

Chapter 1.2

[26] Building Research Establishment. Estimation of thermal and moisture movements and stresses: Parts 1–3. *BRE Digests* 227–229. Garston, Construction Research Communications Ltd, 1979

[27] Ellis B R and Ji T. Floor vibration. Floor vibration induced by dance-type loads: verification. *The Structural Engineer* (February 1994), **72** (3) 45–50

[28] Ji T and Ellis B R. Floor vibration. Floor vibration induced by dance-type loads: theory. *The Structural Engineer* (February 1994), **72** (3) 37–44

[29] Jeary A P. Determining the probable dynamic response of suspended floors. *Building Research Establishment Information Paper* IP 17/83. Garston, Construction Research Communications Ltd, 1983

Further reading

Building Research Establishment. Why do buildings crack? *BRE Digest* 361. Garston, Construction Research Communications Ltd, 1991

Warlow W J and Pye P W. The rippling of thin flooring over discontinuities in screeds. *Modern Plastering* (Autumn 1974), **31**

Chapter 1.3

[30] Anderson B R. The U-value of ground floors: application to building regulations. *Building Research Establishment Information Paper* IP 3/90. Garston, Construction Research Communications Ltd, 1990

[31] Building Research Establishment. *Thermal insulation: avoiding risks.* BRE Report. Garston, Construction Research Communications Ltd, 1994

[32] Department of the Environment and The Welsh Office. *The Building Regulations 1991 Approved Document L: Conservation of fuel and power.* London, The Stationery Office, 1995

[33] Department of the Environment for Northern Ireland. *The Building Regulations (Northern Ireland) 1990 Technical Booklet F: Conservation of fuel and power.* London, The Stationery Office, 1990

[34] Anderson B R. U-values for basements. *Building Research Establishment Information Paper* IP 14/94. Garston, Construction Research Communications Ltd, 1994

Further reading

Building Research Establishment. Standard U-values. *BRE Digest* 108. Garston, Construction Research Communications Ltd, 1975 (revised 1991)

Building Research Establishment. Heat losses through ground floors. *BRE Digest* 145. Garston, Construction Research Communications Ltd, 1972

Chapter 1.4

[35] British Standards Institution. Code of practice for installation of sheet and tile flooring. *British Standard* BS 8203:1987. London, BSI, 1987

[36] Building Research Establishment. Drying out buildings. *BRE Digest* 163. Garston, Construction Research Communications Ltd, 1974

[37] Building Research Establishment. Design of timber floors to prevent decay. *BRE Digest* 364. Garston, Construction Research Communications Ltd, 1991

[38] Building Research Establishment. Damp-proofing solid floors. *BRE Digest* 54. Garston, Construction Research Communications Ltd, 1971

[39] British Standards Institution. Specification for mastic asphalt for building and civil engineering (limestone aggregate). *British Standard* BS 6925:1988. London, BSI, 1988

[40] British Standards Institution. Code of practice for protection of buildings against water from the ground. *British Standard* BS CP 102:1973. London, BSI, 1973

[41] British Standards Institution. Specification for materials for damp-proof courses. *British Standard* BS 743:1970. London, BSI, 1970

[42] Department of the Environment and The Welsh Office. *The Building Regulations 1991 Approved Document C: Site preparation and resistance to moisture.* London, The Stationery Office, 1992

Further reading

Building Research Establishment. Domestic floors: construction, insulation and damp-proofing. *BRE Good Building Guide* GBG 28, Part 1. Garston, Construction Research Communications Ltd, 1997

Building Research Establishment. Domestic floors: repairing or replacing floors and flooring – magnesite, tiles, slabs and screeds. *BRE Good Building Guide* GBG 28, Part 4. Garston, Construction Research Communications Ltd, 1997

Chapter 1.5

[43] Cox S J and O'Sullivan E F. *Building regulation and safety.* Building Research Establishment Report. Garston, Construction Research Communications Ltd, 1994

[44] Health and Safety at Work etc Act 1974. Chapter 37. London, The Stationery Office, 1974

[45] Construction (Working Places) Regulations. London, The Stationery Office, 1966

[46] Construction (Lifting Operations) Regulations. London, The Stationery Office, 1961

[47] Construction (General Provisions) Regulations. London, The Stationery Office, 1961

[48] Davis K and Tomasin K. *Construction safety handbook.* London, Thomas Telford Services, 1990

[49] British Standards Institution. Stairs, ladders and walkways. *British Standard* BS 5395:Parts 1–3:1977–85. London, BSI, 1977–85

[50] Harper F C, Warlow W J and Clark B L. The forces applied to the floor by the foot in walking. *Building Research Station National Buildings Study Research Paper* No 32. London, The Stationery Office, 1967

[51] Pye P W. *A brief review of the historical contribution made by BRE to slip research. Slipping – towards safer flooring.* Shrewsbury, RAPRA Technology Ltd, 1994

[52] Road Research Laboratory. Instructions for using the portable skid-resistance tester (2nd edition). *RRL Road Note* 27. London, The Stationery Office, 1969

[53] Greater London Council. Slip resistance of floors, stairs and pavings. *GLC Bulletin* No 43 (2nd Series), Item No 5. London, GLC, 1971

[54] British Standards Institution. Screeds, bases and in-situ floorings. Code of practice for polymer modified cementitious wearing surfaces. *British Standard* BS 8204:Part 3: 1993. London, BSI, 1993

[55] British Standards Institution. Screeds, bases and in-situ floorings. Code of practice for terrazzo wearing surfaces. *British Standard* BS 8204:Part 4: 1993. London, BSI, 1993

[56] British Standards Institution. Screeds, bases and in-situ floorings. Code of practice for mastic asphalt underlays and wearing surfaces. *British Standard* BS 8204: Part 5: 1994. London, BSI, 1994

[57] British Standards Institution. Artificial sports surfaces. Methods of test. Methods for determination of person/surface interaction. *British Standard* BS 7044:Part 2: Section 2.2:1990. London, BSI, 1990

[58] The UK Slip Resistance Group. *The measurement of floor slip resistance. Guidelines recommended by the UK Slip Resistance Group.* Shrewsbury, RAPRA Technology Ltd, 1996

[59] Department of the Environment and The Welsh Office. *The Building Regulations 1991 Approved Document B: Fire Safety.* London, The Stationery Office, 1992

[60] Building Research Establishment. *Radon: guidance on protective measures for new dwellings.* BRE Report. Garston, Construction Research Communications Ltd, 1992

[61] Building Research Establishment. *Construction of new buildings on gas-contaminated land.* BRE Report. Garston, Construction Research Communications Ltd, 1991

[62] Building Research Establishment. Minimising noise from domestic fan systems and fan-assisted radon mitigation systems. *BRE Good Building Guide* GBG 26. Garston, Construction Research Communications Ltd, 1996

[63] Commission of the European Communities, Directorate General for Science. Research and Development. Guidelines for ventilation requirements in buildings. *European Concerted Action: Indoor Air Quality and its Impact on Man Report* No 11. Luxembourg, Office for Publications of the European Communities, 1992

[64] Department of the Environment and The Welsh Office. *The Control of Substances Hazardous to Health Regulations 1988.* Statutory Instrument 1988 No 1657. London, The Stationery Office, 1988

Further reading

British Standards Institution. Code of practice for installation of textile floor coverings. *British Standard* BS 5325:1983. London, BSI, 1983

British Standards Institution. Screeds, bases and in-situ floorings. Code of practice for concrete wearing surfaces. *British Standard* BS 8204:Part 2:1987. London, BSI, 1987

Building Research Establishment. *Construction of new buildings on gas contaminated land.* BRE Report. Garston, Construction Research Communications Ltd, 1991

Department of the Environment, Her Majesty's Inspectorate of Pollution. The control of landfill gas. *Waste Management Paper* No 27 (2nd edition). London, The Stationery Office, 1991

The Concrete Society. Concrete industrial ground floors – a guide to their design and construction. *Technical Report* No 34. Slough, The Concrete Society, 1994

Department of the Environment and The Welsh Office. *The Building Regulations 1991 Approved Document K: Stairs, ramps and guards.* London, The Stationery Office, 1992

Bailey M. The Measurement of Slip Resistance of Floor Surfaces, The Tortus and Pendulum. *Construction and Building Materials* (September 1988), **2** (3)

Chapter 1.6

[65] Hamilton S B. A short history of the structural fire protection of buildings, particularly in England. *Building Research Station National Building Studies Special Report* No 27. London, The Stationery Office, 1958

[66] Harrison H W. *Roofs and roofing. Performance, diagnosis, maintenance, repair and the avoidance of defects (BRE Building Elements series).* London, Construction Research Communications Ltd, 1996

[67] London County Council. London Building Constructional By-laws. No 3790. London, LCC, 1952

[68] Morris W A, Read R E H and Cooke G M E. *Guidelines for the construction of fire-resisting structural elements.* BRE Report. Garston, Construction Research Communications Ltd, 1988

[69] Building Research Establishment. Increasing the fire resistance of existing timber floors. *BRE Digest* 208. Garston, Construction Research Communications Ltd, 1988

[70] British Standards Institution. Fire precautions in the design, construction and use of buildings. Code of practice for firefighting stairs and lifts. *British Standard* BS 5588:Part 5:1991. London, BSI, 1991

[71] British Standards Institution. Fire tests on building materials and structures. Method of determination of the fire resistance of elements of construction (general principles). *British Standard* BS 476:Part 20: 1987. London, BSI, 1987

[72] British Standards Institution. Fire tests on building materials and structures. Methods for determination of the fire resistance of loadbearing elements of construction. *British Standard* BS 476: Part 21:1987. London, BSI, 1987

[73] Read R E H and Adams F C. The role of suspended ceilings in structural fire protection. *Building Research Establishment Information Paper* IP 1/80. Garston, Construction Research Communications Ltd, 1980

[74] The Institution of Electrical Engineers. *Regulations for electrical installations.* Stevenage, IEC, 1995

[75] Fire Precautions Act 1971. London, The Stationery Office, 1971

[76] Loss Prevention Council. *Code of practice for the construction of buildings. Insurers' rules for the fire protection of industrial and commercial buildings.* London, LPC, 1992

[77] Home Office and The Scottish Office. *Fire Precautions Act 1971. Guide to fire precautions in premises used as hotels and boarding houses which require a fire certificate.* London, The Stationery Office, 1991

[78] Home Office and Scottish Home and Health Department. *Guide to fire precautions in existing places of entertainment and like premises.* London, The Stationery Office, 1990

[79] British Standards Institution. Specification for assessment and labelling of textile floor coverings tested to BS 4790. *British Standard* BS 5287:1988. London, BSI, 1988

[80] British Standards Institution. Method for determination of the effects of a small source of ignition on textile floor coverings (hot metal nut method). *British Standard* BS 4790:1987. London, BSI, 1987

Further reading

British Standards Institution. Fire tests on building materials and structures. Methods for determination of the fire resistance of non-loadbearing elements of construction. *British Standard* BS 476:Part 22:1987. London, BSI, 1987

British Standards Institution. Fire tests on building materials and structures. Methods for determination of the contribution of components to the fire resistance of a structure. *British Standard* BS 476:Part 23: 1987. London, BSI, 1987

Chapter 1.7

[81] Barker P, Berrick J and Wilson R. *Building sight.* London, Royal National Institute for the Blind, 1995

[82] Chartered Institution of Building Services Engineers. *Code for interior lighting.* London, CIBSE, 1994

[83] Building Research Establishment. Estimating daylight in buildings: part 2. *BRE Digest* 310. Garston, Construction Research Communications Ltd, 1986

Chapter 1.8

[84] Building Research Establishment. Sound insulation of separating walls and floors. Part 2: floors. *BRE Digest* 334. Garston, Construction Research Communications Ltd, 1988

[85] Building Research Establishment. Improving the sound insulation of separating walls and floors. *BRE Digest* 293. Garston, Construction Research Communications Ltd, 1985

[86] Department of the Environment and The Welsh Office. *The Building Regulations 1991 Approved Document E: Resistance to the passage of sound.* London, The Stationery Office, 1992

[87] Department of the Environment for Northern Ireland. *The Building Regulations (Northern Ireland) 1990 Technical Booklet G: Sound.* London, The Stationery Office, 1990

[88] Department of the Environment for Northern Ireland. *The Building Regulations (Northern Ireland) 1994 Technical Booklet G1: Sound (Conversions).* London, The Stationery Office, 1994

[89] British Standards Institution. Measurement of sound insulation in buildings and of building elements. Field measurements of impact sound insulation of floors. *British Standard* BS 2750:Part 7:1980. London, BSI, 1980

[90] British Standards Institution. Methods for rating the sound insulation in buildings and of building elements. Method for rating the impact sound insulation. *British Standard* BS 5821:Part 2:1984. London, BSI, 1984

[91] Parkin P H, Purkis H J and Scholes W E. Field measurements of sound insulation between dwellings. *National Building Studies Research Paper* No 33. London, The Stationery Office, 1960

[92] Department of the Environment. Sound insulation of new dwellings. *DOE Construction* (1978), (25) 19–20

[93] Sewell E C and Alphey R S. Field measurements of the sound insulation of timber-joist party floors. *Building Research Establishment Information Paper* IP 5/81. Garston, Construction Research Communications Ltd, 1981

[94] Sewell E C and Alphey R S. Field measurements of the sound insulation of floating party floors with a solid concrete structural base. *Building Research Establishment Current Paper* CP 3/79. Garston, BRE, 1979

[95] Sewell E C and Alphey R S. Sound insulation of floating party floors with a solid concrete structural base. *Building Research Establishment Information Paper* IP 9/79. Garston, Construction Research Communications Ltd, 1979

[96] Langdon F J, Buller I B and Scholes W E. Noise from neighbours and the sound insulation of party floors and walls in flats. *Journal of Sound and Vibration* (1982), **83** (2) 171–80

[97] Grimwood C. Complaints about poor sound insulation between dwellings. *Acoustics Bulletin*, July/August 1995, 11–6

[98] Building Research Establishment. The acoustics of rooms for speech. *BRE Digest* 192. Garston, Construction Research Communications Ltd, 1976

Chapter 1.9

[99] Harper F C. The abrasion resistance of flooring materials. *Wear* (1961), (4) 461–78

[100] European Union of Agrément (UEAtc). Directives for the Assessment of Plastics Floorings. *Method of Assessment and Test* No 23. Paris, UEAtc, 1983

[101] European Union of Agrément (UEAtc). UEAtc Directives for the Assessment of Manufactured Plastics Floorings. *Method of Assessment and Test* No 36. Paris, UEAtc, 1987

[102] British Board of Agrément. Assessment of Plastics Floorings. *Information Sheet* No 2. Garston, British Board of Agrément, 1988

[103] British Standards Institution. Guide to durability of buildings and building elements, products and components. *British Standard* BS 7543:1992. London, BSI, 1992

[104] Yates T. Visitor damage to stone floors in historic buildings. *Proceedings of the Heritage and Tourism Conference, Canterbury, Kent, March 1990*

[105] Frommes B. Performance of abrasion machines for flooring materials. *Wear* (1961), (4) 479–94

[106] Warlow W J, Harper F C and Pye P W. The resistance to wear of flooring materials. *Wear* (1967), (10) 89–102

[107] Ministry of Agriculture Fisheries and Food, Department of Agriculture and Fisheries for Scotland and Scottish Home and Health Department. *Dairy floors*. London, The Stationery Office, 1967

[108] Bravery A F, Berry R W, Carey J K and Cooper D E. *Recognising wood rot and insect damage in buildings*. Building Research Establishment Report. Garston, Construction Research Communications Ltd, 1992

[109] Berry R W. *Remedial treatment of wood rot and insect attack in buildings*. Building Research Establishment Report. Garston, Construction Research Communications Ltd, 1994

[110] Building Research Establishment. Reducing the risk of pest infestations: design recommendations and literature review. *BRE Digest* 238. Garston, Construction Research Communications Ltd, 1980

[111] Coad J R and Rosaman D. Failures with site-applied adhesives. *Building Research Establishment Building Technical File* (October 1984), (7) 57-62

[112] British Standards Institution. Classification of adhesives for construction. Classification of adhesives for use with flooring materials. *British Standard* BS 5442: Part 1:1989. London, BSI, 1989

[113] Matthews S L. The role of subsurface radar in investigation and diagnosis of defects in building structures. *Proceedings of the Symposium of the International Council for Building Research Studies and Documentation on Dealing with Defects in Building, Varenna, Italy, 27–30 September 1994*

Chapter 1.10

[114] British Standards Institution. Care and maintenance of floor surfaces. Code of practice for resilient sheet and tile flooring. *British Standard* BS 6263:Part 2:1991. London, BSI, 1991

[115] The Institution of Structural Engineers. *Appraisal of existing structures* (2nd edition). London, SETO Ltd, 1996

Chapter 2.1

[116] Department of the Environment and The Welsh Office. *The Building Regulations 1991 Approved Document A: Structure*. London, The Stationery Office, 1992

[117] Building Research Establishment. Joist hangers. *BRE Good Building Guide* GBG 21. Garston, Construction Research Communications Ltd, 1996

[118] British Standards Institution. Specification for hot applied damp resisting coatings for solums. *British Standard* BS 2832:1957. London, BSI, 1957

[119] British Standards Institution. Structural use of timber. *British Standard* BS 5268:Part 2:1991. London, BSI, 1991

[120] Building Research Establishment. Suspended timber ground floors: repairing rotted joists. *BRE Defect Action Sheet* 74. Garston, Construction Research Communications Ltd, 1986

[121] Building Research Establishment. Suspended timber floors: notching and drilling of joists. *BRE Defect Action Sheet* 99. Garston, Construction Research Communications Ltd, 1987

[122] British Standards Institution. Guide to evaluation of human exposure to vibration in buildings (1 Hz to 80 Hz). *British Standard* BS 6472:1992. London, BSI, 1992

[123] British Standards Institution. Joist hangers. Specification for joist hangers for building into masonry walls of domestic dwellings. *British Standard* BS 6178:Part 1: 1990. London, BSI, 1990

[124] Building Research Station. The design of timber floors to prevent dry rot. *BRS Digest* 1, First series. Garston, Construction Research Communications Ltd, 1948

[125] Department of the Environment and The Welsh Office. *The Building Regulations 1972*. Statutory Instrument 1972 No 317. London, The Stationery Office, 1972

[126] British Standards Institution. Code of practice for control of condensation in buildings. *British Standard* BS 5250:1989. London, BSI, 1989

[127] British Standards Institution. Structural use of timber. Fire resistance of timber structures. *British Standard* BS 5268: Part 4:1978–89. London, BSI, 1978–89

[128] British Standards Institution. Fire tests on building materials and structures. Method for classification of the surface spread of flame of products. *British Standard* BS 476:Part 7:1987. London, BSI, 1987

[129] Raw G J and Oseland N A. Subjective response to noise through party floors in conversion flats. *Applied Acoustics* (1991), **32** (3) 215–32

[130] Utley W A and Cappelen P. The sound insulation of wood-joist floors in timber-frame constructions. *Building Research Establishment Current Paper* CP 46/78. Garston, BRE, 1978

[131] Paxton B H. Impact testing and requirements of chipboard for flooring. *Journal of the Institute of Wood Science* (1980), **8** (5) (Issue 47) 208–13

[132] Building Research Establishment. Ground floors: replacing suspended timber with solid concrete – dpcs and dpms. *BRE Defect Action Sheet* 22. Garston, Construction Research Communications Ltd, 1983

[133] Building Research Establishment. Substructure: dpcs and dpms – specification. *BRE Defect Action Sheet* 35. Garston, Construction Research Communications Ltd, 1983

[134] Building Research Establishment. Substructure: dpcs and dpms – installation. *BRE Defect Action Sheet* 36. Garston, Construction Research Communications Ltd, 1983

[135] Building Research Establishment. Rising damp in walls: diagnosis and treatment. *BRE Digest* 245. Garston, Construction Research Communications Ltd, 1984

[136] Building Research Establishment. Damp-proof courses. *BRE Digest* 380. Garston, Construction Research Communications Ltd, 1993

[137] Building Research Establishment. House inspection for dampness: a first step to remedial treatment for wood rot. *BRE Information Paper* IP 19/88. Garston, Construction Research Communications Ltd, 1988

[138] Building Research Establishment. Dry rot: its recognition and cure. *BRE Digest* 299. Garston, Construction Research Communications Ltd, 1985

[139] Building Research Establishment. Wet rots: recognition and control. *BRE Digest* 345. Garston, Construction Research Communications Ltd, 1989

[140] Building Research Station. The BRE annular rebate plane. *BRS Information Sheet* 3/76. Garston, BRE, 1976

[141] British Standards Institution. Specification for design, installation, testing and maintenance of services supplying water for domestic use within buildings and their curtilages. *British Standard* BS 6700:1987. London, BSI, 1987

Further reading

Department of the Environment. Minimising thermal bridging in new dwellings. A detailed guide for architects and building designers. *Good Practice Guide* 174. London, DOE, 1996

Department of the Environment and The Welsh Office. *The Building Regulations 1991 Approved Document E: Resistance to the passage of sound.* London, The Stationery Office, 1992

Building Research Establishment. Domestic floors: construction, insulation and damp-proofing. *BRE Good Building Guide* GBG 28, Part 1. Garston, Construction Research Communications Ltd, 1997

Building Research Establishment. Domestic floors: assessing them for replacement or repair – timber floors and decks. *BRE Good Building Guide* GBG 28, Part 3. Garston, Construction Research Communications Ltd, 1997

Welsh P, Pye P W and Scivyer C R. *Protecting dwellings with suspended timber floors: a BRE guide to radon remedial measures in existing dwellings.* BRE Report. Garston, Construction Research Communications Ltd, 1994

Chapter 2.2

[142] Building Research Establishment. *BRS Type 4 houses.* BRE Report. Garston, Construction Research Communications Ltd, 1989

[143] Building Research Establishment. *Fidler houses.* BRE Report. Garston, Construction Research Communications Ltd, 1989

[144] Building Research Establishment. *Incast houses.* BRE Report. Garston, Construction Research Communications Ltd, 1989

[145] Building Research Establishment. *Cast rendered no-fines houses.* BRE Report. Garston, Construction Research Communications Ltd, 1989

[146] British Standards Institution. Structural use of concrete. Code of practice for design and construction. *British Standard* BS 8110:Part 1:1985. London, BSI, 1985

[147] British Standards Institution. Specification for steel fabric for the reinforcement of concrete. *British Standard* BS 4483:1985. London, BSI, 1985

[148] British Standards Institution. Specification for carbon steel bars for the reinforcement of concrete. *British Standard* BS 4449:1988. London, BSI, 1988

[149] Bonshor R B and Bonshor L L. *Cracking in buildings.* Garston, Construction Research Communications Ltd, 1996

[150] Department of the Environment and The Welsh Office. *The Building Regulations 1985 Approved Document C: Site preparation and resistance to moisture.* London, The Stationery Office, 1985

[151] Malhotra H L and Morris W A. An investigation into the fire problems associated with wood wool permanent shuttering for concrete floors. *Building Research Establishment Current Paper* CP 68/78. Garston, BRE, 1978

[152] British Standards Institution. Workmanship on building sites. Code of practice for concrete work. Mixing and transporting concrete. *British Standard* BS 8000:Part2:Section 2.1:1990. London, BSI, 1990

[153] British Standards Institution. Workmanship on building sites. Code of practice for concrete work. Sitework with in situ and precast concrete. *British Standard* BS 8000:Part 2:Section 2.2:1990. London, BSI, 1990

[154] British Standards Institution. Testing concrete. Recommendations on the use of electromagnetic covermeters. *British Standard* BS 1881:Part 204:1988. London, BSI, 1988

Further reading

Department of the Environment. Minimising thermal bridging in new dwellings. A detailed guide for architects and building designers. *Good Practice Guide* 174. London, DOE, 1996

Department of the Environment and The Welsh Office. *The Building Regulations 1991 Approved Document E: Resistance to the passage of sound.* London, The Stationery Office, 1992

Building Research Establishment. Domestic floors: construction, insulation and damp-proofing. *BRE Good Building Guide* GBG 28, Part 1. Garston, Construction Research Communications Ltd, 1997

Chapter 2.3

[155] British Standards Institution. Precast concrete masonry units. Specification for precast concrete masonry units. *British Standard* BS 6073:Part 1:1981. London, BSI, 1981

[156] British Standards Institution. Precast concrete masonry units. Method for specifying precast concrete masonry units. *British Standard* BS 6073:Part 2:1981. London, BSI, 1981

[157] Moss R M. Load testing of floor and roof assemblies. *Proceedings of the international Symposium on Re-evaluation of Concrete Structures, Technical University of Denmark, Lyngby, 13–15 June 1988*

[158] Ellis B R, Kerridge B and Osborne K P. A study of the vibration characteristics of shallow floor structures. *Proceedings of the International Colloquium on Structural Serviceability of Buildings, Goteborg, Sweden, June 1993*

[159] Building Research Establishment. Alkali aggregate reactions in concrete. *BRE Digest* 330. Garston, Construction Research Communications Ltd, 1991

[160] British Standards Institution. Screeds, bases and in-situ floorings. Code of practice for concrete bases and screeds to receive in-situ floorings. *British Standard* BS 8204:Part 1:1987. London, BSI, 1987

[161] Building Research Establishment. Assessment of existing high alumina cement concrete construction in the UK. *BRE Digest* 392. Garston, Construction Research Communications Ltd, 1994

[162] Building Regulations Advisory Committee. Report by sub-committee P (high alumina cement concrete) BRAC (75) p 40. London, The Stationery Office, 1975. Addendum No 1: BRAC (75) p 59, 1975; Addendum No 2: BRAC (76) p 3, 1976. London, Department of the Environment

Further reading

Building Research Establishment. *Thermal insulation: avoiding risks.* BRE Report. Garston, Construction Research Communications Ltd, 1994

Department of the Environment. Minimising thermal bridging in new dwellings. A detailed guide for architects and building designers. *Good Practice Guide* 174. London, DOE, 1996

Building Research Establishment. Static load testing: concrete floor and roof structures within buildings. *BRE Digest* 402. Garston, Construction Research Communications Ltd, 1995

Building Research Establishment. Domestic floors: construction, insulation and damp-proofing. *BRE Good Building Guide* GBG 28, Part 1. Garston, Construction Research Communications Ltd, 1997

Currie R J and Crammond N J. Assessment of existing high alumina cement concrete construction in the UK. *Proceedings of the Institution of Structures and Buildings* (February 1994) **104** 83–92

Chapter 2.4

[163] Sutherland R J M. Recognition and appraisal of ferrous metals. *Proceedings of the Symposium on Building Appraisal, Maintenance and Preservation, University of Bath, July 1985*

[164] British Standards Institution. Code of practice for protective coating of iron and steel structures against corrosion. *British Standard* BS 5493:1977. London, BSI, 1977

[165] Harrison H W. *Steel-framed and steel-clad houses: inspection and assessment.* Building Research Establishment Report. Garston, Construction Research Communications Ltd, 1987

Further reading

British Steel General Steels. *Corrosion protection guide for steelwork in building interiors.* British Steel General Steels, 1992

Chapter 2.6

[166] PSA Specialist Services, Directorate of Building and Quantity Surveying Services. Platform Floors (Raised Access Floors): Performance Specification. MOB PF2 PS. Croydon, PSA Specialist Services, 1990

[167] British Standards Institution. Code of practice for fire protection for electronic data processing installations. *British Standard* BS 6266:1992. London, BSI, 1992

[168] Contract Flooring Association. *The CFA guide to contract flooring.* Rickmansworth, Phebruary Publications Ltd, 1991

[169] Fothergill L C and Royle P. The sound insulation of timber platform floating floors in the laboratory and field. *Applied Acoustics* (1991), **33** (4) 249–61

Chapter 3.1

[170] British Standards Institution. Specification for mastic asphalt for building and civil engineering (limestone aggregate). *British Standard* BS 6925:1988. London, BSI, 1988

[171] British Standards Institution. Concrete. Guide to specifying concrete. *British Standard* BS 5328:Part 1:1991. London, BSI, 1991

[172] Pye P W. *Sealing cracks in solid floors: a BRE guide to radon remedial measures in existing dwellings.* Building Research Establishment Report. Garston, Construction Research Communications Ltd, 1993

[173] Building Research Establishment. Concrete: part 1 – materials. *BRE Digest* 325. Garston, Construction Research Communications Ltd, 1987

[174] Building Research Establishment. Concrete: part 2 – specification, design and quality control. *BRE Digest* 326. Garston, Construction Research Communications Ltd, 1987

[175] Building Research Establishment. Fill. Part 1: classification and load-carrying characteristics. *BRE Digest* 274. Garston, Construction Research Communications Ltd, 1983

[176] Building Research Establishment. Fill. Part 2: site investigation, ground improvement and foundation design. *BRE Digest* 275. Garston, Construction Research Communications Ltd, 1983

[177] Building Research Establishment. Hardcore. *BRE Digest* 276. Garston, Construction Research Communications Ltd, 1983

[178] Building Research Establishment. Assessment of damage in low-rise buildings with particular reference to progressive foundation movement. *BRE Digest* 251. Garston, Construction Research Communications Ltd, 1981

[179] Building Research Establishment. Mini-piling for low-rise buildings. *BRE Digest* 313. Garston, Construction Research Communications Ltd, 1986

[180] British Standards Institution. Particleboard. Specification for wood chipboard. *British Standard* BS 5669:Part 2: 1989. London, BSI, 1989

[181] Anderson B R. U-values of uninsulated ground floors: relationship with floor dimensions. *Building Services and Engineering Research and Technology* (1991), **12** (3) 103–5

[182] Anderson B R. The U-value of solid ground floors with edge insulation. *Building Research Establishment Information Paper* IP 7/93. Garston, Construction Research Communications Ltd, 1993

[183] Anderson B R and Robinson M J. The U-value of solid and suspended floors with foundations of low density masonry units. *Masonry International* (1995), **8** (3) 79–81

[184] British Standards Institution. Code of practice for protection of structures against water from the ground. *British Standard* BS 8102:1990. London, BSI, 1990

[185] British Standards Institution. Code of practice for design and installation of damp-proof courses in masonry construction. *British Standard* BS 8215:1991. London, BSI, 1991

[186] Building Research Establishment. Sulphate and acid resistance of concrete in the ground. *BRE Digest* 363. Garston, Construction Research Communications Ltd, 1991

[187] Walton P L. Effects of alkali-silica reaction on concrete foundations. *Building Research Establishment Information Paper* IP 16/93. Garston, Construction Research Communications Ltd, 1993

[188] Collins R J. Case studies of floor heave due to microbiological activity in pyritic shales. *Proceedings of the Symposium on Microbiology in Civil Engineering, Silsoe College, Bedford, 3–5 September 1990*

[189] Building Research Station. Colliery shale as hardcore or filling. *BRS Digest* 84, First series. London, The Stationery Office, 1956

[190] Building Research Station. Questions and answers. Lifting of concrete floors. *BRS Digest* 97, First series. London, The Stationery Office, 1957

[191] Eldridge H J. Concrete floors on shale hardcore. *Building Research Station Current Paper (Design Series)* 30. Garston, BRE, 1964

[192] British Standards Institution. Screeds, bases and in-situ floorings. Code of practice for concrete wearing surfaces. *British Standard* BS 8204:Part 2:1987. London, BSI, 1987

[193] The Concrete Society. Concrete industrial ground floors – a guide to their design and construction. *Technical Report* No 34. Slough, The Concrete Society, 1994

[194] British Standards Institution. Workmanship on building sites. Code of practice for cement/sand floor screeds and concrete floor toppings. *British Standard* BS 8000:Part 9:1989. London, BSI, 1989

Further reading

Anderson B R. The influence of edge insulation on the steady-state heat loss through a slab-on-ground floor. *Building and Environment* (1993), **28** (3) 361–7

Building Research Establishment. *Thermal insulation: avoiding risks.* BRE Report. Garston, Construction Research Communications Ltd, 1994

Building Research Establishment. Substructure: dpcs and dpms – specification. *BRE Defect Action Sheet* 35. Garston, Construction Research Communications Ltd, 1983

Building Research Establishment. Substructure: dpcs and dpms – installation. *BRE Defect Action Sheet* 36. Garston, Construction Research Communications Ltd, 1983

Building Research Establishment. Damp-proofing solid floors. *BRE Digest* 54. Garston, Construction Research Communications Ltd, 1971

Building Research Establishment. Floor screeds. *BRE Digest* 104. Garston, Construction Research Communications Ltd, 1973

Building Research Establishment. Sheet and tile flooring made from thermoplastic binders. *BRE Digest* 33. Garston, Construction Research Communications Ltd, 1971

Building Research Establishment. Domestic floors: construction, insulation and damp-proofing. *BRE Good Building Guide* GBG 28, Part 1. Garston, Construction Research Communications Ltd, 1997

Building Research Establishment. Domestic floors: assessing them for replacement or repair – concrete floors, screeds and finishes. *BRE Good Building Guide* GBG 28, Part 2. Garston, Construction Research Communications Ltd, 1997

Mastic Asphalt Council and Employers Federation. *Flooring handbook.* Haywards Heath, MACEF, 1983

British Cement Association. *Concrete on site* (series of 11 booklets). BCA, Crowthorne, 1993

Anderson B R. U-values for basements. *Building Research Establishment Information Paper* IP 14/94. Garston, Construction Research Communications Ltd, 1994

Department of the Environment. Minimising thermal bridging in new dwellings. A detailed guide for architects and building designers. *Good Practice Guide* 174. London, DOE, 1996

Chapter 3.2

General reading

Department of the Environment. Minimising thermal bridging in new dwellings. A detailed guide for architects and building designers. *Good Practice Guide* 174. London, DOE, 1996

Building Research Establishment. Domestic floors: construction, insulation and damp-proofing. *BRE Good Building Guide* GBG 28, Part 1. Garston, Construction Research Communications Ltd, 1997

Building Research Establishment. Domestic floors: assessing them for replacement or repair – concrete floors, screeds and finishes. *BRE Good Building Guide* GBG 28, Part 2. Garston, Construction Research Communications Ltd, 1997

Chapter 3.3

General reading

Department of the Environment. Minimising thermal bridging in new dwellings. A detailed guide for architects and building designers. *Good Practice Guide* 174. London, DOE, 1996

Chapter 4.1

[195] British Standards Institution. Specifications for building sands from natural sources. *British Standard* BS 1199 and 1200:1976. London, BSI, 1976

[196] British Standards Institution. Specification for aggregates from natural sources for concrete. *British Standard* BS 882:1992. London, BSI, 1992

[197] British Standards Institution. Specification for test sieves. *British Standard* BS 410:1986. London, BSI, 1986

[198] Warlow W J and Pye P W. The rippling of thin flooring over discontinuities in screeds. *Building Research Station Current Paper* CP 94/74. Garston, Building Research Establishment, 1974

[199] British Standards Institution. In-situ floor finishes. Metric units. *British Standard* BS CP 204:Part 2:1970. London, BSI, 1970

[200] Pye P W and Warlow W J. A method of assessing the soundness of some dense floor screeds. *Building Research Station Current Paper* CP 72/78. Garston, Building Research Establishment, 1978

[201] British Standards Institution. Determination of in situ crushing resistance of floating levelling screeds. *British Standard* BS DD 230:1996. London, BSI, 1996

Further reading

Perkins P H. Floor screeds: recommendations for cement-sand and lightweight screeds. *Cement and Concrete Association Advisory Note* 48.023. Slough, C&CA, 1974

Barnbrook G. Laying floor screeds. *Cement and Concrete Association Construction Guide* 48.046. Slough, C&CA, 1979

P W Pye. BRE screed tester: classification of screeds, sampling and acceptance limits. *Building Research Establishment Information Paper* IP 11/84. Garston, Construction Research Communications Ltd, 1984

Building Research Establishment. Domestic floors: assessing them for replacement or repair – concrete floors, screeds and finishes. *BRE Good Building Guide* GBG 28, Part 2. Garston, Construction Research Communications Ltd, 1997

Building Research Establishment. Domestic floors: repairing or replacing floors and flooring – magnesite, tiles, slabs and screeds. *BRE Good Building Guide* GBG 28, Part 4. Garston, Construction Research Communications Ltd, 1997

Thompson E and Lilley A. Distribution of cement in floor screeds. *National Builder* (April 1982)

Warlow W J and Pye P W. Floor screeds; bakery floors; synthetic resin flooring. Movements leading to the cracking and curling of floor screeds. *Building Research Establishment Current Paper* CP 28/74, Paper 1. Garston, Construction Research Communications Ltd, 1974

Chapter 4.2

[202] British Standards Institution. Code of practice for sound insulation and noise reduction for buildings. *British Standard* BS 8233:1987. London, BSI, 1987

Further reading

British Standards Institution. Method for determination of in situ crushing resistance of floating levelling screeds. *British Standard* BS DD 230:1996. London, BSI, 1996

Chapter 4.3

[203] British Standards Institution. Specification for unbacked flexible PVC flooring. Homogeneous flooring. *British Standard* BS 3261:Part 1:1991. London, BSI, 1991

[204] British Standards Institution. Specification for sheet linoleum (calendered types) cork carpet and linoleum tiles. *British Standard* BS 810:1950. London, BSI, 1950

[205] British Standards Institution. Specification for linoleum and cork carpet sheet and tiles. *British Standard* BS 6826: 1987. London, BSI, 1987

Chapter 4.4

General reading

Building Research Establishment. Domestic floors: repairing or replacing floors and flooring – magnesite, tiles, slabs and screeds. *BRE Good Building Guide* GBG 28, Part 4. Garston, Construction Research Communications Ltd, 1997

Chapter 4.6

General reading

Department of the Environment and The Welsh Office. *The Building Regulations 1991 Approved Document E: Resistance to the passage of sound.* London, The Stationery Office, 1992

Chapter 5.1

[206] Harper F C and Stone P A. Floor finishes for factories. *Building Research Station Factory Building Studies* No 3. London, The Stationery Office, 1959

Further reading

British Standards Institution. Screeds, bases and in-situ floorings. Code of practice for concrete wearing surfaces. *British Standard* BS 8204:Part 2:1987. London, BSI, 1987

Chapter 5.2

General reading

British Standards Institution. Screeds, bases and in-situ floorings. Code of practice for polymer modified cementitious wearing surfaces. *British Standard* BS 8204:Part 3: 1993. London, BSI, 1993

Federation of Resin Formulators and Applicators. Polymer flooring guide. *Application Guide* No 6. Aldershot, FeRFA

Chapter 5.3

General reading

British Standards Institution. Screeds, bases and in-situ floorings. Code of practice for concrete wearing surfaces. *British Standard* BS 8204:Part 2:1987. London, BSI, 1987

British Standards Institution. In-situ floor finishes. Metric units. *British Standard* BS CP 204:Part 2:1970. London, BSI, 1970

Chapter 5.4

General reading

British Standards Institution. Screeds, bases and in-situ floorings. Code of practice for terrazzo wearing surfaces. *British Standard* BS 8204:Part 4: 1993. London, BSI, 1993

British Standards Institution. In-situ floor finishes. Metric units. *British Standard* BS CP 204:Part 2:1970. London, BSI, 1970

Chapter 5.5

[207] British Standards Institution. Testing of resin and polymer/cement compositions for use in construction. Method for measurement of compressive strength. *British Standard* BS 6319:Part 2:1983. London, BSI, 1983

[208] British Standards Institution. Testing of resin and polymer/cement compositions for use in construction. Methods for measurement of modulus of elasticity in flexure and flexural strength. *British Standard* BS 6319:Part 3:1990. London, BSI, 1990

[209] British Standards Institution. Testing of resin and polymer/cement compositions for use in construction. Method for measurement of bond strength (slant shear method). *British Standard* BS 6319: Part 4:1984. London, BSI, 1984

[210] British Standards Institution. Testing of resin and polymer/cement compositions for use in construction. Method for measurement of tensile strength. *British Standard* BS 6319:Part 7:1985. London, BSI, 1985

[211] British Standards Institution. Testing of resin and polymer/cement compositions for use in construction. Methods for determination of creep in compression and in tension. *British Standard* BS 6319:Part 11:1993. London, BSI, 1993

[212] British Standards Institution. Testing of resin and polymer/cement compositions for use in construction. Methods for measurement of unrestrained linear shrinkage and coefficient of thermal expansion. *British Standard* BS 6319:Part 12:1992. London, BSI, 1992

Further reading

Warlow W J and Pye P W. Osmosis as a cause of blistering of in situ resin flooring on wet concrete. *Magazine of Concrete Research* (September 1978), **30** (104)

Building Research Advisory Service. Defects in epoxy resin flooring. *Building Technical File* (July 1986), **14**

Pye P W. Osmosis in flooring. *Contract Flooring Journal* (April 1989), **2** p38

Federation of Resin Formulators and Applicators. Osmosis in flooring. *Technical Report.* Aldershot, FeRFA

Pye P W. Blistering of in situ thermosetting resin flooring by osmosis. *Proceedings of the International Colloquium on Industrial Floors, Stuttgart, January 1995*

Federation of Resin Formulators and Applicators. Synthetic resin flooring. *Application Guide* No 4. Aldershot, FeRFA, 1991

Chapter 5.7

[213] British Standards Institution. Specification for mastic asphalt (limestone fine aggregate) for roads, footways and pavings in building. *British Standard* BS 1447:1988. London, BSI, 1988

Chapter 5.8

[214] British Standards Institution. Specification for materials for magnesium oxychloride (magnesite) flooring. *British Standard* BS 776:Part 2:1972. London, BSI, 1972

Further reading

British Standards Institution. In-situ floor finishes. Metric units. *British Standard* BS CP 204:Part 2:1970. London, BSI, 1970

Chapter 6.1

[215] British Standards Institution. Code of practice for installation of textile floor coverings. *British Standard* BS 5325:1983. London, BSI, 1983

Chapter 6.4

[216] British Standards Institution. Specification. Backed flexible PVC flooring. Needle-loom felt backed flooring. *British Standard* BS 5085:Part 1:1974. London, BSI, 1974

[217] British Standards Institution. Specification. Backed flexible PVC flooring. Cellular PVC backing. *British Standard* BS 5085:Part 2:1976. London, BSI, 1976

[218] British Standards Institution. Specification for semi-flexible PVC floor tiles. *British Standard* BS 3260:1969. London, BSI, 1969

Chapter 6.5

[219] British Standards Institution. Resilient floor coverings – Semi-flexible polyvinyl chloride tiles – Specification. *British Standard* BS EN 654:1997. London, BSI, 1977

Chapter 6.6

[220] British Standards Institution. Specification for solid rubber flooring. *British Standard* BS 1711:1975. London, BSI, 1975

[221] British Standards Institution. Specification for electrically conducting rubber flooring. *British Standard* BS 3187: 1978. London, BSI, 1978

[222] British Standards Institution. Sheet and tile flooring. Cork, linoleum, plastics and rubber. *British Standard* BS CP 203. London, BSI

Chapter 6.7

[223] British Standards Institution. Specification for thermoplastic flooring tiles. *British Standard* BS 2592:1973. London, BSI, 1973

Chapter 7.1

[224] British Standards Institution. Specification for clay tiles for flooring. *British Standard* BS 1286:1974. London, BSI, 1974

[225] British Standards Institution. Code of practice for tile flooring and slab flooring. *British Standard* BS CP 202:1972. London, BSI, 1972

[226] British Standards Institution. Wall and floor tiling. Code of practice for the design and installation of ceramic floor tiles and mosaics. *British Standard* BS 5385: Part 3:1989. London, BSI, 1989

[227] British Standards Institution. Wall and floor tiling. Code of practice for tiling and mosaics in specific conditions. *British Standard* BS 5385:Part 4:1992. London, BSI, 1992
[228] British Standards Institution. Ceramic floor and wall tiles. Method for determination of scratch hardness of surface according to Mohs. *British Standard* BS 6431:Part 13:1986. London, BSI, 1986
[229] British Standards Institution. Ceramic floor and wall tiles. *British Standard* BS 6431:Parts 1–23:1983–86. London, BSI, 1983–86
[230] British Standards Institution. Specification for adhesives for use with ceramic tiles and mosaics. *British Standard* BS 5980:1980. London, BSI, 1980
[231] British Ceramic Tile Council and British Institute of Cleaning Science. *The cleaning of ceramic tiles.* Stoke on Trent, British Ceramic Tile Council, 1989
[232] Yates T, Davison S and Martin W. The care and conservation of medieval tile pavements at English Heritage sites (chapter in *Architectural ceramics: their history, manufacture and conservation*). London, James and James (Science Publishers) Ltd, 1996
[233] British Standards Institution. Workmanship on building sites. Code of practice for wall and floor tiling. Ceramic tiles, terrazzo tiles and mosaics. *British Standard* BS 8000:Part 11:Section 11.1:1989. London, BSI, 1989

Further reading

Building Research Establishment. Domestic floors: repairing or replacing floors and flooring – magnesite, tiles, slabs and screeds. *BRE Good Building Guide* GBG 28, Part 4. Garston, Construction Research Communications Ltd, 1997

Chapter 7.2

[234] British Standards Institution. Precast concrete flags, kerbs, channels, edgings and quadrants. Specification. *British Standard* BS 7263:Part 1:1994. London, BSI, 1994
[235] British Standards Institution. Wall and floor tiling. Code of practice for the design and installation of terrazzo tile and slab, natural stone and composition block floorings. *British Standard* BS 5385:Part 5:1994. London, BSI, 1994

Chapter 7.4

General reading

British Standards Institution. Specification for terrazzo tiles. *British Standard* BS 4131: 1973. London, BSI, 1973
British Standards Institution. Wall and floor tiling. Code of practice for the design and installation of terrazzo tile and slab, natural stone and composition block floorings. *British Standard* BS 5385:Part 5:1994. London, BSI, 1994

Chapter 8.1

[236] Armstrong F H. Timbers for flooring. *Forest Products Research Bulletin* No 40. London, The Stationery Office, 1957
[237] Princes Risborough Laboratory. Flooring and joinery in new buildings. *PRL Technical Note* No 12. Garston, Building Research Establishment, 1966
[238] Forest Products Research Laboratory. Hardwoods for industrial flooring. *FPRL Technical Note* No 49. London, The Stationery Office, 1971
[239] British Standards Institution. Code of practice for flooring of timber, timber products and wood based panel products. *British Standard* BS 8201:1987. London, BSI, 1987
[240] Princes Risborough Laboratory. The movement of timbers. *PRL Technical Note* No 38. Garston, Building Research Establishment, 1969
[241] Building Research Establishment. *Handbook of hardwoods.* BRE Report. Garston, Construction Research Communications Ltd, 1972
[242] Building Research Establishment. *A handbook of softwoods.* BRE Report. Garston, Construction Research Communications Ltd, 1977

Further reading

Building Research Establishment. Domestic floors: assessing them for replacement or repair – timber floors and decks. *BRE Good Building Guide* GBG 28, Part 3. Garston, Construction Research Communications Ltd, 1997
British Standards Institution. Wooden flooring. *British Standard* CP 209:Part 1: 1963. London, BSI, 1963

Chapter 8.2

General reading

Building Research Establishment. Domestic floors: repairing or replacing floors and flooring – wood blocks and suspended timber. *BRE Good Building Guide* GBG 28, Part 5. Garston, Construction Research Communications Ltd, 1997
Building Research Establishment. Design of timber floors to prevent decay. *BRE Digest* 364. Garston, Construction Research Communications Ltd, 1991

Chapter 8.3

[243] British Standards Institution. Wooden flooring. *British Standard* BS CP 209:Part 1:1963. London, BSI, 1963

Further reading

Building Research Establishment. Domestic floors: assessing them for replacement or repair – timber floors and decks. *BRE Good Building Guide* GBG 28, Part 3. Garston, Construction Research Communications Ltd, 1997
Building Research Establishment. Domestic floors: repairing or replacing floors and flooring – wood blocks and suspended timber. *BRE Good Building Guide* GBG 28, Part 5. Garston, Construction Research Communications Ltd, 1997
British Standards Institution. Code of practice for flooring of timber, timber products and wood based panel products. *British Standard* BS 8201:1987. London, BSI, 1987

Chapter 8.5

[244] British Standards Institution. Particleboard. Specification for oriented strand board (OSB). *British Standard* BS 5669:Part 3:1992. London, BSI, 1992
[245] British Standards Institution. Specification for fibre building boards. *British Standard* BS 1142:1989. London, BSI, 1989
[246] British Standards Institution. Particleboard. Specification for cement bonded particleboard. *British Standard* BS 5669:Part 4:1989. London, BSI, 1989
[247] Building Research Establishment. Selecting wood-based panel products. *BRE Digest* 323. Garston, Construction Research Communications Ltd, 1987
[248] British Standards Institution. Plywood. Specification for classification of resistance to fungal decay and wood borer attack. *British Standard* BS 6566:Part 7: 1985. London, BSI, 1985
[249] British Standards Institution. Plywood. Specification for bond performance of veneer plywood. *British Standard* BS 6566:Part 8:1985. London, BSI, 1985
[250] Building Research Establishment. Plywood. *BRE Digest* 394. Garston, Construction Research Communications Ltd, 1994
[251] British Standards Institution. Particleboard. Code of practice for the selection and application of particleboards for specific purposes. *British Standard* BS 5669:Part 5:1993. London, BSI, 1993
[252] Building Research Establishment. Oriented strand board. *BRE Digest* 400. Garston, Construction Research Communications Ltd, 1994

[253] Dinwoodie J M and Paxton B H. Cement-bonded particleboard. *Building Research Establishment Information Paper* IP 14/92. Garston, Construction Research Communications Ltd, 1992

[254] Rodwell D F, Dinwoodie J M and Paxton B H. Fibre building board: types and uses. *Building Research Establishment Information Paper* IP 12/91. Garston, Construction Research Communications Ltd, 1991

[255] Building Research Establishment. Suspended timber floors: chipboard flooring – specification. *BRE Defect Action Sheet* 31. Garston, Construction Research Communications Ltd, 1983

[256] Dinwoodie J M. Wood chipboard: recommendations for use. *Building Research Establishment Information Paper* IP 3/85. Garston, Construction Research Communications Ltd, 1985

[257] British Standards Institution. Classification of non-structural adhesives for joining of wood and derived timber products. *British Standard* BS EN 204:1991. London, BSI, 1991

Further reading

Building Research Establishment. Wood chipboard. *BRE Digest* 373. Garston, Construction Research Communications Ltd, 1992

British Standards Institution. Particleboard. Specification for wood chipboard. *British Standard* BS 5669:Part 2: 1989. London, BSI, 1989

Building Research Establishment. The structural use of wood based panels. *BRE Digest* 423. Garston, Construction Research Communications Ltd, 1997

Index

Floors and flooring is systematically structured to enable the reader seeking particular information to identify quickly the parts of the book relevant to his or her search. The broad structure of chapter and sub-chapter titles will be seen in the Contents list on page iii.

Chapter 0, the introduction to *Floors and flooring*, describes the performance of floors in buildings in the UK, sources of information and data, and a brief review of industry problems.

Chapter 1 and its sub-chapters are self-explanatory.

Chapters 2 and 3 describe the main types of floors; section headings within the sub-chapters, which mainly deal with materials forming the structure, broadly link to the functions shown as sub-chapters in Chapter 1. Much of the structure of sub-chapters is standardised, with section headings drawn from a standard list (see below). Not all of these section headings are used in every sub-chapter, and some have been modified to fit particular circumstances.

Chapter 3 deals with substrates and Chapters 4 – 8 are concerned with finishes.

Section headings in sub-chapters of Chapters 2–8

Characteristic details
- Basic structure
- Bearing details
- Non-loadbearing abutments
- Services

Main performance requirements and defects (not necessarily in the order shown)
- Choice of materials for structure
- Strength and stability
- Dimensional stability, deflections etc
- Thermal properties
- Control of dampness and condensation
- Fire
- Sound insulation
- Durability
- Maintenance

Work on site
- Storage and handling of materials
- Restrictions due to weather conditions
- Workmanship
- Inspection (in panel)

To avoid what would otherwise be considerable repetition of text (because a lot of the information given for one type of floor will apply to one or more other floors), many sections refer the reader to the same sections in other earlier sub-chapters.

Using the Index

The Index excludes words and expressions that are already presented in the list of contents or the list of section headings. Therefore the reader should undertake his or her search in the following order.

1 list of contents (pages iii and iv)
2 list of standard section headings (above)
3 Index (starts page 296)

Where words and expressions which appear in the list of contents and/or the list of section headings also appear in other contexts, they may be shown in the Index.

Page references to captions to illustrations are shown in bold.